MEDICAL USES
OF STATISTICS

2nd Edition

MEDICAL USES
OF STATISTICS

2nd Edition

Edited by
JOHN C. BAILAR III
FREDERICK MOSTELLER

NEJM Books
Boston, Massachusetts

Library of Congress Cataloging in Publication Data

Medical uses of statistics, 2nd ed.
Includes bibliographical references and index.
1. Medical statistics. 2. Clinical medicine — Research — Statistical methods. I. Bailar, John C. (John Christian), 1932- . Mosteller, Frederick, 1916- . III. New England Journal of Medicine. [DNLM: 1. Research — methods — collected works. 2. Statistics — collected works.
WA 950 M489]
Ra409.M43 1992 610'.72—dc20
DNLM/DLC for Library of Congress
ISBN 0-910133-36-0 (soft)

Printed in the United States of America
10 9 8 7 6 5 4 3 2

This book was produced by the Electronic Production Department of the Massachusetts Medical Society. The text was printed on recycled, acid-free paper.

Design by Emily Stuart.

To
Kenneth S. Warren
and
Kerr L. White
for their many successful efforts
to improve the quantitative support
of research in medicine and health

CONTENTS

CONTRIBUTING AUTHORS

John C. Bailar III, M.D., Ph.D., Department of Epidemiology and Biostatistics, Faculty of Medicine, McGill University, and U.S. Office of Disease Prevention and Health Promotion, Department of Health and Human Services

Jayne Berrier, M.A., Clinical Trials Unit, Mount Sinai Medical Center

Elisabeth Burdick, M.S., Department of Health Policy and Management, School of Public Health, Harvard University

Thomas C. Chalmers, M.D., Department of Health Policy and Management, School of Public Health, Harvard University, and Clinical Trials Unit, Mount Sinai School of Medicine

L. Joseph Charette, Ph.D., Westfield, New Jersey

Graham A. Colditz, M.D., D.P.H., Channing Laboratory, Harvard Medical School

Fernando Delgado, M.S., Columbia, South America

Rebecca DerSimonian, D.Sc., National Institute of Child Health and Human Development, Bethesda, Maryland

Christl Donnelly, M.S., Department of Biostatistics, School of Public Health, Harvard University

John D. Emerson, Ph.D., Department of Mathematics, Middlebury College

Howard S. Frazier, M.D., Department of Health Policy and Management, School of Public Health, Harvard University

Jennie A. Freiman, M.D., New York City, New York

Katherine Godfrey, Ph.D., Cambridge, Massachusetts

Katherine Taylor Halvorsen, D.Sc., Department of Mathematics, Smith College

Hossein Hosseini, Ph.D., Digital Equipment Corporation, Irvine, California

Joseph A. Ingelfinger, M.D., Department of Medicine, Carney Hospital

Jerome P. Kassirer, M.D., Editor-in-Chief, *New England Journal of Medicine*

Roy R. Kuebler, Ph.D., Department of Biostatistics, School of Public Health, University of North Carolina, Chapel Hill (Deceased)

Stephen Lagakos, Ph.D., Department of Biostatistics, School of Public Health, Harvard University

Philip W. Lavori, Ph.D., Department of Psychiatry and Human Behavior, Brown University

Thomas A. Louis, Ph.D., Division of Biostatistics, School of Public Health, University of Minnesota

Bucknam McPeek, M.D., Department of Anesthesia, Massachusetts General Hospital

Lincoln E. Moses, Ph.D., Department of Statistics, Stanford University

Frederick Mosteller, Ph.D., Department of Health Policy and Management, School of Public Health, Harvard University

Daniel Pagano, B.A., Clinical Trials Unit, Mount Sinai Medical Center

Stephen G. Pauker, M.D., Department of Medicine, New England Medical Center

Marcia Polansky, D.Sc., Department of Biometrics and Computing, Hahnemann University

Dinah Reitman, M.P.S., Clinical Trials Unit, Mount Sinai Medical Center

Henry S. Sacks, Ph.D., M.D., Clinical Trials Unit, Mount Sinai Medical Center

Harry Smith, Jr., Ph.D., Department of Biomathematical Sciences, Mount Sinai School of Medicine (Retired)

James H. Ware, Ph.D., Department of Biostatistics, School of Public Health, Harvard University

Daniel Zelterman, Ph.D., Division of Biostatistics, School of Public Health, University of Minnesota

PREFACE

The first edition of this book, published over five years ago, found favor with a gratifyingly large number of readers and was widely praised as a unique contribution to its field. The Preface to the first edition, reprinted in almost its entirety, describes the book's origins and purposes. This second edition builds on the strengths of the first, extending its scope to new topics, while revising and updating treatment of many of the old ones and replacing a few of the original chapters with entirely new material.

The result is a slightly longer book, but I believe it is even better and more useful than its predecessor. The general philosophy and organization remain the same, but the range of subjects is broader and the overall treatment more comprehensive. Every effort has been made to achieve a readable and interesting text that explains the important ideas behind current medical uses of statistics without burdening the reader with the technical details of mathematical manipulations.

I found this new edition more interesting and accessible than the first. I trust readers will enjoy it as much as I did.

Arnold S. Relman, M.D.
Editor-in-Chief Emeritus, *New England Journal of Medicine*

PREFACE TO THE FIRST EDITION (1986)

No one who reads the current medical literature, and certainly no one who performs clinical studies these days, can be unaware of the growing importance of statistics. Sound clinical research, as well as the ability to understand published results of research, increasingly depends on a clear comprehension of the fundamental concepts of statistical design and analysis.

This book is the fruit of an idea that originated in 1977, in conversations with John Bailar and Frederick Mosteller of the Department of Biostatistics of the Harvard School of Public Health. Convinced that the readers of the *New England Journal of Medicine* needed a clearer idea of how statistical techniques were being applied in current clinical studies, my editorial colleagues and I (including most prominently our former Deputy Editor, Dr. Drummond Rennie) suggested to Bailar and Mosteller that they organize a study of the research papers published in recent volumes of the *Journal* (and some other important medical journals), to determine what statistical methods were actually being used. We also asked them to tell us whether the methods were appropriately applied and how their use might be improved, and we asked them to do so in simple language that would be understood even by readers who had no education in biostatistics.

With the aid of a generous grant from the Rockefeller Foundation, Bailar and Mosteller, assisted by a host of colleagues at Harvard and elsewhere, set out to do just that. Their work was greatly helped by encouragement from Dr. Kenneth Warren, Director of the Division of Health Sciences, and Dr. Kerr White, Special Projects Officer at the Rockefeller Foundation.

The result, in my view, has been spectacular. First of all, they carried out a survey of statistical practice in the *New England Journal* and a few other journals, demonstrating the frequency with which different types of statistical methods were applied and identifying the need for improvement in the selection and use of these methods. In addition, the group produced a series of articles on a wide range of statistical subjects, drawn from the insights gained during their survey of actual practice.

All together, more than 30 papers have come from this project so far. Some have appeared in the *Journal* as part of our "Statistics in Practice" series. A dozen or so have been published in other journals or as book chapters. Still others have been reserved for first publication in this book.

There are many books on biostatistics, but there are two unique and important characteristics of this one that I believe set it apart. First of all, as already noted, it is based on current usage, and it is concerned with improving that usage. Unlike most standard textbooks, this book takes an empirical, practical approach. It does not simply use examples from the literature to illustrate didactic points; it carefully surveys what clinical investigators are actually doing with statistical methods, as revealed mostly in the pages of the *Journal*. It tells readers what they need to know to understand those methods, and it points out ways in which medical writers can make their reporting of methods and results more informative and their analyses of data more useful.

Secondly, the orientation of this book is toward an understanding of ideas — when and why to use certain statistical techniques. There are many textbooks that explain statistical calculations but few or none that attempt, as this one does, to get behind the calculations and tell what they are all about. This book does not concern itself with the mechanics of statistical computation. There are no instructions on how to perform calculations, and there are few mathematical formulas. The emphasis here is on explaining the purpose of the statistical methods, so that the general reader will have a better understanding of the strategy to be employed and the alternatives that need to be considered. Most chapters, however, cite other "how-to" textbooks of statistics, to which readers may refer for detailed explanations of the mathematical calculations.

The authors have striven to write in a straightforward style, as unencumbered by biostatistical jargon as possible. Their object has been to make this book understandable to almost anyone who has a nodding acquaintance with biomedical research and an elementary grasp of numerical concepts. How well they have succeeded only the reader can judge, but, as an amateur myself, I have found their writing lucid and readable. I should think that most medical students and physicians — even those with no formal statistical education — would agree.

I should note here that this book constitutes one of the *Journal*'s first ventures in book publishing. We hope it meets the standards of quality we have always tried to maintain for the *Journal*, and that it will find favor with a broad cross-section of physicians and students.

Arnold S. Relman, M.D.
Editor, *New England Journal of Medicine*

ACKNOWLEDGMENTS

We had a great deal of cooperation in preparing the second edition of this book. A number of authors from the first edition provided updates or rewrites of their articles, while another group of authors prepared new chapters to improve and augment our original work.

Dr. Arnold Relman, who helped initiate and develop the original project in 1977, generously repeated his essential role for this new edition. He assisted us at each step — selecting topics, reorganizing and revising chapters to focus on the problems facing clinicians and medical investigators, and providing detailed critiques of each chapter to promote precision and ready comprehension.

Marie McPherson handled the heavy effort of managing the manuscripts and dealing with both authors and publishers. We also had statistical and manuscript support from Elisabeth Burdick and Bruce Kupelnik.

At the Publishing Division of the Massachusetts Medical Society, Rob Stuart guided us through copyediting and proofreading with the help of Tommie Richardson and Robert Dall. Emily Stuart designed this new edition, while Martha Soule and Sioux Waks lent their desktop publishing skills to its production.

Many others contributed to the first edition of this book, and we thank them once again, including faculty and students of the Harvard School of Public Health, numerous visitors, and the Rockefeller Foundation, as well as our past and present publisher, the *New England Journal of Medicine.*

We are indebted to them all.

ORIGINS OF CHAPTERS

Chapter 1.   Substantially revised, expanded, and updated from an article originally published in the *New England Journal of Medicine* (1985; 312:890-7).

Chapter 2.   Condensed and revised from the original publication in the Carolina Environmental Essay Series (1988; No. 9), Institute for Environmental Studies, University of North Carolina at Chapel Hill. Printed with permission of the publisher.*

Chapter 3.   Updated from the original article published in the *New England Journal of Medicine* (1983; 309:709-13).

Chapter 4.   Updated from the original article published in the *New England Journal of Medicine* (1983; 309:1291-9).

Chapter 5.   Updated from the original article published in the *New England Journal of Medicine* (1984; 310:24-31).

Chapter 6.   Updated from the original article published in the *New England Journal of Medicine* (1984; 311:156-62).

Chapter 7.   Updated from the original article published in the *New England Journal of Medicine* (1984; 311:705-10).

Chapter 8.   Updated from the original article published in the *New England Journal of Medicine* (1984; 311:1482-7).

Chapter 9.   Originally published in the *New England Journal of Medicine* (1987; 316:250-8).*

Chapter 10. This article was written for this edition of this book. The version prepared for the first edition was based heavily on material in Ingelfinger JA, Mosteller F, Thibodeau LA, Ware JH. What are *P* values? In: *Biostatistics in Clinical Medicine.* New York: Macmillan Publishing Co., Inc. 1983:160-76. Printed with permission of the publisher.*

Chapter 11. Revised and expanded from an article published in the *New England Journal of Medicine* (1985; 313:1629-36).

Chapter 12.   Slightly expanded from an article published in the *New England Journal of Medicine* (1985; 313:1450-6).

Chapter 13.   Originally published in the *New England Journal of Medicine* (1984; 311:442-8).

Chapter 14.   This article was written for this edition of this book.*

Chapter 15.   This article was written for this edition of this book.*

Chapter 16.   This article is slightly modified from its original appearance in the *Annals of Internal Medicine* (1988; 108:266-73). Printed with permission of the publisher.*

Chapter 17.   Updated from the original article published in the *New England Journal of Medicine* (1982; 306:1332-7).

Chapter 18.   Slightly revised from the original published in *Statistics in Medicine* (1984; 3:1-5). Copyright© 1984 by John Wiley. Printed with permission of the publisher.

Chapter 19.   Updated from the original article published in the *New England Journal of Medicine* (1978; 299:690-4).

Chapter 20.   Updated from an article written for the first edition of this book.

Chapter 21.   This article was written for this edition of this book.*

Chapter 22.   This article was written for this edition of this book. It replaces an article in the first edition.*

Chapter 23.   This article was prepared for this edition of this book. It brings up to date information from an article published in the *New England Journal of Medicine* (1987; 316:450-5).*

*Indicates a chapter new to this edition or completely rewritten for this edition.

INTRODUCTION

Biomedical research investigators and clinicians who are keeping up with the literature often find a need to update their preparation in statistical concepts, methods, and techniques. Like other fields, statistics advances rapidly, and even those with strong previous training may wish to freshen their information. Although readers can find review and didactic papers on specific statistical methods in textbooks or journals, they may have nowhere to turn for overviews of the field. This book surveys the state of the art of statistical applications in clinical research and illustrates good and poor uses of methods. With few exceptions, we stress the concepts underlying statistics rather than its more technical how-to-do-it aspects. Most of our examples are from the pages of the *New England Journal of Medicine*. Chapters carried over from the first edition have been updated, and two have been totally rewritten. Seven chapters are new to this edition. Four have been replaced. We now have 23 rather than 20 chapters. The origins of each chapter are explained in detail on pages xx-xxi.

These 23 chapters are divided into 5 broad sections. Each chapter stands alone and can be read as a separate work. Section I opens with a chapter (Statistical Concepts) on the larger concepts of statistics. That chapter surveys some of the central ideas that underlie statistical methods and techniques — ideas that guide all statistical work. These broad concepts are important even when no numbers appear in a research article: users of statistical methods should not think of numerical techniques (such as estimation or special methods of testing hypotheses) as the main ideas in statistics, while leaving the big ideas unrecognized and neglected. A new chapter (Statistical Thinking) extends the concepts in the first chapter, and illustrates with three examples how the practicalities of real life often make the uncertainty associated with statistical inferences much larger than the usual formulas for confidence limits would report. Some reasons are unrecognized influences on the data, errors in critical assumptions about relative magnitudes, and uncertainty about generalizing results in complicated situations, such as moving from data acquired from animal experiments to future human experience. The next chapter (Statistical Analysis in the *Journal*) tells how often various statistical procedures were used in the *Journal* during a two-year period. A brief follow-up notes new emphasis on some procedures. Once the big ideas of statistics are well in mind, one needs to know about applications; this third chapter on frequency of usage should provide practical guidance to persons planning a program of study, whether they are instructors developing courses or interested readers pursuing their own education.

Section II deals with a major statistical area — the design of investigations of treatment and prevention. In four volumes of the *Journal*, we found only three different basic research designs for such studies; these approaches are described and evaluated, with numerous examples, in Chapters 4, 5, and 6. More detailed comment on simple reporting of experience with a series of cases is then offered in Chapter 7 (The Series of Consecutive Cases). A broader classification of the statistical designs in all Original Articles published in those same four volumes is presented in Chapter 8 (A Classification for Research Reports). The three standard methods of investigating treatment and prevention (discussed in Chapters 4, 5, and 6) are presented together with some discussion of their difficulties and of precautions that can be taken to improve their strength.

The five chapters in Section II on the design of investigations give the reader an overview of the main methods of studying the effect of an intervention. Specialized methods that are not treated in the book, such as "balancing," have been described by Shapiro and Louis[1] and in more detail in the references they cite. Chapter 19 (The Importance of Beta), described below, also discusses the design of investigations, especially with regard to sample size.

Although an investigation must start with a study design, analysis becomes the important feature after the data are in. Section III describes some special topics in data analysis. A new Chapter 9 (Decision Analysis) explains how situations involving uncertainty and multiple options can be analyzed to develop algorithms for the treatment of single patients or to decide on programs of screening or treatment policies. Systematic analysis of situations and bringing together data from many sources, including benefits and losses, human as well as fiscal, can achieve realistic appraisals of uncertainty and ultimately improve therapeutic policies. Chapter 10 (*P* Values) discusses the meaning of *P* values, which are the usual means of expressing the results of tests of significance. *P* values are widely used but often misunderstood. The chapter explains their underlying assumptions, and why *P* values have a straightforward meaning only in the presence of likely alternative hypotheses. It is most important to understand the strengths of *P* values in terms of achieving objectivity, as well as their weaknesses in terms of

decisions or policy. Therefore, this chapter deals with both uses and mis-
uses. This chapter was originally based on Chapter 7 of *Biostatistics in
Clinical Medicine*,[2] but the medical literature has reported so much discussion about *P* values since the first edition that Chapter 10 has been completely rewritten.

Section III then turns to five specific sets of techniques. Two of these deal
with major problems in statistics — simple linear regression (Chapter 11)
and comparisons of several group means (Chapter 12). They make clear
that investigators must have in mind specific questions about a set of data
before they can make a rational choice of analytic methods. Both chapters
discuss the problems that commonly occur and the steps needed to take
account of them. They are intended to help investigators to look below the
surface of a mechanical recitation of numerical results.

Categorical or classification data are widely used in biomedical work,
but the appropriate methods of analysis depend on whether the categories
have a natural ordering (Chapter 13, Ordered Categories). When ordered
categories are studied by means appropriate for unordered data, investigators may lose large amounts of information. Chapter 13 gives some notion
of the effect of ordered categories and explains a little about the general
methods used to handle order. The authors of this chapter discovered
during their study that some of the popular computer packages used to
exploit the additional strength of order contained errors. The authors have
communicated with the key workers in computer software and believe these
problems have been corrected.

The increased use of survival analysis in the clinical literature has caused
us to add a discussion of the analysis of failure-time data (Chapter 14,
Survival Data). Such analysis must account for the fact that not all subjects
in an investigation will have experienced the key event such as death or
stroke by the time the analysis must be made. The widely used Kaplan–
Meier method of estimating survival distributions and the Cox proportional-hazards model are explained.

Our update of Chapter 3 showed that many *Journal* articles use contingency tables to describe patients under study and to analyze the consequences of treatment. Thus a new Chapter 15 (Contingency Tables)
explains the notions behind the 2×2 contingency table, including odds
ratios, Fisher's exact test for contingency tables, and the paradoxes that arise
when tables are collapsed. The much used technique of logistic regression,

including multiple logistic regression, is introduced to extend the ideas of regression to situations where the outcome variable (a proportion) must lie in the range from 0 to 1.

Once an investigation has been executed, the results must be conveyed. Investigators may be helped by five chapters (16-20) in Section IV on reporting results. When faced with the masses of numbers generated by any large quantitative study, one must consider what part of the background and results to present. Chapter 16 (Guidelines for Statistical Reporting) gives the investigator some general ideas about what to offer readers and what to keep in one's notebooks. The chapter expands on the brief statistical guidelines given as the *Uniform Requirements for Manuscripts Submitted to Biomedical Journals* distributed by the International Committee of Medical Journal Editors. It gives advice about 15 specific issues that frequently arise in preparing a clinical paper containing numerical data.

Details of what to publish may vary a lot from one kind of study to another. Chapter 17 (Reporting on Methods) presents an empirical study of whether certain items were reported in a set of 67 papers on randomized clinical trials published in four leading general medical journals. Of 11 simple, important aspects of statistical design and analysis in each of these papers, only about 56 percent were clearly reported. A second problem facing scientists is how to get the most out of their statistical consultants. As explained in Chapter 18 (Statistical Consultation), effective collaboration with a statistical consultant requires good communication in both directions. The statistician must be an integral part of the scientific team, and other team members must understand and contribute to the common statistical effort. The authors discuss how to facilitate efforts in both directions.

Because statistical power is not widely understood, we asked Freiman et al. for permission to republish as Chapter 19 their landmark paper on power (The Importance of Beta), which appeared in the *Journal* in 1978. Besides explaining the idea, this chapter shows that many published papers with low power may have missed sizable medical improvements.

Chapter 20 (Writing about Numbers) describes some common but easily avoided perils to those whose experience is primarily in working with words rather than numbers. It offers some conventions and rules about reporting numerical data.

Section V deals with two important, related problems, medical technology assessment and meta-analysis. The new Chapter 21 (Medical Technology Assessment) explains why assessing medical technologies is important to the clinician, the patient, and society. It describes some common problems and methods used to find out about the efficacy and safety of medical technologies and the process of judging when new drugs, devices, and procedures should be brought into the medical system.

Often the reader of the literature must assemble information about a particular topic from many papers. This assembly must go beyond the methods of the usual review of the literature to a more formal integration of quantitative information from different reports. Such formal quantitative reviews are called meta-analyses. Chapters 3, 12, 17, and 19 all contain meta-analyses. Each quantitatively combines the results from several papers to summarize data and answer scientific questions. Chapter 22 (Combining Results) describes the various features of the research synthesis carried out by meta-analysts, illustrates the variety of methods used, and explains what a reader should be looking for in appraising a meta-analysis. So much has occurred in this field since the first edition that this chapter has been entirely rewritten. To give some idea of the developments in the field we have also included Chapter 23 by Sacks et al. (Meta-Analyses of Randomized Control Trials), completely revised and based on new data. This article emphasizes some of the methods being used and gives the reader some notions of the quality-scoring used to appraise progress in clinical investigations.

Most chapters provide references to additional material on how to carry out statistical techniques. One of the many useful and practical introductions to statistical methodology for the clinician is the textbook by Ingelfinger et al.[2]

John C. Bailar III
Frederick Mosteller

REFERENCES

1. Shapiro SH, Louis TA, eds. Clinical trials: issues and approaches. New York: Marcel Dekker, 1983.
2. Ingelfinger JA, Mosteller F, Thibodeau LA, Ware JH. Biostatistics in clinical medicine. 2nd ed. New York: Macmillan, 1987.

MEDICAL USES
OF STATISTICS

2nd Edition

SECTION I

Broad Concepts and Analytic Techniques

1
~

STATISTICAL CONCEPTS FUNDAMENTAL TO INVESTIGATIONS

Lincoln E. Moses, Ph.D.

ABSTRACT   Statistics is a body of methods for learning from experience. Clinical research often draws on statistical methods, and a good understanding of their rationale is therefore important for clinicians as well as research investigators. This chapter examines the underlying logic of statistical methods as applied to clinical research. The discussion focuses on four key concepts: operational definition, the precise specification of terms and procedures; the infinite-data case, a way of considering what conclusions might be reached if the study were so large that statistical variation was negligible; probabilistic thinking, which focuses on the resemblance to be expected between the study's outcome and what an infinitely large study would show; and induction, the process of reaching conclusions about future cases on the basis of the data in the present study. The design of an investigation needs to take these concepts into account. The publication reporting the study should disclose fully and clearly how the study was done and analyzed and how the authors interpret the results.

S tatistics may be defined as a body of methods for learning from experience — usually in the form of numbers from many separate measurements showing individual variations. But because many qualitative matters of clinical interest, such as alive or dead, improved or worse, and male or female, can be presented as counts, rates, or proportions, the scope of statistical reasoning and methods is surprisingly broad. Nearly all scientific investigators find that their work sometimes presents statistical problems that demand solution; similarly, nearly all readers of research reports find that understanding a study's reported results often requires an understanding of statistical issues and of the way in which the investigators have addressed those issues.

Even more striking than the range of clinical studies where statistical issues arise is the importance of a few statistical concepts that apply to many different types of studies. This chapter presents and discusses four of these broad concepts.

The first key concept is *operational definition*. To learn from experience, we must first be able to state what that experience is. Labels are insufficient for this purpose. "Stage II disease" can have different meanings in different clinical settings. "Suicide" rates are likely to be very different in jurisdictions that do and do not demand the presence of a suicide note before applying the term. A statistic reports the outcome of some measurement process; unless we specify that process, we cannot know the meaning of the statistic. It is this kind of specification that is meant by the term operational definition.

Before they consider finite sets of data, statisticians usually find it valuable to consider what conclusions might be reached if the data set were infinitely large, so that statistical variation was negligible. In thinking about this *infinite-data case*, they pose these questions: If we had a very large quantity of data of the kind under consideration, would the data answer our questions? How would we analyze that infinite-data set to explore and reveal its meaning? Could we change some feature of the data-gathering process to make that body of data more useful or informative?

Any actual study produces only a finite body of data, which can be regarded as approximating the infinite-data set. *Probabilistic thinking,* which focuses on the closeness of that approximation, takes account of the number of observations and makes use of such statistical concepts as bias and variability. Its premise is that when the laws of probability are known to have governed the acquisition of data, then statistical inferences have the force of logical consequences of these laws.

Statistical inference, or *induction*, is ordinarily — perhaps always — a two-stage process. First, we must ask how well the data reflect what we would learn from an infinite body of data collected in the same way. What we hope to discover is how greatly chance may have distorted the resemblance of our finite-data set to its corresponding infinite-data set. The issue raised by this question is sometimes labeled *internal validity*. A second question follows: If the data had been collected instead in a somewhat different way (e.g., by including patients younger than 55, by considering patients from community hospitals as well as teaching hospitals, or without excluding patients with diabetes), how closely might the data from our sample resemble the infinite-data case corresponding to such a modification? This question raises the issue sometimes labeled *external validity*. Internal validity is primarily a statistical issue; external validity can be evaluated only with the help of expertise and judgment in areas outside statistics.

The next four sections of this chapter discuss these four key concepts in detail and provide examples of their importance in clinical research. Two sections follow, examining the effects of all four concepts on study design and statistical reporting.

OPERATIONAL DEFINITION

Many medical investigations follow a characteristic pattern: the investigator imposes one or more treatments on subjects of certain kinds, observes and perhaps compares outcomes, and then tries to reach conclusions about the effects of the treatments. The specific meaning of such a study grows out of the answers to a host of questions about the patients, the treatments as actually applied, the outcomes, and how the outcomes were assessed. For these answers to be accurate, they must faithfully take into account the actual procedures used in the study, and they must be precise and specific.

DESCRIPTION OF TERMS

Reports of laboratory investigations typically include specific accounts of equipment, procedures, and materials. An operational account of a clinical investigation is equally necessary, but often more demanding. A statement that patients have "disease A, Stages II and III" sounds definite enough, but the diagnosis of disease A may be somewhat tricky. We need to know how that diagnosis was made. What criteria were applied? How were the patients assessed? Were all cases assigned stages by the same person, team, or committee? If not, then how were disease stages determined? How reproducible is the staging? For instance, were any cases staged twice and blindly? If laboratory or microscopic confirmation was required, what is the effect of leaving out subjects with disease A who did not have that confirmation?

Measured characteristics present analogous demands. "Cardiac output" may be one thing if measured by angiography, another if assessed from blood gases. Blood pressure can vary greatly, depending on the state of the subject, the person who measures, and the device used. It is important to know how a measurement was made and whether the same method was used for all subjects. An often useful way to dispel ambiguity about the measurement of some elusive yet important variable is to employ a stand-ard well-known instrument for the purpose; examples include the New York Heart Association Index of Cardiac Function, the Karnofsky Scale for disability in cancer patients, and the Mini Mental State Examination for cognitive function in the elderly.

Treatments are often not what investigators believe and intend them to be, and careful operational definition can require subtle distinctions. Drug A administered by mouth in a syrup also includes the syrup. (A series of deaths in the early days of sulfanilamide treatment attests tragically to this fact.[1]) A medicine prescribed is not necessarily a medication actually used. The analgesic pill has both its active ingredient and its function as a placebo to relieve the patient's pain. An office procedure comprises both the proce-dure and the visit, with whatever effects on well-being each may entail. Here we see highlighted the need for carefully devising (and operationally defining) any control treatment.

PHASES IN A STUDY

A comparative trial of treatments typically comprises several sequential phases: determination of a patient's eligibility for the study, the patient's entry into the study, assignment of treatment, the care itself (using the as-signed treatment and any adjuvant treatments), evaluation of the patient's outcome (perhaps after a follow-up interval), statistical analysis of the data (including the information on this patient and others) and reporting. Fair comparison of treatments can be difficult if at any of these phases knowl-edge of the treatment assigned to the patient influences other aspects of the process. Thus, if the decision to enroll each patient in the trial is made with the knowledge of the treatment the next patient will receive, then the op-portunity for constructing noncomparable treatment groups is ample. If evaluation of subjective end points is made by observers who know which treatment the patient received, then another potential source of bias exists (hence the value of double-blind studies). Different follow-up periods for different treatment groups may conceal some mortality (or longevity), to the advantage of one treatment or the other. Careful planning of the processes at each phase can reduce the risk that knowledge of treatment assignment or results may lead to contamination of the conclusions.

INTEGRITY OF OPERATIONAL DEFINITION

When we move from studies where the investigators impose treatments to those where they simply observe different groups or similar groups in different epochs, the problems are likely to be markedly more difficult to solve. For example, the record may not always contain sufficient information for the operational definition of crucial matters. In such a case, the conclusions of the study rest heavily on assumptions about the undefined terms and procedures, along with data about those that are adequately defined.

Concern for the integrity of operational definitions leads investigators to take important precautions in well-conducted studies. Identification of disease stages and laboratory analysis may be checked by introducing, blindly, occasional standard specimens. Samples of study records may be checked against clinical records. Visits and audits by personnel from a center that is charged with responsibility for quality control may be routinely conducted in multicenter studies. All such steps have the purpose of ensuring the proper operational definition of patients' characteristics, treatments actually applied, evaluation of outcomes, and record-keeping methods.

THE INFINITE-DATA CASE

In the planning phase of a study, few questions are more useful to consider than this: What could we learn from an unlimited amount of data obtained in the same way that we are planning to obtain ours in this study? Careful consideration of this question can lead to dropping a study, to improving it, or simply to clarifying issues of procedure and analysis, as in the examples discussed below.

APPROPRIATE SUBJECTS, CONTROLS, AND DATA

In earlier days medical students were sometimes used as volunteers to assess the risks of side effects from prospective new drugs. If a drug is intended to treat a disease affecting mainly elderly patients, and if younger subjects are expected to be used in a clinical trial, this question should arise before the study begins: What could unlimited data about the responses of healthy 25-year-olds tell us about the incidence of, say, nausea and vomiting in 70-year-old sick patients who will take this drug? The question is a good one to pose before data collection begins. Even though the answer to the question may be obscure, its obvious importance may lead to changing the investigational approach. Thus, giving thought to the infinite-data case can clarify what groups of subjects are appropriate for what aspects of the study at hand.

Such thinking can also help to define appropriate controls. In one case, a promising method for directly dissolving a clot in patients during the first two hours after a heart attack was to be studied. The initial proposal was to apply the new method in all eligible patients and to use as controls those patients who arrived more than two hours, but less than eight hours, after a heart attack; this control group would be treated with current standard therapy. An infinite supply of data gathered in this way could at best resolve whether it was better to receive the new therapy within two hours or the standard therapy after two hours. Not even an infinitely large study could determine whether the new method was better than the standard one, either in the first two hours or in the next six.

Even an infinitely large study will not provide information about questions for which data are not collected. Thinking about analyzing the data as if they were already in hand and infinitely abundant can point both to unnecessary information that should not be collected and to key items of information that must be gathered.

STATISTICAL RELATIONSHIPS AND REGRESSION

Laws of physics such as Ohm's Law, Newton's Laws of Motion, and Einstein's famous $E = mc^2$ allow one to calculate exactly the value of one variable that must accompany the stated value of another. But in medicine and everyday life such relationships are rare; instead we see "statistical relationships" that may hold true on average, but not case by case. Thus, tall people tend to be heavier than short people; older children tend to be taller than younger ones. Higher doses of drug usually produce larger effects. A useful way to make this idea of a statistical relationship more amenable to quantitative treatment is the concept of regression. Think of two variables x (dose) and y (response). We define the regression of y upon x to be the curve that depicts at each value of x (dose) the *average* value of y (response) for those elements of the population having that value of x (receiving that dose). Now, though individual variability still attends the pair of variables x and y, there is a well-defined single curve relating the average of one variable to stated values of the other.

This idea of regression is far reaching, and has broad applicability. Generalizations to more than two variables lead to the concept of *multiple regression*.

THE LIMITS OF ASSOCIATIONS

An observational study with infinite data can definitely demonstrate the presence of an association between two variables, such as lung cancer and smoking, without resolving questions of cause and effect. For example, an extensive study of adult men might show strong and roughly equal positive associations between height and weight and between girth and weight. We must draw on other information to support the proposition that by increasing the weight of a man we will increase his girth but not his height. To establish cause and effect typically demands recourse to knowledge outside the particular study.

When an experiment is carried out, treatments are imposed and subsequent events are followed; these procedures make causal inference much more direct, but dependence on outside knowledge is unlikely to be wholly absent. The point would be quickly illustrated by an experiment in which subjects were given large drinks of whiskey and water, rum and water, or brandy and water and all showed signs of intoxication. It is "outside knowledge" that supports the conclusion that the effect was not due to the "common factor," water.

CONFOUNDING VARIABLES

In a study discussed earlier in this section, the time elapsed after a heart attack and the method of therapy were confounded. Two variables are said to be confounded in a study if they appear in such a pattern that their separate effects cannot be distinguished. A common, often subtle, and sometimes ruinous form of confounding occurs when the personal choice of a patient (or physician or other key participant) can affect either side of a treatment comparison. The polio-vaccine trials (involving 2 million children) provide a surprising illustration. In that study the incidence of polio was clearly lower among unvaccinated children whose parents refused permission for injection than among children who received the placebo after their parents gave permission.[2] As it turned out, families who gave permission differed from those who did not in ways that were related to susceptibility to poliomyelitis.

Personal choice also acts as an enemy of easy inference in questions of drug compliance. Studies with clofibrate[3] showed that subjects who took 80 percent or more of the prescribed dose had substantially lower mortality than subjects with poorer drug compliance; this evidence seemed to indicate that the drug was beneficial. But the same difference in mortality was observed between high- and low-compliance subjects whose medication was the placebo. Drug compliance, a matter of personal choice, was for some reason related to mortality in the patients in this study. Had there not been a placebo group, the confounding between the quantity of the drug actually taken and unknown factors related to survival might have gone unnoticed, and the reasoning "more drug, lower mortality; therefore, the drug is beneficial" might have gone unchallenged. As these examples suggest, consideration of the infinite-data case, before the study begins, should include efforts to identify points where personal choice may be confounded with variables under study.

"EXHAUSTING EXPERIENCE"

To think about the infinite-data case is to consider what could be learned from an infinitely large study of the kind contemplated. A closely related question is what could be learned by exhausting experience of the kind that the study will sample. Occasionally, this exhaustion of the data would involve only a finite set of observations. Thus, a sample of the current opinions of pediatricians in the United States on confidentially furnished contraceptive information for teenagers corresponds not to an infinite-data case but, rather, to the finite collection of the opinions of all pediatricians in the country on this topic. We could avoid concern about such special cases by speaking of the *all-possible-data case*. Some statistical writings use the term *population* to capture the ideas discussed here under the rubric of the infinite-data case.

A sometimes troublesome point is illustrated by the following example, which deals with motorcycle accident fatalities and helmet laws.[4] In Colorado, in the period 1964 – 1968, when the state had no law requiring helmets, there were 74 fatal motorcycle accidents (an annual rate of 6.3 per 10,000 registered motorcycles); in the period 1970 – 1976, after the enactment of a helmet law, there were 248 such accidents (an annual rate of 4.6 per 10,000 registered motorcycles); in the period 1978 – 1979, after the helmet law had been repealed, there were 137 deaths (an annual rate of 6.1 per 10,000). Since these figures include all the fatal motorcycle accidents in Colorado during those years, should we regard this information as itself exhausting experience?

Most statisticians would say no. They might say, for instance, that the deaths observed in these periods could be thought of as random outcomes of complex probabilistic processes; we happen to have relatively brief peeks at these processes; each death rate we have observed indicates the average level of risk per registered motorcycle in its period, but we must doubt that any of them exactly reflects that average risk. There is clearly a role for the concept of an infinite-data case in thinking about this problem. By observing indefinitely long periods (under unchanging conditions) *with* a helmet law and also *without* a helmet law, we might in principle learn the exact relationship between a helmet law and the risk of motorcycle accident death. Our actual finite data tell us, uncertainly, about that infinite-data case and about the actual level of risk in the three periods observed.

Overall, then, thinking about the data to be acquired in a study as if they were already in hand and as abundant as desired can identify problems, opportunities, and fruitful questions early enough to help most studies and profitably to abort some.

PROBABILISTIC THINKING

When unpredictable variation is large enough that it may affect conclusions, then probabilistic thinking is likely to be helpful. In principle, the
laws of physics plus a lot of elaborate instrumentation could permit us to
treat the result of rolling a die as a deterministic matter, but at our usual
practical level of analysis, that outcome is a chance matter to be regarded as
probabilistic. Two similar patients with the same disease may have different
outcomes. Perhaps that, too, is in principle deterministic (though much
more complex), but at our usual level of analysis it is better regarded as
probabilistic.

With an infinite number of observations we would learn, in the problem
with the die, the probabilities that it would come to rest showing 1, 2, 3, 4, 5,
or 6. (If it is a fair die, these probabilities will all be one sixth.) In more complicated situations, the infinite data would show the probabilities of more
complex kinds of outcomes. Thus, if two diagnostic tests, A and B, are used
and the outcomes measured are survival to one year or death in the same
period, then the infinite-data case would answer all questions of this form:
What is the probability of survival to one year for patients with an A score
between a_1 and a_2 and a B score between b_1 and b_2? With more variables
under study, the description of the possibilities grows rapidly more complex. The actual finite data can at best provide approximate answers to questions that could be answered precisely from the infinite-data set. A principal
objective of probabilistic thinking is to appraise the closeness of that approximation by drawing on the finite data themselves for the appraisal.

Attention typically is focused not on the entire probability distribution
but on particular aspects of it. This idea becomes more concrete if we consider, briefly, some especially important elements of probabilistic thinking.

SAMPLE MEANS AND STANDARD DEVIATIONS

A statistic is a number computed from the observations in a sample. The
sample mean (the familiar average, learned in grade school) is a statistic
that tells about the "general size" of the sample's observations. Different
samples drawn from the same infinite-data case will have somewhat different sample means, so any one sample mean must be thought of as only a
probabilistic approximation of the mean that would be found if the full in-

finite-data case could be examined. How closely the sample mean approximates the infinite-data mean is a major concern of probabilistic thinking.

The standard deviation is a statistic that describes the degree of variation among the individual observations in the sample. If all had the same value, the standard deviation would be zero; the farther apart from one another (and from their mean) the individual observations are, the larger the standard deviation is. If the standard deviation of some sample is very small, then the sample average closely represents every individual value, whereas a large standard deviation tells us that this is not so. As a general rule, when random samples have small standard deviations, then sample means are more likely to be close to the all-data mean than when standard deviations are large. This principle points to the key role of the standard deviation in probabilistic thinking. It is an intuitively appealing principle; it says that if individual random observations are in close agreement, their average is likely to serve as a good estimate of the all-data mean.

RANDOM VARIATION AND THE SIZE OF SAMPLES

It is a mathematical fact that in all cases of practical interest the differences among random samples of size n from a single population tend to be smaller when n is greater. Moreover, many important sample statistics from random samples are also increasingly similar with larger sample sizes. Therefore, the random variation of statistics can be reduced by using larger samples, so long as the processes of acquiring the data are not degraded in the effort to enlarge the sample. (There are important practical issues here. Doubling the sample size by allowing a particle counter to operate for twice as long is one thing. But doubling the sample size of a clinical study by extending its intake period from two years to four years may be quite another: quality control may be harder to maintain, changes in personnel and shifts in the patient population may be more likely, the treatment itself may change, and so on.)

BIAS

If all our observations use one instrument and it is out of adjustment, so that all readings are too high by four units, then our data have a bias of +4 units. A large sample size will neither increase nor decrease this bias; with sufficiently large samples, this bias will stand out (falsely) as a numerical fact having almost no sampling error. We can think of bias as a numerical discrepancy between the mean of some statistic from our intended infinite-data case and the mean for our actual infinite-data case. Of course, no one intends to use an instrument that gives readings four units too large, but if that happens, then the actual infinite-data case will incorporate this bias, and the sample data will approximate the biased figure.

Bias may be the most difficult problem in quantitative research. It can enter a study in many and subtle ways. Articles have been written on biases that affect clinical research.[5] The use of treatment and control groups can help sometimes; if all readings are four units too large (though not known to be), then in the difference between the mean for the treatment group and the mean for the controls, the bias cancels out. On the other hand, if the treatment group's measurements were made on instrument I and the control group's measurements on instrument II, any instrument bias would be confounded with treatment; large sample size would not reduce this bias.

Bias can enter not only through equipment but in follow-up, identification of disease stages, treatment, history taking, record keeping, and patient responses. Patients referred from different sources often differ materially, and if different referral streams enter different treatment groups then comparisons of treatments will be biased. Alertness on the part of the investigators and symmetry between treatment groups in all operational respects are the primary weapons for combating bias (hence the value of randomization and blinding). Bias must be fought in planning the study, in its execution, and in the analysis of the data.

INDUCTION

The very notion of learning from experience carries with it the idea that the experience comes from one set of instances and the learning is to be applied to other (often future) instances. This is the philosopher's problem of induction. Probabilistic thinking helps us to approach this problem. Sometimes it is possible to draw a sample of n individuals from a population so that every possible sample of size n has the same probability of being chosen; this is called a simple random sample. The laws of probability are directly applicable and enable us to make definite statements about the whole population on the basis of the statistics computed from the one random sample we have drawn. (More complicated schemes of sampling often have powerful advantages, but we don't treat them in this chapter.)

INFERENCE FROM A PROBABILITY SAMPLE

Two things need to be said about these inferences. First, they necessarily take a probabilistic form, such as "It is a 99-to-1 bet that the population mean lies between 20.6 and 21.3" or "It is a 19-to-1 bet that the population mean is negative." Other forms for stating statistical inferences exist; they also include a numerical statement about the population or infinite-data case and quantification of the degree of confidence with which that statement can be made. Second, these inferences depend for their integrity on the actual imposition of the probabilistic mechanism (e.g., a table of random numbers) whose mathematical properties then provide the inference. If two treatment groups have not been constructed as random subsets of a single group of eligible patients but instead have simply been found to be similar in some ways, then one may still go through the motions of statistical methods, but the conclusions are no longer logical consequences of the laws of probability. They are less direct, dependent on ad hoc assumptions, and altogether less reliable.

The Health Interview Survey of the National Center for Health Statistics exemplifies probabilistic reasoning from a finite sample (of about 40,000 households annually) to the whole population of the United States. That survey uses *probability sampling,* a modest extension of random sampling; all possible samples of size n have known (but not equal) probabilities of being drawn, and the statistical analysis uses that information.

A random or probability sample from a population is often subdivided, for instance, into males and females or smokers and nonsmokers; these subsamples are also random or probability samples of the males in the population, of the females, and so on.

COMPARISON OF DIFFERENTLY TREATED GROUPS AS FOUND

Often statistical inference is attempted under circumstances that are much more adverse than comparing two subsamples of a random sample from a population. Results from a therapy used in recent years are frequently compared with results from the therapy used earlier (*historical con-*

trols; see Chapter 6 of this book). Or results from operation I applied in hospital A are compared with results from operation II applied in hospital B. Inference is difficult and hazardous in such cases because many influential interfering variables may be systematically different in the populations furnishing the data. In principle, the difficulties could be reduced — or perhaps even eliminated — by successfully carrying out three steps: identifying the interfering variables; finding the values of these variables in the two samples; and adjusting appropriately for the values of these interfering variables.

Unfortunately, it is rare for an investigator to have an effective command of any one of these three steps. Indeed, identifying all of the important interfering variables alone is likely to be a more complex issue than whether operation I is better than operation II. Attempting to make the measurements required to find the values for the variables may greatly complicate and burden the study, increase costs, and multiply opportunities for error. Adjusting for interfering variables depends heavily on external information that may, in fact, not be available. Successfully carrying through these three steps is unlikely to be feasible; hence, valid comparison of differently treated groups is typically difficult, though sometimes possible.

Clarification of the relation between lung cancer and cigarette smoking, for instance, was achieved through a host of studies stretching over years. For each study that found an adverse effect of smoking, it was possible to suggest biases that had not been controlled, thus casting doubt on the conclusions. By 1964 so many kinds of studies had been done — some free of one kind of bias, some of another — that consensus was reached that heavy cigarette smoking increased the risk of lung cancer. (Ultimately, the increase in risk was recognized to be about 10-fold.) This example demonstrates that induction in the absence of an applicable probability model is possible, but that in those circumstances it can be difficult, slow, and hard to defend against even inappropriate criticism.

COMPARING TREATMENTS IN RANDOMLY CONSTITUTED GROUPS

Symmetry was mentioned earlier as a device for controlling bias. The randomized study enforces symmetry by taking a suitable group of subjects and dividing them at random (e.g., with a table of random numbers) into two subgroups, one to receive treatment I and the other to receive treatment II. What does the randomization achieve? It ensures that (except for treatment) both subgroups have the same infinite-data case. Both are in this sense alike with regard to all relevant "other" variables — even though some of the important ones might be unrecognized and even though none may have been measured. The randomization does not ensure that the two subgroups will themselves be identical (any more than any two finite random samples from the same population will agree with each other or with their infinite-data case); rather, it ensures that no bias favoring one treatment or the other can operate. Equally important, randomization leads to a precisely quantified measure of the uncertainty arising from the differences that do occur between subgroups. Because of this, the randomization rigorously justifies statistical inference about the comparative efficacy of the treatments in the group of subjects studied. The method forgoes attempting to untie the Gordian knot of interfering variables but, instead, cuts through it in one stroke.

Readers of the Surgeon General's 1964 report, *Smoking and Health*,[6] will find among the 392 references cited in the chapter on cancer seven key observational (nonrandomized) studies published in 1954 (their references 38, 84, 138, 158, 175, 268, and 365). In that same year, 1954, the polio-vaccine trial, a randomized double-blind placebo-controlled experiment, definitively demonstrated protection by the vaccine, which cut the polio rate by a factor of about 2.5. That effect, much smaller than a factor of 10, was established once and for all by the study. When the randomized experiment is feasible, it can be very powerful indeed.

Not all studies can be done by assigning treatments to randomly constituted subgroups. First, conditions to be investigated may not be assignable at all. For example, people choose whether or not to smoke, and they are born with attributes such as their blood type. Second, there may be ethical obstacles; if patients can definitely be expected to benefit more from treatment I than from treatment II, that should bar their being assigned to treatment II. (Thus, the initial "suitable group" should consist of patients for whom both treatments are equally appropriate, insofar as is known.) Third, resources may be inadequate. Because irregular practices or systematic mistakes can make an experiment meaningless, great care is given to quality-control measures in clinical trials, and care often entails expense.

INTERNAL AND EXTERNAL VALIDITY

The sample survey using probability sampling provides specific inferences about the whole population at the time of the survey; these conclusions are logical consequences of the laws of probability, and we say that
they have high internal validity. If we apply these conclusions next year or
to a somewhat different, though similar, population, we move away from a
statistical inference to a different kind of inference, one that draws on our
knowledge of the phenomena under investigation — smoking habits, in-
come, health status, or other factors. In this case, we must be concerned
with the external validity of the conclusions.

These two kinds of inference (first, from the data to the corresponding
infinite-data case and, second, from that infinite-data case to wider situations) arise not only with sample surveys but in clinical studies. The 1954
polio trial involved randomized comparison of placebo and vaccine in children in grades one, two, and three. Generalization of its conclusions to
older and younger children belongs to virology and immunology, not to
statistics.

How a study is carried out can affect its external validity. One study
might compare two treatments in a large body of patients in a narrow age
range in a particular large city. A different study, using the same number of
patients, might compare those two treatments in several locations and
might use wider age limits. The second study would pose fewer problems
of external validity, though this gain would be won at a cost in complexity
of statistical analysis; the second study, moreover, would require additional
quality-control measures to ensure compliance with the protocol and adherence to standards. Investigators often must balance the demands of internal validity, including narrowly selected groups of subjects for study,
against the demands of external validity, where broad representation of potential future subjects may be important.

In summary, induction is the process of reaching conclusions from one
set of instances that can then be applied to other instances. The relationship can be especially direct if a probability sample is drawn from a population with the aim of describing the whole population. But even in such a
straightforward situation, interpretation may be difficult, simply because
the infinite case itself could rarely, if ever, settle questions about cause and
effect. Comparison of treatments in differently established groups raises

thorny questions, both theoretical and practical; inference in such cases is likely to be error-prone, difficult, and slow. Comparing treatments in randomly chosen subsets of a suitable group of subjects ensures that, except for treatment, both groups have the same infinite-data case; thus, bias, even from unrecognized interfering variables, is ruled out. If the suitable group of subjects is defined too narrowly, problems of external validity may prove troubling.

STUDY DESIGN

Statistical considerations merit attention at the time of planning a study fully as much as they inevitably claim attention at the time of analysis. Four statistical issues deserve particular care during study design: preventing bias, ensuring efficiency (a high information yield per observation), ensuring the integrity of the process (quality control), and determining the size of the sample.

BIAS

Prevention of bias requires identifying possible interfering variables and then ensuring that their uncontrolled effects will not favor one treatment or another. Such variables must first be listed. Then any variable on the list can be coped with in one of two ways: its effects can be controlled — for example, the design might call for an equal age distribution in each treatment group; or its effects can be made symmetrical through randomization (which may or may not be followed by statistical adjustments, perhaps by regression methods). The key is to prevent any influential variable from becoming confounded with the effects of treatment.

EFFICIENCY

A subfield of statistics, called statistical design of experiments, is largely devoted to devising ways to collect data that reduce statistical variation. One device exploits knowledge about interfering variables by applying regression methods (analysis of covariance). Another chooses closely similar subsets of the eligible population within which to compare the treatments; comparison of treatments in identical twins epitomizes this approach. The crossover design goes further, using both treatments in every subject, but results may be invalid if certain conditions do not apply to the physiologic processes under study.[7] Statistical design of experiments has other devices in its kit as well, and may produce substantial economies in time, money, or other resources; it can also produce substantial improvements in both internal and external validity.

INTEGRITY OF THE PROCESS

Ensuring that the operational definitions really do apply in the data-collection process is necessary for a successful study. The process of operational definition, discussed earlier in this chapter, must be planned, and its execution needs to be provided for in the study protocol.

SAMPLE SIZE

How many observations should we make? People often bring this question to a statistician (usually, fortunately, before the study begins). For this question to be answered, more detail is needed; it resembles the question, How much money should I take when I go on vacation? One needs to ask, How long a vacation? Where? With whom?

Three questions must be answered before the sample size can be determined: How variable are the data? How precise an answer is required? How much confidence do you want to have in the answer? These questions are worth asking even if the sample size will be determined by the budget or the time available to do the study. Sometimes a planned study is dropped because sample size analysis shows that the study has little chance of providing a useful answer under the existing constraints of time or budget. Thus, if eligible subjects must have a disease D, with low prevalence, it may not be realistic to undertake the study if, say, 20 years would be needed to accumulate the required number of cases of this rare disease.

Another place that prevalence enters into consideration at the stage of study design is in screening. If one wishes to identify cases of disease D by applying a test to unselected members of the population, the practicality of the enterprise depends on the relation between the test's false-positive rate, r, and p, the prevalence of the disease in the population. A little reflection shows that if the prevalence p is 1 in 1000 and the false-positive rate r

is 1 in 100, then there will be about 10 false positives turned up by the test for each true positive turned up. Those testing positive then are 10 to 1 bets *not* to have the disease. The time to consider this possible difficulty is at the outset — the design stage. If p were 0.15 and r were 0.01 that same test might be quite serviceable, since true positives would outnumber false positives (among those who had positive tests) by about 15 to 1.

STATISTICAL REPORTING

Good statistical reporting aims for full, clear disclosure of what was done, what resulted, how the authors interpret the results, and why.

DESCRIBING THE DATA ACQUISITION

The operational definitions of all terms should be clear. The reader should not be left with doubts about how the subjects were selected, how they were assigned to treatments, what treatments were applied, how outcomes were measured, or in what order these steps were taken. DerSimonian et al.[8] reviewed reports of clinical trials published in four leading medical journals. Their criteria for full reporting required that the papers answer these six questions: What were the eligibility criteria for admission to the study? Did admission to the study occur before treatment was allocated? Was treatment allocated at random? What method of randomization was used? Were patients blind to the treatment they received? Were outcomes assessed by persons who were blind to treatment? All these are crucial in presenting the results of clinical trials.

REPORTING THE DATA

The data should be summarized clearly in tables, in words, in charts (frequently it is desirable to use all three). Missing data and extreme observations, both those retained and those excluded, should be explicitly reported. It is desirable to publish the raw data, though often there are obstacles — such as editorial policies — to doing so. At a minimum, the raw data should be carefully labeled and filed, not melded into the next analysis or merged with new cases as they arrive. If raw data are not published, they can sometimes be made available through archives — perhaps locally, through a university library or computer center, or nationally, through the National Technical Information Service, for government-supported studies, or through such an agency as the National Auxiliary Publication Service or the Interuniversity Consortium for Political and Social Research, which, despite its name, now reaches into the fields of health and medicine. Good reporting will also include an explanation of the quality-control measures applied, the methods used for follow-up, and the auditing measures employed.

DESCRIBING THE ANALYSES

The reader should be told not only what was done and how, but also what happened. Summary statistics should have the aim of revealing information to the reader. The methods of statistical analysis should be explained. The best way to do this is topic by topic — the same way that the analysis is usually carried out. This kind of reporting promotes understanding, specificity, and incidentally, ease of writing. Sometimes a published paper lists statistical procedures in the Methods section — for instance, "We used chi-square, the t-test, the F test, and Jonckheere's test." This style of reporting, if applied to the recipes used at a banquet, would list all the ingredients in all the dishes together and report the use of stove, mixer, oven, meat grinder, eggbeater, and double boiler.

QUESTIONS OF TWO KINDS

In addition to showing the data or generously detailed summaries of them, the statistical analysis should state each of the principal questions that motivated the study and discuss what light the data shed on those questions. (This is not the same thing at all as reporting just the results that are statistically significant.) To lend clarity to both significant and nonsignificant results, it is wise to use confidence intervals when feasible and to report the power of statistical tests.[9] Interesting statistical results that arise out of studying the data (rather than out of studying the principal questions that motivated the study) are necessarily on a different, and somewhat ambiguous, logical footing. A study may enable the investigator to look at scores or even hundreds of interesting questions involving different subgroups and kinds of response. Investigating all these questions, either systematically or implicitly, by careful scanning, will inevitably turn up some "interesting outcomes." But this inevitability serves as a warning that the particular interesting findings may not hold up under repetition. It is usually wise to regard such outcomes with considerable reserve, more as hypotheses than as established facts. It is especially important to be candid about the nature and amount of "data dredging" that has accompanied the analysis.

Any doubts about this message may be quelled by considering a thought experiment. In the first phase we might assess a person's literacy in English by asking that he or she spell any five-letter English word. In the second phase we might program a computer to use a random number generator, and produce a succession of letters of the alphabet chosen with probabilities for the various letters that correspond to their frequencies in written English. Such a text will surely contain some correctly spelled five-letter English words. (The sample at hand as this is being written shows "sites" as the 148th through 152nd letters.) It would be foolish to attribute any meaning to such a result. In a parallel vein, if only chance is at work one must expect among 1000 significance tests to find 50 (5% of 1000) to be of a size judged as significant at $P \leq 0.05$. Thus, post hoc "significant findings" demand candor when they are reported and great reserve when they are interpreted.

Good statistical reporting can help users address the issue of external validity. The Hypertension Detection and Follow-up Program Cooperative Group reported comparative five-year mortality rates under two treatment regimens for patients with high blood pressure.[10] In a second paper,[11] the group explored how their basic finding held up in subgroups defined by age, by race and sex, and by years of follow-up. Such information can help readers to gauge the probable applicability of the study's basic result, that stepped care resulted in better survival experience than referred care.

Finally, good statistical reporting includes a discussion of the limitations and possible weaknesses of the study; this step can help to forestall unsound critical attacks on a study and to focus attention on any real problems that deserve further consideration.

REFERENCES

1. Sollman TH. A manual of pharmacology and its applications to therapeutics and toxicology. Philadelphia: WB Saunders, 1957:129.

2. Meier P. The biggest public health experiment ever: the 1954 field trial of the Salk poliomyelitis vaccine. In: Tanur JM, Mosteller F, Kruskal WH, et al., eds. Statistics: a guide to the unknown. 3rd ed. Pacific Grove: Wadsworth & Brooks/Cole, Advanced Books & Software, 1989:3-14.

3. Coronary Drug Project Research Group. Influence of adherence to treatment and response of cholesterol on mortality in the Coronary Drug Project. N Engl J Med 1980; 303:1038-41.

4. Krane S. Motorcycle crashes, helmet use and injury severity: before and after helmet law repeal in Colorado: Symposium on Traffic Safety Effectiveness (Impact) Evaluation Projects, May 29-31, 1981, Chicago, 1981:330 (Table 1). (Conducted by National Safety Council under contract no. DTNH22-80-C-01564).

5. Sackett DL. Bias in analytic research. J Chronic Dis 1979; 32:51-63.

6. Smoking and health: report of the advisory committee to the Surgeon General of the Public Health Service. Washington, D.C.: Government Printing Office, 1964:235-57. (Public Health Service publication no. 1103).

7. Brown BW Jr. Statistical controversies in the design of clinical trials — some personal views. Contr Clin Trials 1980; 1:13-27.

8. DerSimonian R, Charette LJ, McPeek B, Mosteller F. Reporting on methods in clinical trials. N Engl J Med 1982; 306:1332-7. [Chapter 17 of this book.]

9. Freiman JA, Chalmers TC, Smith H Jr, Kuebler RR. The importance of beta, the type II error, and sample size in the design and interpretation of the randomized controlled trial: survey of 71 "negative" trials. N Engl J Med 1978; 299:690-4. [Chapter 19 of this book.]

10. Hypertension Detection and Follow-up Program Cooperative Group. Five-year findings of the Hypertension Detection and Follow-up Program. I. Reduction in mortality of persons with high blood pressure, including mild hypertension. JAMA 1979; 242:2562-71.

11. Idem. Five-year findings of the Hypertension Detection and Follow-up Program. II. Mortality by race, sex and age. JAMA 1979; 242:2572-7.

2

SOME USES OF STATISTICAL THINKING*

John C. Bailar III, M.D., Ph.D.

ABSTRACT   Many researchers have become skilled in the application of statistical concepts and methods appropriate for describing and summarizing the uncertainty of their results given some probability model and its assumptions; P values and confidence limits are especially important in this context. However, these measures do not account for deviations from the presumed (often implicit) model or for other nonrandom sources of error or uncertainty about the findings.

The need for a broader concept of statistical thinking is illustrated by three examples of unexpected uncertainty in research. These involve studies of postmenopausal estrogens and cardiovascular disease (illustrating the role of unrecognized influences on the data), the relation between dose of a carcinogen and its effects (illustrating error in a critical assumption), and analyses of the carcinogenic potential of Tris (illustrating uncertainty in generalizing results).

Actual uncertainty and error in empirical studies may well exceed the uncertainty estimated from the data in hand and the statistical model used for estimation and testing. A substantially broader view of statistics and statistical thinking could do much to improve quantitative analysis in science.

B oth scientific conclusions and technical decisions must include realistic appraisals of the likelihood that they are wrong; it usually matters how, and how much, they are wrong; and we commonly underestimate the degree of error in both decisions and conclusions. The scientific method gives special attention to uncertainty, the word uncertainty being used rather broadly to cover all actual or possible sources of error, recognized or not, as well as the probabilities of error of different

*From the Carolina Environmental Essay Series, Institute for Environmental Studies, University of North Carolina at Chapel Hill. Modified and reprinted with the permission of the publisher.

types and sizes. This differs a little from common usage, in which the word uncertain reflects our recognition that we may be wrong in some way, as, for example, when we recognize that our information is inadequate or that we may have misinterpreted some observation. I have chosen to broaden the term to include unrecognized problems because the broader concept needs a name, and no other term seems more appropriate. Three examples will illustrate what I mean by uncertainty resulting from unrecognized problems. These examples are drawn from the fields of environmental hazards and medical research, but the concepts cover the whole range of academic disciplines, as well as everyday life.

Many scientists and users of science are already familiar with statistics in the formal development of a hypothesis and its negation (often designated *alternative* and *null* hypotheses), the calculation of *P* values and confidence limits, and their interpretation. These matters are important and indeed often critical to scientific progress, but so are many other, less widely understood aspects of statistics. These include the initial development of research questions and strategies, devising and testing detailed protocols, the optimum use of such checks on bias as randomizing and blinding, measuring and maintaining the quality of data, selecting the most appropriate methods of analysis, generalizing results to populations or outcomes not directly studied, and many other things. Training, practice, and sometimes expert consultation about these additional facets of statistical thinking can often enhance research investigations and promote progress in science.

UNCERTAINTY ABOUT CONFOUNDERS

My first example of unrecognized sources of uncertainty comes from a pair of papers published side by side in the *New England Journal of Medicine* in 1985.[1,2] They dealt with an important medical problem: the incidence of certain cardiovascular diseases in women who had been taking postmenopausal hormones versus the incidence in those who had not. One study found that the incidence in women taking hormones was about twice as high as in the control group; the other found that it was only half as high. The estimate of relative risk in each paper had rather narrow confidence limits that excluded the no-effect point. Both papers seemed to be technically sound. Both had been reviewed and approved for publication, and they even had one reviewer in common. Neither study assigned treatments to patients by randomization, but both included substantial and careful consideration of differences in known risk factors between subjects and controls.

Either study, taken alone, would have been convincing. Then why did they disagree so sharply? I do not know, but said in an accompanying editorial, "[T]he results of these studies (and by implication the results of countless other observational studies) are subject to a great deal more variability than is captured in the usual kind of statistical tests and confidence limits," and that "[s]uch problems would lead to the improper calculation of error probabilities and confidence limits."[3] In other words, our calculated P values tend to be too low, and our confidence limits tend to be too narrow, especially when applied to observational studies.

Of course, research studies commonly disagree in reported conclusions, and often in their data as well. For example, handbooks of physics often publish several different estimates of the same physical constant, with confidence limits that do not overlap. Some of this lack of agreement may come from unrecognized study-to-study variation in routine procedures or operational definitions, some from differences in the subjects or other experimental materials, and some from other sources. But if the finest laboratories of physics cannot agree on the measurement of, say, the speed of light, which has been well researched for a century or more, we should not be surprised when we find disagreement in fields of research with much less experimental control.

Much of the difficulty comes because the usual statistical measures of uncertainty, P values and confidence limits, reflect only that part of the uncertainty that can be attributed to random variability within the context of the specific (often tacit) statistical model used in the analysis. Common assumptions that may fail and lead to incorrect results are that different observations are statistically independent and that the data are normally distributed. For example, year-to-year differences may matter, but if all the measurements have been taken in a single year, this source of variation will be concealed from the investigator. Choice of the wrong model and other errors in assumptions, biases in the data, and a host of other problems generally increase "real" uncertainty and for mathematical reasons never decrease it. Thus, to the extent that we rely only on P values and confidence limits, we have too much confidence in our scientific inferences.

The effects of postmenopausal hormones on the incidence of cardiovascular disease still have not been fully elucidated, but may reflect the substantial reduction in dose between the time of treatment of the subjects of Wilson et al.[1] and those of Stampfer et al.[2] It seems likely that the apparent conflict in results also arises in part from unresolved differences between the study populations, research methods, and end points — that is, from confounders that are either unrecognized or have larger effects than authors, editors, and most other observers have recognized. This example also illustrates that uncertainty may sometimes be helpful. If future research identifies the critical confounders, we may have a way to tell which women are at high risk of cardiovascular complications from added estrogens and which are not. Thus, study of the uncertainty in the broad, unqualified, and incompatible inferences from these two studies might advance clinical medicine as well as medical science.

We were fortunate in this instance that the two papers were submitted to the same journal at about the same time, that both were judged worthy of publication, and that the editor saw the conflict in findings as a way to bring attention to some general problems in our methods of drawing scientific inferences. Readers of the scientific literature will remember countless other such divergences, where apparently good studies produced incompatible results, but we are not often treated to having them published side by side.

UNCERTAINTY FROM UNRECOGNIZED ERROR IN A CRITICAL ASSUMPTION

My second example of uncertainty in quantitative results has to do with the shape of the dose–response curve for a carcinogenic chemical. Most forms of cancer occur at some low rate (called *background*) even when there is no specific chemical exposure, and the critical scientific issue is the size of any additional risk caused by very small exposures. Because of the limited sensitivity of experiments of feasible size, direct observations at low dose levels are not generally useful, and one must make a reasoned inference from effects at higher levels.

For effects other than cancer, many dose–response curves appear to have a threshold. For example, small amounts of carbon dioxide in the air we breathe are innocuous (indeed they can be beneficial), but larger amounts are toxic and even lethal. Our metabolic systems can handle low salt in-takes, but higher intakes appear to produce hypertension in susceptible persons. A little aspirin, or pesticide, or alcohol may have no serious acute effects on function, but moderately larger amounts can have large effects.

Figure 1 shows some possible dose–response curves. For each curve, the response (cancer, for example) is shown to be small (the "background"

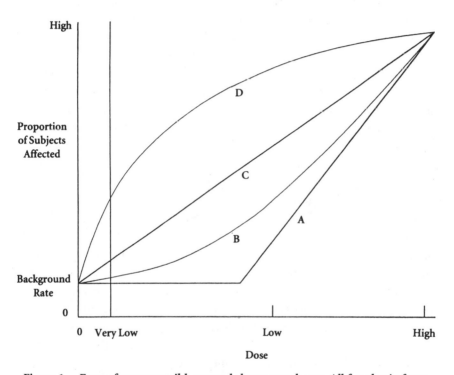

Figure 1.   Four of many possible general shapes are shown. All four begin from the same small "background" effect in unexposed animals and increase to the same high rate of effects at a high dose. Shape A illustrates a threshold, with no effect at all at the lowest doses followed by a sharp rise at higher doses. B shows responses below the straight line between effects at the lowest (zero) and highest doses, with little effect at the lowest doses (almost a threshold) but with each in-crease in dose causing a larger increase in effect than previous increases of the same size. C shows a straight-line dose–response relationship from zero to high dose. D shows a dose–response curve above the fitted straight line, as appears to occur for many carcinogens.

rate) at zero exposure; each curve also shows the same response at a "high" dose, but they differ at intermediate points. Curve A shows a true threshold; there is no response until the dose approaches the level designated "low" in the figure. Curve B has no threshold, but each intermediate point is below the straight line between the response rates at zero and high dose. Curve C is the straight line, and D is entirely above the line. Clearly, risk at the "very low" dose depends a lot on the true curve, but there is no direct way to tell which of these forms, if any, is closest to the truth. Research studies suggest that carcinogens may sometimes have a threshold, but that more often they may not; that there is some risk from even the smallest exposure.

For a time, carcinogenesis experts — in industry, regulatory agencies, and universities — did assume, however, that whatever the true relation between exposure and carcinogenesis, the response at low doses would be no greater than proportional to that at higher doses. In other words, they believed that one could determine the risk at some high exposure, and assume with confidence that if exposure to the chemical were reduced, say, 100-fold, the carcinogenic risk would also be reduced at least 100-fold, perhaps more, and possibly even to background (curves A and B in Figure 1). Thus, a straight-line model (curve C in the figure) was considered "conservative" of the public health, and risk assessment based on scaling-down effects at high doses was thought to set an upper bound on risk at much lower doses. (In some related approaches, this presumed maximum is considered to be an exponential curve, which bends down as doses increase and responses approach 100 percent, but at low doses this is close to a straight line.) A few dose–response curves seemed to flatten out well before the 100 percent response level, as in curve D, but this pattern was considered uncommon.

The assumption that the straight-line model was ordinarily conservative was thus widely used, though there were plausible biologic reasons for thinking that low doses might at times be more efficient than higher doses in terms of producing more cancers per unit of exposure. Table 1 lists some of these.[5] Extensive data available from the National Toxicology Program were used to actually test the straight-line model at moderate to high doses.[5] The straight-line model (curve C) was fitted to data at zero dose (control) and the highest tested dose, and it was found to underestimate risks at intermediate doses (generally at half the highest dose) almost as often as it overestimated risks.

Figure 2 shows experimental data for the induction of one kind of liver cancer by vinyl chloride. Data were collected for female rats fed 0, 2400, 6000, and 10,000 parts per million (ppm) in the diet. These doses are in the range that might be studied in routine investigations, and the top part of the figure shows that the best fitting line for results at 2400 and 6000 ppm (dashed line) is well within the range of uncertainty in the data (expressed as vertical "error bars" in the figure). This fitted line is an exponential curve, but, as noted above, it is rather close to a straight line. Results at 10,000 ppm might have been considered irrelevant or erroneous. The upper 95 percent confidence limit on the fitted exponential curve (solid line) is also shown; it is rather close to the best-fitting (dashed) line.

The bottom part of Figure 2 shows additional data at much lower doses; it also shows the low-dose range of the upper 95 percent confidence limit (solid line) from the upper part of the figure. It is apparent that actual risks at doses up to, say, 50 ppm were many times higher than estimated from the fitted model. Low exposure levels such as these are generally most relevant to human risk, and human risk might have been seriously underestimated if we had not had the additional low-dose data.

These extra data were available because the dose–response relations of vinyl chloride had been recognized as a problem needing special study. But is this failure to follow a straight-line (or exponential) model common? Table 2 lists some other chemicals for which the data show a risk at an intermediate dose that is close to that at the highest dose; risk assessments for each of these must be suspect because we cannot determine, from what is here, how rapidly the same dose–response curve rises at low exposures or where it has largely leveled out. For example, in the study of 1,2-dibromo-

Table 1. Some Plausible Reasons for Expecting a Straight-Line Model to Give Nonconservative Estimates of Risk.*

1.	Dose related mortality from other causes
2.	Direct interference of experimental methods with the causation of cancer
3.	Saturation of enzyme systems
4.	Population heterogeneity in susceptibility to the carcinogen
5.	Intervening mechanisms, such as cell killing at high doses
6.	Carcinogens with high-dose cancer-suppressive effects

* From Bailar et al.[5] Reprinted with permission of the publisher.

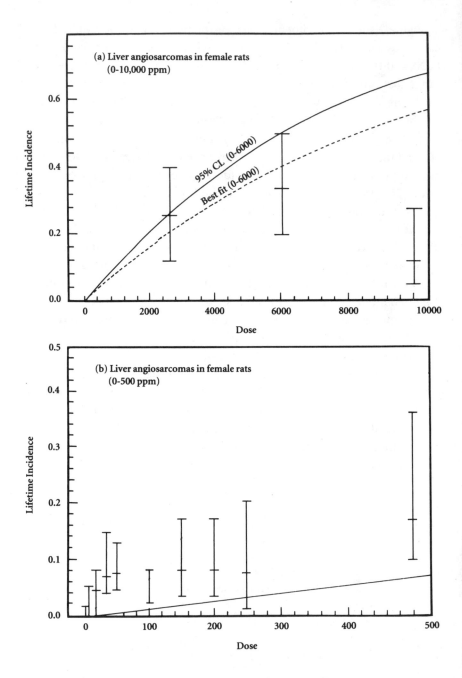

ethane, no unexposed animals developed squamous cell cancer of the oral cavity or gastrointestinal tract, but the incidence of these cancers was 80 percent at the middle-dose level and no higher at the highest dose. We have no way to estimate the risk at 1/10, or 1/100, or 1/1000 of the middle dose except to say that it probably does not exceed that found at the middle dose.

The data in Table 2 were selected post hoc from many thousands of similar sets of observations from the U. S. National Toxicology Program. These estimates of lifetime incidence are based on rather small experimental groups (often 50 to 60 animals), so they are subject to considerable random variation, and by chance alone a few chemicals might give results like those in Table 2. Also, these examples were chosen because they showed a particular kind of result, and so their individual P values arc almost meaningless. (When statistical tests are calculated under the wrong assumptions, they may look reasonable but be very seriously misleading.) The basic question, then, is not whether this large set of results includes some apparent violations of the assumed conservatism of the straight-line model, but whether the number of apparent violations in the entire data set is greater than can be attributed to chance.

My colleagues and I devised a statistical procedure to examine this,[5] such that when the straight-line model holds exactly, a test statistic is about equally likely to take on any value between 0 and 1. (The test statistic is formally a one-tailed P value, but it is not used for testing any hypotheses.) If the straight-line model is truly conservative, many values of the test statistic will be close to 1.00, while if it underestimates low-dose risks, there will be many values near 0.00. If the straight-line model is sometimes too conservative, sometimes too generous, and sometimes right on the mark, we

Figure 2.   Observed rates of liver angiosarcoma in female rats at high doses (top) and low doses (bottom), with 95 percent confidence limits above and below the observed data. The top panel also shows the best fitting exponential model (effectively the straight-line model at low doses) and its upper 95 percent confidence bound. The bottom panel shows similar data at much lower doses, and the low-dose section of the 95 percent upper confidence bound (solid line) is shown in the top panel. All of the low-dose best estimates, and most of the whole confidence ranges on the observed data, lie above the line fitted to the high-dose data. Originally from Maltoni et al.[4] Reproduced here from Bailar et al.[5] Reprinted with permission of the publisher.

Table 2. Examples of Cancer Rates Higher Than Forecast by Linearity in Dose–Response Studies of Animal Carcinogenesis.*

Chemical	Sex	Species and Strain*	Tumor Type	Route	Adjusted Lifetime Cumulative Incidence Dose			P Value†
					None	Mid	High	
4,4′-methylene-dianiline.2HCL	M	b	Hepatocellular carcinoma	Water	0.22	0.66	0.70	0.025
1,2-dibromoethane	F	o	Oral, GI squamous carcinoma	Gavage	0.00	0.80	0.80	0.0002
1,4-dioxane	F	o	Fibroadenoma	Water	0.05	0.46	0.47	0.024
1,3-butadiene	F	b	Hemangiosarcoma	Inhalation	0.00	0.38	0.41	<0.00001
Carbon tetrachloride	M	b	Hepatocellular carcinoma	Gavage	0.07	0.98	0.98	0.008
Dimethylvinyl chloride	M	b	Oral, GI squamous carcinoma	Gavage	0.00	0.86	0.90	0.004
Cytembena	M	f	Mesothelioma, osteosarcoma	Peritoneal injection	0.05	0.74	0.68	0.00003
1,5-naphthalene-diamine	F	f	Endometrial stromal polyp	Food	0.07	0.28	0.27	0.04
Iodinated glycerol	M	f	Plasmacytic leukemia	Gavage	0.31	0.72	0.76	0.04
2-methyl-aziridine	F	c	Skin, breast adenoma	Gavage	0.00	0.81	0.83	0.02

* Species and strain: b-B6C3Fl mouse; f-Fischer 344 rat; o-Osborne Mendel rat; c-Charles River rat.
† One-tailed exact P comparing the mid-dose result to the value obtained from a linear model.

Data from Bailar et al.[5] Reprinted with permission of the publisher.

will see a mixture of these outcomes: a peak near 0.00, another peak near 1.00, and an even spread of values between the extremes. Figure 3 shows that this is just what happened, and that the straight-line model seriously underestimated risks almost as often as it overstated them. Also, the under-estimates may be larger (on a proportionate scale) than the overestimates, which cannot exceed 100 percent. Use of the straight-line model to esti-mate carcinogenic risk appears to be dangerous to the public health.

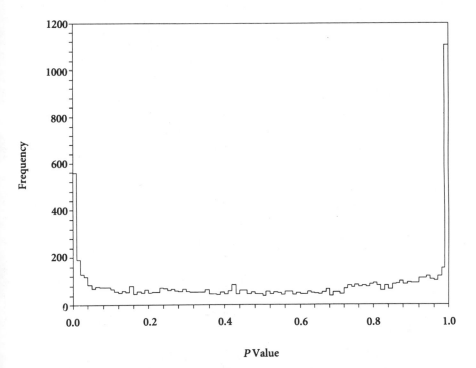

Figure 3.   One-tailed *P* values for the fit of data to the middle dose of observed dose–response data, rounded to each hundredth from 0.00 to 1.00. If the expo-nential model described each true dose–response curve exactly, random varia-tion in the actual data would produce *P* values here that would be uniformly dis-tributed between 0.00 and 1.00. In actuality there was a substantial excess of very low *P* values, indicating dose–response curves above the fitted straight line, such as Curve D in Figure 1, and an even larger excess of high values, corresponding to curves such as A and B in Figure 1. From Bailar et al.[5] Reprinted with permission of the publisher.

There is again much more uncertainty in a broad range of scientific inferences than was commonly recognized. The more general lesson here is that we can all make assumptions that seem perfectly reasonable, but are wrong. Widespread acceptance of some technical assumption does not guarantee truth or even a good approximation of truth.

UNCERTAINTY IN GENERALIZING RESULTS

My third example comes from the field of cancer risk assessment.[6] I shall refer to a set of four estimates of the cancer hazard presented by a compound commonly known as Tris.[7-10] The four risk assessments were all done within a period of about six weeks in 1977, and all four sets of investigators had access to essentially the same information.

Tris was introduced about 1972 as a flame retardant in fabrics, especially in fabrics used to make pajamas for infants. Later, however, the preliminary results of an animal study showed that Tris fed at high doses to mice and rats was a potent carcinogen. Thus there were questions about the risk of sending children to bed in Tris-treated sleepwear. Table 3 shows the experimental results for the four cancer sites where the increase in cancer incidence was judged to be statistically significant in at least one species and

Table 3. Cancer Incidence after Lifetime Feeding with Tris.*

Tris: Parts per Million in the Diet	Ratios of Number of Animals with Cancers to Number Treated, by Site of Cancer							
	Kidney		Stomach		Liver		Lung	
	M	F	M	F	M	F	M	F
	Mice							
0	0/55	0/55	0/51	2/52	28/55	11/54	12/55	4/55
500	4/50	2/50	10/48	14/48	30/49	23/49	18/44	9/50
1000	14/50	2/50	13/50	22/49	23/50	33/50	24/50	17/50
	Rats							
0	0/54	0/54						
50	25/55	4/55						
100	29/55	10/55						

*   Data from Brown et al.[7]

one sex. (About 60 animals of each sex and species were tested at each dose, but a few died early and in some instances not all organs could be examined, so denominators vary.) Large and statistically significant increases were found in the frequency of cancer of the kidney in male mice and in both sexes of rats, the stomach and lung in both sexes of mice, and the liver in female mice. The increase in kidney cancer in rats is especially striking because the level of exposure was only one-tenth of that in mice but the cancer frequency was higher.

The four risk assessments all focused on kidney cancer, using the data in Table 3. Thus, differences in assessed risk must result from differences in how the data were extrapolated from this rodent bioassay to human beings. The size of the risk is critical here, because Tris was, beyond question, effective in reducing the flammability of woven fabrics, and almost as certainly reduced the number of injuries and deaths of children at risk of burns. Substitutes for Tris were available, but they did not have the same desirable chemical and physical properties.

Several kinds of problems arise almost immediately in trying to estimate the human risks of a carcinogen from a laboratory experiment such as that summarized in Table 3. Some of the largest, well recognized by risk assessors, are:

- Extending results from animals (mice and rats, in this case) to humans, who may respond in different ways;

- Extending results from the high doses in the experiments to what were expected to be much lower doses in infants;

- Extending results from feeding (in the animals) to skin exposure (in humans), though some Tris might also be ingested by infants who chew on their pajamas;

- Extending results from a lifetime of exposure in rodents to exposure of humans over a period of perhaps one to two percent of their lifetimes.

While each of these is a problem in toxicology or carcinogenesis, each is also a problem in statistics because it affects overall uncertainty about the meaning of the data for the specific problem at hand, and hence affects the scientific and statistical inferences that may be drawn from the data.

It should be no surprise that different scientists, all quite competent in this sort of work, should make different assumptions, approximations, and guesses about these and other critical uncertainties. These differences are reflected in what the risk assessors presented as their best estimates of the lifetime risk of kidney cancer per million infants exposed: Brown, Schneiderman, and Chu[7] estimated 52 per million (upper 95 percent confidence level, and based on the mouse data); Harris[8] estimated 7 to 6300; Hooper and Ames[9] estimated 17,000; and the Consumer Product Safety Commission (CPSC) staff[10] estimated 160 (averaged over males and females).

The most critical points of difference among the risk analyses concerned the amount of Tris that would be absorbed from treated garments and the relation between risk to animals and risk to humans at similar exposure levels. These involved a host of subsidiary questions such as the decline in availability of Tris as garments are washed repeatedly, how many garments a child would wear, and the effects of age at exposure and duration of exposure. The nature of other differences among the risk assessors is illustrated by some of the decisions made by Hooper and Ames that:

- Twelve Tris-treated garments would be worn during the first year of life;
- Each garment had 5000 cm^2 of surface area, half of which would be in contact with the child at any one time;
- The amount of Tris available for absorption per cm^2 could be determined by the amount removed from polyester fabric (the fabric most commonly used) by solvents; for this they used a number halfway between the amounts of Tris removed from "high" surface Tris garments and those classified as "low," with adjustment for the amount removed by washing weekly for one year and calculated that 30 micrograms/cm^2 of Tris was available for absorption; and
- Six percent of this available Tris is absorbed through the skin and one percent by mouthing the garment.

From such estimates and calculations, Hooper and Ames estimated that a child would absorb 70 mg/kg/year. They then assumed that cancer risk in children and in rats would be the same in terms of mg/kg/year. The other risk assessors went through similar processes but made other assumptions:

for example, that the animal–human conversion should be in terms of surface area rather than weight; that the risk to humans should reflect the small fraction of the normal life span during which they were exposed (perhaps 2 percent, rather than the 100 percent in animals); or that mice would provide a closer estimate of human risk than rats.

Each group of risk assessors had a lively appreciation of the uncertainties in its estimate, and the problems were discussed at length, both in the risk assessment documents and in formal testimony before the CPSC. But errors were discussed in terms of a single order of magnitude — 10-fold — rather than the substantially larger differences actually existing among the four estimates — roughly 2½ orders of magnitude (from 52 to 17,000) or even 3½ orders (from 7 to 17,000). Nor can the skeptical observer even be sure that these ranges bound the true value. If each "best estimate" is equally likely to fall above or below the true value, the chance that four independent estimates would all be too low is 1 in 16, or 6.25 percent. These estimates were not independent — the risk assessors communicated freely with each other, as we would want them to do, and they all had access to the same data — so the chance that all four are too low may be a good bit larger than 6.25 percent.

Despite these uncertainties, the control of the potential hazards of Tris was a success of risk assessment methods. By 1977, when Tris was recognized as a possible human carcinogen, an improved range of other flame retardants had been developed for use on clothing, and there was no evidence to show that they were hazardous. The CPSC commissioners found that even the lowest estimate of cancer risk of Tris — 7 per million over a lifetime, or about 20 to 25 cancers per year if all infants in the United States wore Tris-treated sleepwear for their first year — was too high. As the CPSC commissioners moved to regulate the use of Tris, the industry itself voluntarily stopped using it on clothing and on fabric meant for clothing. These four quantitative health risk assessments differed in many ways, including their conclusions about the size of the risk, but they contributed to the same decision.

USING STATISTICAL CONCEPTS

These three examples differ in obvious and important ways, but they have a few critical similarities. Each involves a serious and, in retrospect, obvious problem in statistical thinking; these problems were not always recognized by the investigators. Each required the interpretation of a body of scientific and technical data that was large, complex, known to have some defects, and not precisely targeted at the question of present interest. Further, each of the three was subjected to extensive statistical analysis by knowledgeable people. And in each, in the end, the uncertainty generated by random variability was essentially trivial in relation to other uncertainties:

- With postmenopausal estrogens, the reconciliation of apparently incompatible results is not complete, but appears to involve important but unrecognized (perhaps unrecognizable) differences between study groups.

- With carcinogen dose–response curves, general acceptance of a critical assumption introduced substantial uncertainty. The assumption was not validated against the data until 1988 (when it was found wanting), despite ready opportunities to do so.

- With Tris, the risk assessors had to make some judgment about the size of the added risk, and to make it as useful as possible, despite huge uncertainties that were acknowledged by all parties.

The problems were apparent in these three examples because we had access to data, from the same or other sources, showing contrary findings. That is not common in science, despite the general notion that the scientific method involves independent replication. Strict replication is in fact rare in science; professional rewards such as advancement and tenure, as well as the sociology of science, produce substantial pressures to refute, but none to replicate. In my work as statistical consultant for the *New England Journal of Medicine,* I reviewed close to 400 papers each year. Nearly all of these had passed the rigorous standards of outside peer reviewers, and had also been judged by the *Journal* staff to be likely to meet other standards for publication, including scientific importance, interest to readers, ultimate relevance to medical practice, length, quality of writing, and balance of fields covered by the *Journal.* About half of the papers that came to me for statistical review were eventually published in the *Journal.* How many of these 400 papers, judged to be of high quality, would pass the test of independent confirmation? How many are really free of major, unrecognized problems in basic assumptions, or study design, or quality of data, or interpretation of findings, or generalization to other subjects? I do not know, but as time passes and as I see more and more scientific work that appears to be sound but is overturned by newer (and sometimes better) studies, some patterns seem to emerge. Many of the problems are what I regard as fundamentally statistical, and many of these statistical problems are subject to considerable amelioration. All of these problems involve what I call statistical thinking, as did the three examples above; that is, they involve concepts, definitions, procedures, and ways of looking at scientific processes and data that focus on drawing inferences from information that is inevitably subject to error. To the extent that information (not just the data from a particular experiment) is subject to error, the inferences are uncertain, and statistical thinking is the best way to anticipate, detect, measure, reduce, and otherwise manage such uncertainty.

We have some reason to be pleased with our long and successful history of educating ourselves and others in some basic statistical matters. That success has not only been helpful in itself; it has been an essential base for moving forward. For example, statistical methods have led to substantial advances in many scientific fields, though the basic assumptions that underlie those methods are rarely satisfied exactly and the results are often misinterpreted or misused. We can now move on to the next level of sophistication, where uncertainty (and the statistical treatment of uncertainty) is even more important and generally different from the routine.

Statistical thinking, as embodied in the best practice of applied statistics, is the art and science of inference, as opposed to subject matter and technical content. Because the sources of uncertainty are so varied, statistics must concern itself with the full range of scientific activity, and indeed statistics can have much to offer at each step, from framing the initial questions for study, to designing an investigation and setting a protocol, to the collection and processing of data, to formal analysis, and ultimately to the drawing of generalizations to subjects and groups not studied.

Statistics, or inference, is the machinery of the scientific method. It is the cogs and wheels, the belts and pulleys, the presses and drills and furnaces that are left when we take away the specifics of problem and discipline and data. A few examples of these very general statistical tools and concepts are randomization, the special role of prior hypotheses, blind assessment of outcomes, and replication. This kind of statistics, like mathematics, can be studied in its own right, and again like mathematics, it can be applied to an enormous and varied range of problems.

I am indebted to several colleagues for thoughtful and helpful comments, especially Drs. Marvin Schneiderman, Frederick Mosteller, Richard N. L. Andrews, and David Bates; and to the Editor of *Risk Analysis* for permission to reprint Tables 1 and 2 and Figures 2 and 3.

REFERENCES

1. Wilson PWF, Garrison RJ, Castelli WP. Post-menopausal estrogen use, cigarette smoking, and cardiovascular morbidity in women over 50: the Framingham study. New Engl J Med 1985; 313:1038-43.

2. Stampfer MJ, Willett WC, Colditz GA, et al. A prospective study of post-menopausal estrogen therapy and coronary heart disease. New Engl J Med 1985; 313:1044-9.

3. Bailar JC. When research results are in conflict. New Engl J Med 1985; 313:1080-1.

4. Maltoni C, Lefemine G, Gilberti A, et al. Vinyl chloride carcinogenicity bioassays (BT project). Epidemiologie Animale et Humaine. Proceedings of the 20th meeting of Le Club Cancerogenese Clinique, Paris, November 10, 1979.

5. Bailar JC, Crouch EAC, Shaikh R, Spiegelman D. One-hit models of carcinogenesis: conservative or not? Risk Analysis 1988; 8:485-97.

6. Needleman J, Burney B, McGiness JM, Bailar JC. Methodologic challenges in risk assessment (in press).

7. Brown C, Schneiderman M, Chu K. Estimation of human lifetime carcinogenic risk from exposure to Tris. Memorandum to the U.S. Consumer Product Safety Commission, March 21, 1977.

8. Harris RH. Estimating the cancer hazard to children from Tris-treated sleepwear. Memorandum to the U.S. Consumer Product Safety Commission. March 8, 1977.

9. Hooper NK, Ames BN. Letter to John Byington, Chair, U.S. Consumer Product Safety Commission. March 21, 1977.

10. Bayard SP. Preliminary analysis of Tris-induced human lifetime risk to cancer based on data from the NCI lifetime animal study and BBS best estimate of lifetime human exposure. Letter to Robert M. Mehir. March 17, 1977.

3

~

USE OF STATISTICAL ANALYSIS IN THE NEW ENGLAND JOURNAL OF MEDICINE

John D. Emerson, Ph.D., and Graham A. Colditz, M.D., D.P.H.

ABSTRACT   A survey of the statistical methods used by authors of the 760 research and review articles in Volumes 298 through 301 (1978-9) and the 115 Original Articles in Volume 321 (1989) of the *New England Journal of Medicine* reveals varied uses of statistics. A reader who is conversant with some simple descriptive statistics (percentages, means, and standard deviations) has full statistical access to 58 percent of the articles. Understanding *t*-tests increases this access to 67 percent. The addition of contingency tables gives complete statistical access to 73 percent of the articles. Familiarity with each additional statistical method gradually increases the percentage of accessible articles.

Original Articles use statistical techniques more extensively than other articles in the *Journal*, and those in Volume 321 (1989) make greater use of statistical methods than did articles appearing a decade earlier.

Research studies based on a longitudinal design make heavier use of statistics than do those using a cross-sectional design. The tabulations in this study should aid clinicians and medical investigators who are planning their continuing education in statistical methods and faculty who design or teach courses in quantitative methods for medical and health professionals.

In this study we report on the frequency of use of statistical techniques in the *Journal* to answer such questions as the following: Will knowledge of a few elementary statistical techniques, such as chi-square and *t*-test analyses, assist readers in understanding the statistical content of a high percentage of research articles in the *Journal?* Which additional techniques are used most often and therefore could be added most profitably to the statistical background of readers? To aid clinicians and medical investigators who are continuing their own education, as well as persons designing courses in biostatistics for physicians, we report on some components of the statistical content of all research and review articles in the *Journal* in 1978 and 1979. We also examine the statistical content of one more volume of the *Journal* from 1989.

METHODS

This study analyzes Volumes 298 through 301 (January 1978 through December 1979) of the *Journal* and includes all articles identified in the table of contents as Original Articles, Special Articles, Medical Progress articles, Medical Intelligence articles, and Seminars in Medicine of the Beth Israel Hospital, Boston. One person screened the 760 articles identified from these volumes and selected 218 Original Articles for detailed review. The 218 articles were selected for their potential value, not only for this study, but also in other projects conducted by our Study Group for Statistical Methods in the Biomedical Sciences. Bailar and his colleagues[1] have reported on their related study of research design, which examined these articles. At least two statisticians independently reviewed each of the 218 articles for their use of statistical methods and completed a checklist documenting research design and statistical content.

The remaining 542 articles were reviewed more briefly for the presence or absence of specific statistical procedures and techniques. The reviewer read the Methods section and all tables, and scanned other sections of the articles for the pertinent information. For each article, the presence or absence of statistical methods in each of 21 categories was recorded and entered in coded form into a computer. All reviewers were faculty members, postdoctoral fellows, or graduate students trained in both the quantitative sciences and statistical applications to medical research.

CATEGORIES OF STATISTICAL METHODS

Table 1 lists the 21 categories of statistical methods used and gives a brief description of their content. Although most categories in Table 1 need no further description, a statement of the criteria used to define the categories requiring a judgmental decision follows.

The first category in Table 1 applies to articles that contain no statistical methods other than percentages, means, standard deviations, standard errors, or histograms. This category contains precisely those articles that lack the statistical content belonging to the other 20 categories. This is the only category defined so that it cannot overlap with any of the others.

The "multiple-comparison" category includes any methods that adjust significance levels or P values when several statistical comparisons are made. Chapters 10 and 12 of this book discuss these and related issues in some detail.[2,3] "Cost–benefit analysis" requires a direct quantitative comparison of costs with benefits.

"Power" is broadly defined; we looked for an indication that sample size was determined, in part, by a consideration of the probability of falsely accepting a null hypothesis. A post hoc analysis of power is acceptable, as are other, more formal analyses. Significance tests or confidence limits, although useful, did not qualify as treatments of power. Chapter 19 of this book describes the importance and interpretation[4] of power in medical research design.

Whenever an article contained an examination or analysis of data in a scale other than that of the raw data (for example, square roots or logarithms), we indicated that a "transformation" was used. We did not count the replacement of raw data with ranks or categories as transformations.

The last category ("other") includes specialized methods, each used in a single article, that did not fit into the defined categories. For example, Lang et al.[5] examined the modeling of plasma glucose and insulin concentrations in human beings by fitting a sine curve to the "smoothed" time series. Another article[6] used the Kolmogorov–Smirnov test for goodness of fit.

Occasionally, a particular method, although not used by the authors themselves, was mentioned when citing the work of others. We attributed such methods to the article concerned, because we believed that an understanding of the method in question would enhance the statistical understanding of the article we reviewed.

QUALITY CONTROL

For the 218 articles receiving detailed review, any discrepancies between the two independent reviewers were discussed and resolved. The accuracy of the coding of the group of 542 articles that received a briefer review was determined by an independent examination of 96 of these articles. Discrepancies were found for 15 items in 11 of these articles. Another careful

Table 1.   Statistical Procedures Used to Assess the Statistical Content of Articles.

Category	Brief Description
No statistical methods or descriptive statistics only	No statistical content, or descriptive statistics only (e.g., percentages, means, standard deviations, standard errors, histograms)
Contingency tables	Chi-square tests, Fisher's exact test, McNemar's test
Multiway tables	Mantel–Haenszel procedure, log-linear models
Epidemiologic statistics	Relative risk, odds ratio, log odds, measures of association, sensitivity, specificity
t-tests	One-sample, matched-pair, and two-sample *t*-tests
Pearson correlation	Classic product-moment correlation
Simple linear regression	Least-squares regression with one predictor and one response variable
Multiple regression	Includes polynomial regression and stepwise regression
Analysis of variance	Analysis of variance, analysis of covariance, and *F* tests
Multiple comparisons	Procedures for handling multiple inferences on same data sets (e.g., Bonferroni techniques, Scheffé's contrasts, Duncan multiple-range procedures, Newman–Keuls procedure)
Nonparametric tests	Sign test, Wilcoxon signed-rank test, Mann–Whitney test
Nonparametric correlation	Spearman's rho, Kendall's tau, test for trend
Life table	Actuarial life table, Kaplan–Meier estimate of survival
Regression for survival	Includes Cox regression and logistic regression
Other survival analysis	Breslow's extension of Kruskal–Wallis, log rank, proportional hazards models
Adjustment and standardization	Pertains to incidence rates and prevalence rates
Sensitivity analysis	Examines sensitivity of outcome to small changes in parameters of model or in other assumptions
Power	Loosely defined, includes use of the size of detectable (or useful) difference in determining sample size
Transformation	Use of data transformation (e.g., logarithms), often in regression
Cost–benefit analysis	Combining estimates of cost and health outcomes to compare policy alternatives
Other	Anything not fitting above headings; includes cluster analysis, discriminant analysis, and some mathematical modeling

reading of these articles indicated that all but two errors involved overlooking a procedure.

ORIGINAL ARTICLES

Of the 760 articles in this study, the *Journal* classified 332 as Original Articles. To relate the study design to the statistical methods used, we partitioned these papers into longitudinal (prospective or retrospective[1,7]) and cross-sectional types. Longitudinal studies aim to elucidate a state of nature that changes or may change. Cross-sectional studies focus on phenomena that are thought to be static over the period of interest.

We extended this study by reviewing a second set of 115 Original Articles from Volume 321 (January through June, 1989). The authors of this chapter independently reviewed the Methods sections and all tables and scanned other parts of the articles for pertinent information about the presence or absence of specific statistical techniques. Each Original Article received a single review, and the results were recorded on a coding form. We did not classify these articles by study design.

ANALYSIS OF TECHNIQUES USED

We determined the frequency of the individual statistical methods in the 760 *Journal* articles. In addition to performing the simple quantification (number and percentage of articles using a method), we assessed how much a reader's acquaintance with all the quantitative techniques in an article would improve with an increased statistical repertoire. In trying to obtain a definite measure, we were handicapped by the lack of a natural order for learning these methods. For the analysis we chose the order that maximally increased the percentage of articles for which a reader would be acquainted with all the statistical techniques reported if he or she learned one more method. We began this analysis by assuming that the reader had no knowledge of statistical techniques. The order was thus determined by the data gathered. This ordering, though useful, intellectually reasonable, and empirically based, is nevertheless arbitrary. In particular, it ignores the fundamental role of broad statistical concepts such as experimental design, variation, and bias in determining the extent of a reader's statistical understanding. Moses[8] discusses the role of these issues in Chapter 1 of this book. Furthermore, it may not be the best order for learning about the statistical methods.

RESULTS

Table 2 shows the frequency of statistical methods used in Volumes 298 through 301 of the *Journal*. Under the assumptions outlined above, we analyzed the impact of increased statistical knowledge on the accessibility of *Journal* articles. We analyzed the data in three ways. First, we gave the number and the percentage of articles that used the method — e.g., for the item "contingency table," 112 articles used the method, or 15 percent of the 760 articles reviewed. This figure of 112 also represents 10 percent of the 1120 "article-method uses"; to determine this number, we counted the number of distinct method categories found in each article and then added these counts together over all 760 articles. Although 179 articles used *t*-tests, 113 of these also used methods farther down in Table 2. The remaining 66 articles would become fully accessible to readers unfamiliar with statistical techniques beyond *t*-tests. Similarly, if one knew descriptive statistics and *t*-tests and then added contingency tables to one's statistical repertoire, one would add 42 (551 − 509) articles, or 6 percent (from the fourth column of figures, 73 − 67 = 6), to the number of articles in which one had access to all the statistical methods. The figure of 6 percent is less than 15 percent because some articles used topics farther down on the list, and so would not become completely accessible by merely adding knowledge of contingency tables.

We use the phrase *statistical access* in a narrow and specific sense; we mean access to the information transmitted when the authors report that they used, say, a *t*-test or a life-table method in their statistical analyses. By *complete access* we mean that a hypothetical reader of the *Journal* is familiar with all the statistical methods used in a particular article. We acknowledge that full statistical understanding of an article depends on far more than a passing acquaintance with the techniques employed in the article; an understanding of many related concepts in statistics is often required.

If a reader with no prior knowledge of statistical techniques were to learn the first seven methods listed in Table 2 (descriptive statistics, *t*-tests, analysis of contingency tables, nonparametric tests, epidemiologic methods, correlation, and simple regression), he or she would be acquainted

Table 2. Statistical Content and Accessibility of *New England Journal of Medicine* Articles.

Procedure	Articles Containing Methods no. (%)	Accumulated by Article no. (%)	Accessibility by Article-Method (%)
No statistical methods or descriptive statistics only	443 (58)	443 (58)	(40)
t-tests	179 (24)	509 (67)	(56)
Contingency tables	112 (15)	551 (73)	(66)
Nonparametric tests	45 (6)	571 (75)	(70)
Epidemiologic statistics	39 (5)	585 (77)	(73)
Pearson correlation	55 (7)	598 (79)	(78)
Simple linear regression	37 (5)	621 (82)	(81)
Analysis of variance	33 (4)	636 (84)	(84)
Transformation	26 (3)	650 (86)	(87)
Nonparametric correlation	15 (2)	662 (87)	(88)
Life table	24 (3)	674 (89)	(90)
Multiple regression	19 (3)	686 (90)	(92)
Multiple comparisons	13 (2)	698 (92)	(93)
Other methods	17 (2)	708 (93)	(94)
Adjustment and standardization	13 (2)	718 (95)	(96)
Multiway tables	12 (2)	728 (96)	(97)
Power	13 (2)	737 (97)	(98)
Other survival analysis	11 (1)	747 (98)	(99)
Regression for survival	6 (1)	753 (99)	(99)
Cost–benefit analysis	6 (1)	758 (100)	(100)
Sensitivity analysis	2 (0)	760 (100)	(100)
Totals:			
Article-method uses	1120		
Articles		760	

with techniques necessary for complete statistical access to 82 percent of the 760 *Journal* articles. These methods also cover 81 percent of all 1120 article-method uses of statistical methods. Addition of each further method adds only a minor proportion of the 760 articles to the total accessible to the reader.

Of these 760 articles, 317 used a total of 677 methods other than those in the descriptive-statistics category, or slightly more than 2 of these methods per article. Because more than 75 percent of these articles were Original Articles, we analyzed these papers to relate study design to methods used. Table 3 shows the frequencies of statistical methods in the Original Articles found in Volumes 298 through 301. The results of their categorization by study design are presented on the right-hand side of Table 3 as deviations from the group percentage, following the residual-analysis approach of Ehrenberg,[9] a standard technique in the analysis of variance. In the first category, "no statistical methods or descriptive statistics only," overall usage is 27 percent. Among the 332 Original Articles, prospective studies yield a deviation of -11, which represents the frequency for this study design (16 percent) minus the overall percentage ($16 - 27 = -11$). Similar calculations apply for the other columns. As indicated by the residual of $+18$ for descriptive statistics, cross-sectional studies have a higher relative frequency of reliance on only descriptive methods.

EXTENDING THE FINDINGS

To extend our investigation of the use of statistical analysis in the *Journal* beyond 1978-79, we reviewed the 115 Original Articles in Volume 321 (July through December, 1989). Table 4 summarizes these findings using the same ordering of method categories as in Table 3. The percentages in the second columns of Table 3 and Table 4 are comparable, because both tables focus on the Original Articles in their respective volumes.

Although the survey of 115 articles in a single volume of the *Journal* is modest in size and scope, we can compare our findings with those from the more extensive review of Original Articles in Volumes 298 through 301. We do not compare these percentages with those for the 760 articles in Table 2, because fewer than half of those were Original Articles and because the method categories of Table 2 are slightly more refined than those used in Tables 3 and 4.

In 1989, only 12 percent of Original Articles used no statistical methods or descriptive statistics only. Among the method categories with gains since 1978-79 in the percentage of Original Articles reporting their use are: Contingency tables (from 27% to 36%), Nonparametric tests (from 11% to 21%), Epidemiologic statistics (from 9% to 22%), Analysis of variance

Table 3.   Statistical Content of 332 Original Articles, According to Study Design.

Procedure	Original Articles (*n* = 332)	Prospective (*n* = 182)	Retrospective (*n* = 18)	Cross-sectional (*n* = 132)
	no. (%)	deviation from overall %*		
No statistical methods or descriptive statistics only	91 (27)	−11	−16	+18
t-tests	147 (44)	+4	−11	−4
Contingency tables	91 (27)	+9	+23	−16
Nonparametric tests	38 (11)	+3	+6	−5
Epidemiologic statistics	33 (9)	0	+40	−5
Pearson correlation	40 (12)	0	−1	0
Simple linear regression	28 (8)	+1	+3	−1
Analysis of variance	25 (8)	+1	−8	−1
Transformation	23 (7)	0	−1	0
Nonparametric correlation	13 (4)	0	+4	+1
Survival methods†	36 (11)	+7	+4	−11
Multiple regression	15 (5)	+1	+1	−1
Multiple comparisons	11 (3)	+2	−3	−1
Adjustment and standardization	9 (3)	+2	−3	−3
Multiway tables	12 (4)	+1	+8	−3
Power	10 (3)	+1	+3	−2
Cost–benefit analysis	3 (1)	0	−1	0
Other methods	12 (3)	0	−3	+1
Total article-method uses	637			

* Deviations from overall percentage were calculated by subtracting the overall percentage of Original Articles (e.g., 27 in row one) from the percentage of articles in each study design using the procedure (e.g., descriptive statistics only). For prospective studies, 16 percent used descriptive statistics only, giving 16 − 27 = −11. Similarly, the use in retrospective studies (11 percent) minus the overall percentage (27 percent) gives a residual of −16. Cross-sectional studies used only descriptive statistics in 45 percent of studies and so yield a residual of +18 (45 − 27 = 18).

† Survival methods here include the original categories of life table, regression for survival, and other survival methods.

(from 8% to 20%), and Multiple regression (from 5% to 14%). The most substantial change is the increase in Original Articles using Survival methods — from 11% to 32% during the 10-year span of our work.

Table 4.   Statistical Content of 115 Original Articles in Volume 321 (1989) of the
New England Journal of Medicine.

Procedure	Original Articles ($n = 115$) no. (%)
No statistical methods or descriptive statistics only	14 (12)
t-tests	45 (39)
Contingency tables	41 (36)
Nonparametric tests	24 (21)
Epidemiologic statistics	25 (22)
Pearson correlation	22 (19)
Simple linear regression	10 (9)
Analysis of variance	23 (20)
Transformation	8 (7)
Nonparametric correlation	1 (1)
Survival methods*	37 (32)
Multiple regression	16 (14)
Multiple comparisons	10 (9)
Adjustment and standardization	10 (9)
Multiway tables	11 (10)
Power	4 (3)
Cost–benefit analysis	0 (0)
Other methods	10 (9)
Total article-methods†	311

* Survival methods here include the original categories of life table, regression for survival, and other survival. The numbers of articles for these categories are: life table, 19; regression for survival, 25; other survival analysis, 13.

† When the three categories of survival methods are distinguished, there are 331 article-method uses.

A total of 241 of the 332 Original Articles in Table 3 used $637 - 91 = 546$ methods other than those in the descriptive-statistics category, or 2.3 methods per article. A decade later, 101 of the articles in Table 4 used $311 - 14 = 297$ methods; the rate of use is 2.9. The increase in the rate of usage means that not only do more Original Articles make use of statistical methods but they also use a greater number of methods.

DISCUSSION

Nearly 42 percent of the 760 *Journal* articles originally surveyed relied on some type of statistical analysis beyond descriptive statistics. These articles used *t*-tests most frequently (24 percent of all articles), followed by chi-square analysis of contingency tables (15 percent). Beyond these two large groups, no methods stand out from the rest as being in particular favor among *Journal* authors. The various categories of statistical procedures defined in Table 1 are not highly clustered in the articles surveyed. We are impressed with the varied combinations of methods employed, although this finding does depend on the initial definition of categories.

Worthy of note is the gradual way (1 or 2 percent per technique) in which increased knowledge of statistical techniques adds to the percentage of statistical uses understood and to the percentage of articles in which the reader has access to all the statistical methods used. One might predict that knowledge of half a dozen common statistical procedures would make almost all the articles statistically accessible to the reader. Although the arbitrary choices of the categories and their resulting small sizes contribute to this gradualness, these choices fit many statistical textbooks fairly well. On the basis of these categories, Table 2 shows that 11 techniques are needed to raise a reader's full access level to more than 90 percent of articles, whether for usage or for accumulated accessibility. The categories listed below "life table" account for the 10 percent not already accessible. Although some categories are obviously quite narrowly defined, others such as "multiple regression" and "multiple comparisons" are broad. Furthermore, these latter methods, though cited infrequently in the *Journal*, are widely used in other fields and may also be gaining in use in medical research.

Table 3 indicates that longitudinal studies, both prospective and retrospective, made heavier use of statistical techniques than did articles using cross-sectional designs. The right-hand column, labeled "cross-sectional" studies, almost always shows negative signs, indicating that these categories were used relatively less often. Although cross-sectional studies used proportionally almost as many *t*-tests as did the other types of studies, these articles used contingency-table analysis much less often — in 11 percent (27 − 16) as compared with 27 percent overall. Perhaps because epidemiologists adopt retrospective study designs most often, retrospective studies show a 40 percent excess usage of epidemiologic statistics.

An extension of our study confirms our main findings of frequent and varied usage of statistical methods by *Journal* authors. If there was a trend in the 1980s, it was toward an increased use of newer and more varied statistical techniques. Methods for analyzing survival data gave the largest apparent gains; in particular, we noticed more use of regression techniques for analyzing survival outcomes. More generally, we found that *Journal* authors frequently use techniques of multiple logistic regression for analyzing binary responses.

We cannot conclude with certainty that the apparent shifts in the counts and percentages reflect changes in statistical practice. The observed changes may arise in part from differences in our own research methods. However, these findings further strengthen our main message that *Journal* authors use a wide array of statistical methods and that many Original Articles use varied combinations of these methods. An acquaintance with a few basic statistical techniques cannot give full statistical access to research appearing in the *Journal*.

The use in medicine of a wide range of statistical methods gives clinicians and medical investigators good reason to continue their own education in statistics. The percentages presented in Tables 2, 3, and 4 may help them in identifying the statistical methods they should master. This study reviews statistical techniques from the perspective of a general reader, and no attempt has been made to identify a hierarchy of statistical backgrounds, which may vary for physicians within each specialty. Furthermore, the *Journal* is oriented to clinical medicine, in general, and professional specialization may influence the relative importance of the various statistical methods.

We are indebted to the following members of the Study Group for Statistical Methods in the Biomedical Sciences for their suggestions and encouragement: John Bailar, Roger Day, Rebecca DerSimonian, Karen Falkner, Katherine Godfrey, Samuel Greenhouse, Katherine Taylor Halvorsen, Hossein Hosseini, Philip Lavori, Robert Lew, Thomas Louis, Lincoln Moses, Frederick Mosteller, Kay Patterson, Marcia Polansky, and John Williamson.

REFERENCES

1. Bailar JC III, Louis TA, Lavori PW, Polansky M. A classification for biomedical research reports. N Engl J Med 1984; 311:1482-7. [Chapter 8 of this book.]

2. What are P values? In: Ingelfinger JA, Mosteller F, Thibodeau LA, Ware JH. Biostatistics in clinical medicine. 2nd ed. New York: Macmillan, 1987:151-67.

3. Godfrey K. Comparing the means of several groups. N Engl J Med 1985; 313:1450-6. [Chapter 12 of this book.]

4. Freiman JA, Chalmers TC, Smith H Jr, Kuebler RR. The importance of beta, the type II error, and sample size in the design and interpretation of the randomized controlled trial: Survey of 71 "negative" trials. N Engl J Med 1978; 299:690-4. [Chapter 19 of this book.]

5. Lang DA, Matthews DR, Peto J, Turner RC. Cyclic oscillations of basal plasma glucose and insulin concentrations in human beings. N Engl J Med 1979; 301:1023-7.

6. Ault KA. Detection of small numbers of monoclonal B lymphocytes in the blood of patients with lymphoma. N Engl J Med 1979; 300:1401-5.

7. White C, Bailar JC III. Retrospective and prospective methods of studying association in medicine. Am J Public Health 1956; 46:35-44.

8. Moses LE. Statistical concepts fundamental to investigations. N Engl J Med 1985; 312:890-7. [Chapter 1 of this book is an expanded version of this article.]

9. Ehrenberg AS. Data reduction: analyzing and interpreting statistical data. New York: John Wiley, 1975.

SECTION II

Design

4

~

DESIGNS FOR EXPERIMENTS —
PARALLEL COMPARISONS OF TREATMENT

Philip W. Lavori, Ph.D., Thomas A. Louis, Ph.D.,
John C. Bailar III, M.D., Ph.D., and Marcia Polansky, D.Sc.

ABSTRACT   Clinical trials of medical treatments often compare treated groups with each other or compare one or more treated groups with a separate but concurrent control group. We have examined a consecutive series of 47 such parallel studies reported in the *New England Journal of Medicine* in 1978–1979, including 35 with random assignment to the treated or control group, to discover how this approach is actually used.

A major strength of these studies as a group was the frequent use of randomized treatment assignment. Common problems included lack of sufficient detail about methods of randomization, failure to provide enough detail about patient sources, and insufficient use of multivariate statistical techniques and of statistical modeling.

We emphasize the importance of avoiding bias by balancing prognostic factors when assigning patients to treatments, reducing bias by modeling the influence of prognostic factors on response, and increasing precision by modeling. We also advocate careful consideration of the relevance of a treatment comparison within the study to the external world of clinical practice.

C linical trials of medical treatments often compare two or more groups of patients treated separately but concurrently (including placebo or "no treatment" groups) as part of the same study. We call these trials *parallel* to emphasize their difference from other clinical trials in which patients are their own controls or controls are drawn from historical or other data external to the work reported. Parallel comparisons include most of the usual forms of randomized clinical trials, as well as some nonrandomized studies.

This report identifies features of parallel comparisons of clinical treatments that affect the strength and validity of inferences. Our observations are based on a survey of reports of 47 parallel comparisons that were published in the *New England Journal of Medicine* during 1978 and 1979 (Volumes 298 through 301), of which 35 were randomized clinical trials. We do not make statistical inferences from our study to other journals or other times. Instead, we seek to describe the state of use of this important design in a consecutive series of studies reported in the *Journal*. We give examples from these studies to illustrate the use of this approach, the problems faced by clinical investigators, and the range of available solutions. We also discuss several preventive and corrective measures that affect the validity and strength of the scientific inferences that are drawn from parallel comparisons, and show how the design can sometimes be improved.

Tables 1 and 2 summarize certain design features of the 35 randomized and 12 nonrandomized parallel studies that are discussed in more detail below.

PROGNOSTIC FACTORS OTHER THAN TREATMENT

If patients who are assigned to separate treatment groups differ before treatment in factors that affect prognosis, such as age, extent of disease, or concurrent medical problems, the observed results may be affected by this difference as well as by treatment. The assessment of treatment effects is then biased. Such factors are called covariates, confounding variables, or prognostic factors; we use the term *covariates*. The term *imbalance* refers to differences in the distribution of covariates among the treatment groups. The overall initial advantage of one treatment group over another depends on how the distribution of the covariates differs among the groups.

In 32 of the 35 randomized studies (91 percent) and 7 of the 12 nonrandomized studies (58 percent), the investigators compared treatment groups for several covariates, usually one variable at a time. They used these comparisons to argue (often implicitly) that the pretreatment differences among groups were too small to jeopardize the validity of conclusions about treatment effects. A ubiquitous error was the interpretation of a nonsignificant statistical difference as sufficient evidence that the groups were not substantially different; this practice is based on a misunderstanding of the meaning of statistical significance.

For example, if 50 patients, evenly divided between men and women, are randomly assigned, 25 to treatment A and 25 to treatment B, the investigator has about a 20 percent chance of obtaining an imbalance for sex that is at least as extreme as if there were 10 men and 15 women in one group and the reverse in the other (a 20 percent difference). Although the statistical significance of this imbalance can be calculated as $P = 0.20$, this P value is not informative, since we already know that the null hypothesis of random allocation is true. With 50 patients or 50,000, what matters is the clinical importance of the observed imbalance, not its statistical significance. P values should rarely be used to examine the efficacy of randomization in a specific study. When they are so used, we recommend that they be presented clearly and explicitly as measures of the size of departure from perfect balance, not as tests of hypotheses.

There are three common ways to resolve or reduce the problem of imbalance of covariates. One, randomization (sometimes with stratification or blocking), is designed to prevent serious imbalance, and two others — stratified analysis and model-based adjustment — can reduce imbalance problems when they occur. These methods can often be combined. For example, pairs of patients could be matched on one or more important background variables (a form of stratification) and then a randomly chosen member of the pair is assigned to an experimental treatment, the other to the control (a form of blocking). The issues characteristic of parallel studies arise from the fact that the treatments are compared in separate groups, in contrast to those in crossover studies (see Chapter 5 in this book), in which each treatment is given (in succession) to each patient. In parallel studies, issues of bias, comparability, and sources of variation other than treatment revolve around the assessment and control of the effects of prognostic factors or covariates. Thus, issues concerning covariates dominate in this sample of reports.

RANDOMIZATION

Randomization confers several benefits on a study. It protects the rules for treatment assignment from conscious or unconscious manipulation by investigators or patients. Randomization may include the various forms of "blindness," but when properly performed it always ensures that the assignment to treatment groups is left to an objective and indifferent proce-

Table 1. Some Design Features of 35 Randomized Parallel Studies Reported in the *New England Journal of Medicine*.*

Volume:Page	Subject	Method of Assignment	Covariate Display†	Control of Covariate Effects	Would Modeling Have Helped?
298:243	Chlordecone toxicity	Strat (block?)	Overall	No	Yes
298:289	Recent myocardial infarction	Rand	By treatment	Model	—
298:413	Urinary-tract infection	Rand	No	No	No
298:758	Traveler's diarrhea	Rand	No	No	No
298:763	DES exposure	Rand	By treatment	Model	—
298:809	Cyanide toxicity	Rand	By treatment	No	Yes
298:981	Herpes zoster and cancer	Rand	By treatment	No	Yes
298:1041	Acute illness	Rand	Outcome	Post-strat	Yes
298:1052	Bone-marrow transplantation	Rand	By treatment	Post-strat	Yes
299:53	Threat of stroke	Strat (block?)	Outcome	Model	—
299:157	Hypersensitivity to insect sting	Strat (block?)	By treatment	Post-strat	No
299:561	Umbilical-artery catheterization	Rand	By treatment	Post-strat	Yes
299:570	Urinary-tract infection	Rand	Outcome	Post-strat	Yes
299:624	Pneumonia	Strat (block?)	By treatment	Post-strat	No
299:1151	Lupus nephritis	Rand	By treatment	No	No
299:1201	Oral candidiasis	Rand	By treatment	No	No
299:1261	Ovarian cancer	Block	Outcome	Model	—
300:149	Coronary bypass	Rand	Outcome	Model	—
300:274	Biliary cirrhosis	Rand	By treatment	No	No

Table 1. (Continued)

300:756	Nephropathic cystinosis	Block	By treatment	No	No
300:1074	Minocycline and gonorrhea	Rand	Outcome	Post-strat	Yes
300:1180	Herpes simplex labialis	Rand	Outcome	Post-strat	No
300:1345	Renal transplant	Strat (block?)	By treatment	No	No
301:126	Cryptococcal meningitis	Block	Outcome	Post-strat	Yes
301:225	Trigeminal-nerve surgery	Rand	No	No	No
301:293	Hypertension	Rand	By treatment	No	Yes
301:509	Gonococcal urethritis	Rand	Outcome	Post-strat	No
301:577	Hemodialysis	Rand	By treatment	Model	—
301:687	Advanced cancer	Strat (block?)	By treatment	Model	—
301:743	Myeloma	Strat, block	Outcome	Model	—
301:797	Acute myocardial infarction	Strat (block?)	Outcome	Model	—
301:855	Venous thrombosis	Strat, block	By treatment	Post-strat	No
301:962	Bypass surgery	Rand	Outcome	Post-strat	No
301:1077	Partial gastrectomy	Rand	By treatment	No	No
301:1301	Membranous nephropathy	Strat (block?)	Outcome	Model	—

* Strat denotes stratification, Block denotes blocking, Rand denotes simple randomization, DES denotes diethylstilbestrol, Post-strat denotes post-stratification of analysis by covariates, and Model denotes modeling or adjustment used.

† Outcome indicates that outcome information is given for covariates; By Treatment, for each treatment group; and Overall, for the pooled sample.

dure that cannot be predicted. Randomization is not the only way to achieve such protection, but it is the simplest and best understood way to certify that one has done so. Because of the critical role of randomization, and because some people use the word randomized to mean haphazard, both author and reader are served well by a short, clear statement about how the randomization was done (e.g., table of random numbers) and executed (e.g., by sealed envelopes prepared in advance).

Other assignment methods may leave a residue of uncertainty about the protection of the assignment rules. For example, Ehrenkranz et al.[1] de-

Table 2. Some Design Features of 12 Nonrandomized Parallel Studies Reported in the *New England Journal of Medicine*.*

Volume: Page	Subject	Covariate Display	Control of Covariate Effects	Would Modeling Have Helped?
299:564	Vitamin E and respiratory distress syndrome	By treatment	No	Yes
298:178	Lithium and neutrophils	No	No	No
298:1	Overweight, hypertension	Outcome	Post-strat	Yes
298:229	Recent acute myocardial infarction	By treatment	No	No
298:815	Acute lymphocytic leukemia	No	No	No
298:1429	Vaginitis and normal controls	No	No	No
299:17	Cholesterol and contraceptives	Outcome	Model	—
299:324	Live-born infants	Outcome	Model	—
299:847	Prolactin-secreting pituitary tumors	Outcome	Model	—
299:1321	Renal transplants	No	No	No
300:278	Doxorubicin for cancer	No	Model	—
300:524	Nephritis or vasculitis	Outcome	No	No

* For abbreviations and explanations, see footnote to Table 1.

scribed a nonrandomized (alternate-assignment) comparison of the use of intramuscular vitamin E with no treatment at all for bronchopulmonary dysplasia in neonates. The two groups, each of 20 infants, differed greatly on one-minute Apgar scores; the means ($\pm$ S.E.) of the Apgar scores were 6.2 $\pm$ 0.5 for the vitamin-treated group and 4.6 $\pm$ 0.5 for the control group. The Apgar-score imbalance favored the treatment group and was both medically important and statistically significant. The statistical significance of the difference suggests the possibility that the assignment was not neutral with respect to prognosis. It is possible, although unlikely, that the observed imbalance could have occurred by chance, and if the assignment had been random, that interpretation would have prevailed. Not having used random assignment, the investigators could not certify to others the objectivity of the treatment assignment, no matter how unbiased the assignment may in fact have been.

In the 47 articles we reviewed, the 12 nonrandomized studies were often undertaken in situations in which the assignment of treatments was not controlled by the investigators, so that there was no possibility of random assignment. We believe that the investigators could have randomized in only 2 of the 12 nonrandomized studies (Table 2), and we cannot be sure about these 2 because the published reports may not have mentioned all the constraints on study design. We are not sure how much this relatively widespread use of randomization in parallel studies reflects choice by research investigators and how much it reflects selection of stronger study design by *Journal* referees and editors.

Randomization alone may provide an adequate balance of covariates,[2] especially in larger studies, where the statistical law of large numbers reduces the chance of serious imbalance. Even with simple randomization, small studies are vulnerable to substantial covariate imbalance among treatment groups and to substantial unplanned imbalance in the sizes of the treatment groups themselves.

In designing a randomized study with a small sample size, it may help to add two features to the process of assigning patients: stratification and blocking. The investigator first classifies patients according to one or more important covariates (stratification), then randomly assigns treatment so that predetermined, approximately fixed proportions (often equal) of patients from each stratum receive each treatment (blocking). For example, an investigator may use a randomization scheme that assigns exactly three of each set of six successive male subjects to each treatment, and the same for female subjects. The randomization is then both stratified (by sex) and blocked (in groups of six). If the treatment groups are reconstituted by pooling the strata, they will be automatically balanced for the variables used for stratification with blocking and for other variables related to them.

Blocking alone can be useful, but stratification without blocking is no better than simple randomization in preventing imbalance. Kirkpatrick and Alling[3] used blocking without stratification to ensure that there were equal numbers of patients in each treatment group. In 13 of the 35 randomized clinical trials in our sample, the authors described their use of stratification or blocking.

Graham and Bradley[4] studied the efficacy of chest physiotherapy and intermittent positive-pressure ventilation for patients with pneumonia (Table 3). Before randomly assigning 54 patients to control or treatment groups, they stratified subjects by age, sex, prior antibiotic therapy, and smoking history. A serious imbalance of any of these covariates would have complicated the investigators' analytic and interpretative tasks and weakened the treatment comparison. They guarded against this possibility by randomizing the treatment assignments within the strata defined by these factors. The authors did not describe the particular method of stratification used in their study.

Research reports should describe or provide references for the specific method of stratification that was used, so that readers may judge its adequacy and better interpret the findings. For illustration, we will describe one such method as it might have been used by Graham and Bradley. Since they split age (under or over 45), smoking history (yes/no), sex, and prior antibiotic therapy (yes/no) into two categories each, there are $2 \times 2 \times 2 \times 2 = 16$ possible combinations of the four factors. The number of subjects in each cell may be random, and some combinations may not occur at all; this is not a barrier to stratification and blocking. Persons in these 16 cells must still be allocated to one or the other treatment. In the unlikely situation that all 54 patients had entered at one time, they could have been randomly and nearly evenly allocated to the treatments. In this manner the treatment groups would have had nearly identical distributions of the four covariates.

It is more usual for patients to arrive one at a time and to be placed in treatment groups immediately. This sequential feature can foil an investi-

Table 3.   Results of Randomization in a Study of Treatment for Pneumonia.*

	Control Group	Treatment Group
Total no. of patients	27	27
No. of men	13	14
Age (yr)†	63±3	61±4
No. of smokers	16	17
No. who took antibiotics before admission	5	10

* Data from Graham and Bradley.[4]

† Values are expressed as means ±S.E.M.

gator's attempt to balance the treatment groups for more than one covariate. Table 3 illustrates one possible outcome of such a procedure: three of the four stratified covariates were almost evenly split between the treatment groups; the fourth (prior antibiotic treatment) was somewhat lopsided. The attempt to balance covariates by stratification succeeded only partially. Although the covariate distribution is compatible with random allocation (the difference between the groups is not statistically significant), if prior antibiotic therapy had a clinically large effect on outcome the imbalance could have biased the results. Blocking within cells to minimize imbalance would have produced the most informative data set, but this can be done only when the treatment can be assigned by the investigator.

For a brief discussion of balancing and several references to other techniques, see Louis et al.[5]

STRATIFIED ANALYSIS BASED ON COVARIATES

A second way to deal with imbalance of covariates is to stratify the analysis — that is, to examine relatively homogeneous subgroups separately (often in lieu of stratifying the randomization). An imbalance of covariates can weaken any study, but even a large imbalance need not make a study unusable. For example, Reisin et al.[6] reported a nonrandomized study of the effects of weight loss and drugs on hypertension, in which there were sharp differences of both age and sex between treatment groups (Table 4).

The mean age of patients in their Group I was 28 years, whereas in Groups IIA and IIB it was 46 and 47, respectively. Furthermore, Groups I and IIA had a preponderance of men (67 and 74 percent, respectively), whereas only 38 percent of the patients in Group IIB were men. If response to treatment depends on age or sex, difficulties will arise in interpreting overall results.

Reisin et al. controlled for the imbalance by a combination of statistical analysis and cautious interpretation. The imbalance for age and sex (Table 4) was handled by presenting separate (stratified) tabulations of outcome for each sex and for each age group across the three treatment groups. Table 5 shows that the overall pattern of the largest effect from treatment IIA and the smallest from IIB persists within the separate age and sex categories. This practice of looking at results in each subgroup is called *post-stratification*, since the subgroups can be (and often are) defined after assignment to treatments or even after all follow-up data have been collected.

Such an analysis provides an opportunity to detect strong covariate influences on outcome and furnishes a straightforward check on the results of the study. Although post-stratification offers a valuable adjunct to more elaborate and sophisticated methods of reducing covariate influences, it has the drawback of requiring partition of the subjects into groups that may be too small to provide reliable estimates of treatment effects within strata. Frequently, one cannot satisfactorily interpret the results

Table 4. Distribution of Numbers of Patients According to Treatment Group, Age, and Sex in a Randomized Controlled Trial of Treatments for Hypertension.*

| | Treatment Group | | | |
	I	IIA	IIB	Totals for Age and Sex
Age				
≤44	17	26	12	55
≥45	7	31	14	52
Sex				
Male	16	42	10	68
Female	8	15	16	39
Treatment totals	24	57	26	107

* Data from Reisin et al.[6]

in individual strata or make informative comparisons. As explained below, statistical adjustment for covariates provides a solution to this problem.

A second problem with stratification, especially with post-stratification, is that when several statistical comparisons are made simultaneously, the chance that at least one will be statistically significant is increased. These comparisons need not all be explicit computations; the problem may occur when investigators calculate only one extra P value because they have noticed an oddity in the data, whether or not they have calculated P values for all the other less striking contrasts that they passed over.

ADJUSTMENT

Adjustment in general is the use of weighted averages of results in various strata — a technique also known as direct standardization. Mosteller and Tukey[7] discuss this technique (as well as indirect standardization) in some detail. Weighted averages are less subject to random variations than are results for single strata. The best choice of weights depends on the goals of the analysis.

For example, the data in Tables 4 and 5 show that the imbalance for sex makes direct comparisons of treatment groups difficult, and we may wonder what the results would have been had the three treatment groups had the same sex distribution. For this we use the sex distribution observed in the three groups combined (68 of 107, or 63.5 percent, were male). We estimate from the actual data on treatment in Group I that if this treatment had been applied to our entire combined population, the 63.5 percent who were men would have had a reduction in systolic pressure of 26.3 mm Hg, and the 36.5 percent who were women would have had a reduction of 24.3 mm Hg. The estimated effect over the whole (hypothetical) group would then have been $(26.3 \times 0.635) + (24.3 \times 0.365) = 25.6$ mm Hg. Similar adjusted effects of treatments in Groups IIA and IIB are 39.0 and 4.2 mm Hg, respectively. Comparisons among these three results are now nearly free of the influence of imbalance for sex because they estimate the treatment effect on the same (hypothetical) population that is 63.5 percent male. Results could have been adjusted to any other sex distribution. Although these values are close to the unadjusted values in the top line of Table 5, we have more confidence in their relevance.

Table 5. Mean Reduction in Systolic Blood Pressure for Three Treatment Groups Classified According to Age and According to Sex.*

	Treatment Group		
	I	IIA	IIB
		mm Hg	
Total for group	25.7	37.4	6.9
Age			
≤44	23.5	32.1	5.9
≥45	30.9	41.6	7.7
Sex			
Male	26.3	32.7	0.4
Female	24.3	49.9	10.9

 * Data from Reisin et al.[6]

Adjusted quantities have the stability of averages taken over several small subgroups and help to smooth out random variation from stratum to stratum. Disadvantages of the adjustment method include increased complexity of computation and loss of information about stratum-to-stratum differences.

MODELING

Modeling covariates is a general approach that includes adjustment as a special case. By constructing and validating a mathematical model of the effects of covariates on outcome, the investigator can remove much or most of the effects of covariates from the observed results, leaving nearly pure treatment effects. We will illustrate this approach with the data on the treatment of hypertension in Tables 4 and 5. Focusing on age, we begin by looking for separate estimates of age and treatment effects that combine to produce the observed outcomes. A set of techniques called analysis of variance,[7,8] which are not discussed here, can be used to explain each observation as a sum of terms that apply to sets of related observations plus an error term.

A model commonly used for adjustment represents the response (mean reduction in blood pressure) as a constant plus terms for various covariates. In this problem, this might be the constant plus a term giving the increment or decrement associated with treatment, a term giving the increment or decrement associated with age group, and a remainder term that compensates for lack of perfect fit.

Table 6 displays the structure. The numerical values in Table 6 are derived from standard analysis-of-variance calculations, which we do not explain here. As is apparent, the analysis-of-variance model is constructed so that the influences of a variable (such as age) add up to zero. The age-adjusted difference between Group I and Group IIA is 9.6 (= 13.2 − 3.6, from the treatment terms in Table 6), as compared with the unadjusted difference of 11.7 (= 37.4 − 25.7, from the "Total for group" row in Table 5).

The analysis-of-variance model adjusts the treatment comparison to compensate for the higher proportion of patients in Group I who are young (71 percent) as compared with the proportion in Group IIA who are young (46 percent). If the age distributions in Groups I and IIA had been equal, so would the adjusted and unadjusted treatment differences.

The remainder terms in Table 6 show the part of the observed value that is not explained by the constant, treatment, and age terms. They are com-

Table 6.   The Analysis-of-Variance Model for Treatment and Age in a Randomized Controlled Trial of Treatments for Hypertension.[*]

Treatment Group	Age	Observed Value		Constant Term		Treatment Term		Age Term		Remainder
I	≤44	23.5	=	23.6	+	3.6	−	3.1	−	0.6
I	≥45	30.9	=	23.6	+	3.6	+	3.1	+	0.6
IIA	≤44	32.1	=	23.6	+	13.2	−	3.1	−	1.6[†]
IIA	≥45	41.6	=	23.6	+	13.2	+	3.1	+	1.7[†]
IIB	≤44	5.9	=	23.6	−	16.8	−	3.1	+	2.2
IIB	≥45	7.7	=	23.6	−	16.8	+	3.1	−	2.2

[*]   Data from Reisin et al.[6] The numerical values for Constant Term, Treatment Term, and Age Term are derived from standard analysis-of-variance calculations, which are not explained in this chapter.

[†]   These values should add to zero. The discrepancy is caused by rounding. Note also that the treatment terms for younger (or older) subjects add to zero, and that the age terms (for any treatment) add to zero.

monly called residuals or unexplained deviations and contain whatever structure and variation in the data that are not represented in the model. When the remainder terms are small, the model is considered to fit the data well. Standard analysis-of-variance methods provide a numerical test for the goodness of fit of the model to the data, though this test depends on assumptions that are not always adequately met.

In other circumstances, a model may have more terms (e.g., treatment, age, and sex), or some terms may be multiplied rather than added, or more complex mathematical functions may be introduced. Guiding the investigator in the appropriate choice of a model is an important part of the statistician's contribution to a collaborative research effort. Developing a model that is relatively simple and fits the data well often requires substantial experience.

A second example of modeling covariate influences comes from Harter et al.,[9] who compared the effects of aspirin and placebo on the incidence of thrombi in 44 patients undergoing hemodialysis. Since the total number of thrombi per patient could only increase with time, the duration of follow-up evidently influenced the outcome. The authors implicitly adopted a mathematical model in which the total incidence of thrombi was proportional to the total length of follow-up. According to this model, if treatment had no effect, the average number of thrombi per patient-month should have been about the same for the two treatments. Their data (Table 7), based on substantial numbers of thrombi, showed a nearly threefold difference between the two treatment groups. Although one could instead compare the proportion of patients with one or more thrombi (6 of 19 vs. 18 of 25), the average number of thrombi per patient (14 thrombi among 19 patients vs. 53 among 25), or the average number of thrombi among patients with at least one thrombus (14 among 6 patients vs. 53 among 18), each of these latter figures answers a somewhat different question, each is subject to substantial change as the period of follow-up increases, and each lacks the simple intellectual appeal of the thrombi-per-patient-month model used by the authors.

In a third example, taken from the previously mentioned study by Hastings et al.,[2] the authors relied on post-stratification to control for the effects of an imbalance among covariates. Modeling of the sort described below would have been a more powerful tool. In their comparison of the value of antacid with that of gastric suction in the prevention of bleeding in

Table 7.   Results of a Study of Aspirin as Compared with Placebo in the Prevention of Thrombi.*

Treatment Group	Patients (no.)	Patients with Thrombi (no.)	Thrombi (no.)	Patient-Months	Thrombi/ Patient-Month
Aspirin	19	6	14	87.5	0.16
Placebo	25	18	53	115.2	0.46

* Data from Harter et al.[9]

critically ill patients, the authors post-stratified the results by simply counting the number of risk factors (from a fixed list) that were present in each patient (Fig. 1). Table 8, which is derived from their Figure 2, shows that the risk of bleeding increases with the number of risk factors. The "two-by-two" table for treatment versus outcome, when the data are pooled for all risk categories (Table 8), shows that the investigator detected gastric bleeding in 2 of 51 patients on antacid and 12 of 49 on gastric suction — a significant advantage for antacid treatment (Yates' corrected chi-square = 7.2, $P = 0.008$).

Analyses of the separate strata can be performed, but Gregory and Brown, in a letter commenting on the study,[10] pointed out that the chi-square test is inappropriate for such sparse tables. They suggested using exact tests or "more sophisticated techniques that estimate the relationship of risk factors and antacid prophylaxis to the incidence of bleeding." To do that, one must model probabilities, which range from 0.0 to 1.0. Additive models of probabilities can lead to absurd results, since there is nothing to prevent the model from estimating a probability as either negative or greater than 1. If it is more plausible that risk factors act multiplicatively on the probability, one is led to models that are additive in the logarithm of the probability. Such models, called *log-linear*,[11] have many other advantages, though their proper use may involve substantial computation.

We reanalyzed the data in Table 8 with a log-linear model that takes into account the slight imbalance between treatment groups in numbers of risk factors and found that $P = 0.011$. This P value is less extreme than the (incorrect) value of 0.008 reported for all risk categories combined, but the analysis confirms the advantage of antacid, regardless of risk factors. Re-

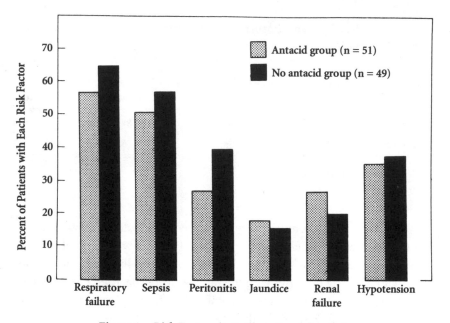

Figure 1.   Risk Factors in Patients Receiving and
Those Not Receiving Antacid Prophylaxis.
The similarity of the two groups is indicated by
the almost equal rate of each risk factor.
Reproduced from Hastings et al.[2] with the permission of the publisher.

analysis also shows a large, positive influence of the number of risk factors
on the probability of bleeding and a similar treatment effect for each stratum. Such conclusions may add to the value of the study and provide useful information for designing subsequent investigations.

The results of modeling are sometimes surprising. Neutra et al.[12] described a complex example of covariate adjustment with modeling in a
nonrandomized, nonexperimental design. About half of a group of 17,080
pregnant women received electronic fetal monitoring during labor. Comparing neonatal death rates in the monitored and unmonitored groups, the
authors reported that fetal monitoring appeared to reduce mortality: the
relative risk for unmonitored neonates was 1.7. Contrary to expectation,
the apparent advantage of monitoring declined when the authors adjusted,
using a log-linear model for prior inherent risk: the relative risk for unmonitored neonates dropped to 1.44. This drop was unexpected because

the prior supposition was that mothers with high-risk pregnancies were more likely to receive monitoring, so that monitoring should have appeared less favorable before adjustment for risk than after.

To derive full benefit from the scope and power of modeling often requires substantial consultation with an experienced statistician, but the alternative to modeling may be to leave valuable results buried in a mass of data, where they lie unused and perhaps even unsuspected. Modeling, in conjunction with simpler methods of analysis, should be considered whenever the investigator has reason to believe that measured factors other than treatment have a substantial influence on outcome.

If the investigator specifies the general form of the model (including the covariates to be included) before the data are analyzed, then a statistician can provide useful indicators of the precision and accuracy of the estimated effects of the treatments and covariates. These tell the investigator how much the estimated effects would vary by chance in repeated experiments and thus provide the basis for hypothesis testing. However, if the in-

Table 8. Relations among Treatment, Number of Risk Factors, and Incidence of Gastric Bleeding in a Study of Antacids as Compared with Gastric Suction to Prevent Gastrointestinal Bleeding.*

Treatment	Risk Factors	Bleeding		% with Bleeding
		yes	no	
	no. present	no. of patients		
Antacid	0–1	0	24	0
	2	0	8	0
	≥3	2	17	11
	Total	2	49	4
Gastric suction	0–1	1	13	7
	2	3	12	20
	≥3	8	12	40
	Total	12	37	24

* Data from Hastings et al.[2]

vestigator first examines the data with the intent of selecting the model (and the covariates) to obtain a good fit, then it is often impossible to assess the variability of the estimated effects. The investigator would know a great deal about the data set in hand but would have little basis for generalizing to other samples or to the general population.

MODELING TO INCREASE PRECISION

Concerns for covariate balance often arise first from the desire to obtain an unbiased assessment of the effects of treatment — that is, to avoid slanting the results one way or another. However, a second important consideration is that covariance adjustment can improve the precision of estimates of effects, even when balance is excellent.

The Veterans Administration Cooperative Study on Antihypertensive Agents[13] provides an example of the uncertainties arising from an incomplete covariate analysis. The study group compared ticrynafen and hydrochlorothiazide for antihypertensive effect at two dose levels. Among the many results, it appeared that ticrynafen at 250 mg (the lower dose) was a less effective antihypertensive than the other three regimens. This finding was not statistically significant, and the authors correctly used the term "suggested" rather than "demonstrated." However, if they had modeled covariates, the increase in precision might have been enough to confirm a real difference, thus promoting the suggested difference to a "finding." We find a clue to this possibility in the observation that the unadjusted difference was of clinical interest to the authors but statistically not significant. The adjustment procedure might instead have indicated the opposite, and the suggestion might even have disappeared because of a reduction in the treatment effect. Either way, both investigator and reader would have profited from a better understanding of the effects of the covariates. Stratified analysis does not offer the same possibility, since the reduction of sample size within strata acts to reduce the degree of precision.

In stratifying, one sacrifices precision to reduce bias; in modeling, one may achieve both a reduction of bias and an increase in precision if the model is appropriate, though of course an inappropriate model may seriously misrepresent the data. It appears to us (without consulting the original data) that in 26 of the 47 parallel studies in our sample, treatment comparisons profited or could have profited from well-chosen covariate adjustment. Adjustment was actually used in 14 of these 26 studies, including one[14] in which the overall result depended heavily on covariate influences. Here, the covariate analysis identified important clinical measures that were useful for prognosis and in the design of a new treatment study. The other 12 studies seemed to have the potential for either dramatic strengthening of a marginally significant conclusion or a substantial increase in the precision of an estimate.

In the remaining 21 of the 47 parallel studies, adjustment to improve precision would have added little; these studies had either definitive differences (e.g., $P<0.001$) in treatment effect or differences so small that it seems to us that no credible covariate adjustment could have increased them to either clinical importance or statistical significance. The authors of none of these 21 studies reported having performed covariate adjustment.

A COMPARISON OF METHODS OF COVARIATE CONTROL

The first defense against the effects of covariates rests with the method of allocating treatments to patients. When samples are large, randomized assignment of patients is likely to provide approximate balance on both measured and unmeasured (even unsuspected) covariates. If, in addition, a few important prognostic factors are known, random allocation with stratification and blocking for one or two factors (rarely more) can help balance those covariates and others related to them. Still, if investigators measure many covariates they are likely to find at least one that is seriously out of balance by pure chance. In these 47 studies post-stratification was often useful in finding and controlling for the influences of covariates, but adjustment was used less often. If the sample size permits and if the imbalance is not severe, division into more homogeneous categories with moderately large numbers can control unwanted covariate effects simply and effectively. We stress the requirement that the investigators choose such covariates before analysis to avoid statistical and other scientific problems associated with the use of covariates that are of interest mainly because the study in hand for analysis suggests that they are influential (see Ware et al., Chapter 10).

In addition to its other virtues, stratified random allocation with blocking makes the final conclusions relatively insensitive to the choice of adjustment methods. Also, even when covariates are in excellent balance, adjusting for their effects on outcome offers a means to refine statistical comparisons by reducing the standard error of such comparisons. Variance reduction requires moderate balance and works best when balance is excellent. Snedecor and Cochran[8] and Mosteller and Tukey[7] provide detailed information on how to perform and interpret covariate adjustments.

Post-stratification and adjustment complement each other; the former offers a conservative method for checking the validity of conclusions in the presence of influential covariates, and the latter offers a way to sharpen the precision of the results and extract more information from the study. We think there should be more of each in the conduct and analysis of future studies.

GENERALIZING RESULTS

So far, we have dealt with methods for ensuring that the analysis of a parallel-treatment study correctly assesses the relative effects of treatment on patients enrolled in that study. The investigator must also consider whether the results can properly be extended to patients outside the study; we refer to this issue as generalizability. The following example illustrates how the idiosyncrasies of a study sample may limit generalization of its results.

Bibbo et al.[15] reported the incidence of breast cancer after 25 years of follow-up of 1361 women who had been subjects in a randomized controlled trial of the effects on pregnancy of diethylstilbestrol. Although the difference in incidence of breast cancer between exposed and unexposed groups was not statistically significant, the authors noted that both the treated and control groups of women over 50 years of age had a significantly higher incidence of cancer than was expected from general-population data ($P<0.01$). It appears that women who were at high risk for breast cancer were more likely to have enrolled in this study. Thus, generalization of these findings to women at low risk would require independent justification.

This study reminds us that selection processes can influence the characteristics of patients in clinical trials to an important degree. Bibbo et al. carefully described the characteristics of their patient sample,[15] but in many other studies there is little effort to place the clinical-trial sample in a larger setting or to learn whether it matches a reference population.

Unless there is random sampling from a well-defined population (not to be confused with random allocation of treatment), the question of generalizability may not be completely resolvable. Nevertheless, a brief discussion of selection criteria and a concise description of the clinically relevant facts about the patients chosen for study can go a long way toward meeting this objective.

Mokrohisky et al.[16] tested the effect of the position of umbilical catheters on complication rates for high-risk infants. Their description of their methods takes the reader step by step through the selection process, starting with this statement: "From September, 1976, through May, 1977, all infants requiring an umbilical-artery catheter at the University of Colorado Medical Center were eligible for the study" and proceeding through initial exclusions, randomization, and subsequent exclusions to the final evaluated group. This description provides the reader with an excellent idea of the patient population to which the study refers. The reader is therefore prepared to judge whether the results in this study can be extended to a specific infant in need of treatment. Not all papers we reviewed have been so informative.

CONCLUSIONS

The issues of generalizability just discussed are not at all unique to parallel comparisons. Any other study design can have similar problems of sample selection, description, and inference for the general population. To judge the strength of inferences from a trial, the reader needs more information about its design and analysis than was presented in most of the reports in our sample. For a discussion of reporting protocol details in randomized trials, see the chapters by Bailar and Mosteller and DerSimonian et al. (Chapters 16 and 17 of this book). Many of their suggestions apply as well to other types of parallel-treatment comparisons.

Many investigators have successfully used the basic tools of design and analysis, including randomization and post-stratification. The wider use of stratified and blocked randomization and of the techniques of adjustment and modeling that we advocate in this chapter will require greater technical sophistication, probably including expert statistical help early in the design of trials. We think that the resulting improvement in power and validity will be worth the extra effort.

REFERENCES

1. Ehrenkranz RA, Bonta BW, Ablow RC, Warshaw JB. Amelioration of bronchopulmonary dysplasia after vitamin E administration: a preliminary report. N Engl J Med 1978; 299:564-9.

2. Hastings PR, Skillman JJ, Bushnell LS, Silen W. Antacid titration in the prevention of acute gastrointestinal bleeding: a controlled, randomized trial in 100 critically ill patients. N Engl J Med 1978; 298:1041-5.

3. Kirkpatrick CH, Alling DW. Treatment of chronic oral candidiasis with clotrimazole troches: a controlled clinical trial. N Engl J Med 1978; 299:1201-3.

4. Graham WGB, Bradley DA. Efficacy of chest physiotherapy and intermittent positive-pressure breathing in the resolution of pneumonia. N Engl J Med 1978; 299:624-7.

5. Louis TA, Mosteller F, McPeek B. Timely topics in statistical methods for clinical trials. Annu Rev Biophys Bioeng 1982; 11:81-104.

6. Reisin E, Abel R, Modan M, Silverberg DS, Eliahou HE, Modan B. Effect of weight loss without salt restriction on the reduction of blood pressure in overweight hypertensive patients. N Engl J Med 1978; 298:1-6.

7. Mosteller F, Tukey JW. Data analysis and regression: a second course in statistics. Reading, Mass.: Addison-Wesley, 1977.

8. Snedecor GW, Cochran WG. Statistical methods. 8th ed. Ames, Iowa: Iowa State University Press, 1989.

9. Harter HR, Burch JW, Majerus PW, et al. Prevention of thrombosis in patients on hemodialysis by low-dose aspirin. N Engl J Med 1979; 301:577-9.

10. Gregory PB, Brown BW Jr. Antacid titration for prevention of gastrointestinal bleeding. N Engl J Med 1978; 299:830-1.

11. Fienberg SE. The analysis of cross-classified categorical data. 2nd ed. Cambridge, Mass.: MIT Press, 1980:95-119.

12. Neutra RR, Fienberg SE, Greenland S, Friedman EA. Effect of fetal monitoring on neonatal death rates. N Engl J Med 1978; 299:324-6.

13. Veterans Administration Cooperative Study Group on Antihypertensive Agents. Comparative effects of ticrynafen and hydrochlorothiazide in the treatment of hypertension. N Engl J Med 1979; 301:293-7.

14. Canadian Cooperative Study Group. A randomized trial of aspirin and sulfinpyrazone in threatened stroke. N Engl J Med 1978; 299:53-9.

15. Bibbo M, Haenszel WM, Wied GL, Hubby M, Herbst AL. A twenty-five-year follow-up study of women exposed to diethylstilbestrol during pregnancy. N Engl J Med 1978; 298:763-7.

16. Mokrohisky ST, Levine RL, Blumhagen JD, Wesenberg RL, Simmons MA. Low positioning of umbilical-artery catheters increases associated complications in newborn infants. N Engl J Med 1978; 299:561-4.

5

CROSSOVER AND SELF-CONTROLLED DESIGNS IN CLINICAL RESEARCH

Thomas A. Louis, Ph.D., Philip W. Lavori, Ph.D.,
John C. Bailar III, M.D., Ph.D., and Marcia Polansky, D.Sc.

ABSTRACT   In crossover studies, each patient receives two or more treatments in sequence, and in self-controlled studies each patient serves as his or her own control. These designs for clinical trials can produce results that are statistically and clinically valid with far fewer patients than would otherwise be required.

Before choosing a crossover design, an investigator should be prepared to demonstrate the absence of carry-over effects, guarantee low drop-out rates, and expect a stable underlying disease. Since a parallel study with covariate adjustment may still be a strong competitor, we recommend careful consideration of the trade-off between power and fragility.

We investigated the use of the crossover design in the 13 crossover studies that appeared in the *New England Journal of Medicine* during 1978 and 1979. We considered the following important features of design and analysis as they applied to these studies: the method for assigning initial treatment (only 7 of 13 studies used random assignment); the determination of when to switch treatments (10 of the 13 used a time-dependent rule, and 1 a less appropriate disease-state–dependent rule; in 2, this point could not be determined); blinding of the crossover point (in only 3 of the 13 studies was the crossover point concealed, but in 4 of the remaining 10 concealment was impossible); assessment of the effects of the order of treatments (included in only 1 of the 13 studies); and the use of at least minimally acceptable statistical analysis (11 of the 13 studies had such an analysis).

We also report briefly on 28 additional studies of a single treatment each. In these studies each patient served as his or her own control before or after treatment or both. The scientific issues were much the same as in crossover studies except that self-controlled comparisons of treatments tended to be less precisely designed and conducted and to focus on clinical problems and patient groups that are especially difficult to study.

In any clinical trial the investigator studies the responses to treatment of a sample of patients for the purpose of inferring which treatment to prefer in a wider group or population. Chapter 4 discussed issues bearing on the strength of scientific inferences from parallel comparisons of clinical treatment — in such studies each patient receives only one treatment and responses in one group are compared with those in another.[1] This chapter extends the discussion to crossover studies — those in which each patient receives two or more treatments in sequence, and outcomes in the same patient are contrasted. One of the treatments may be a placebo or no treatment. We also comment more briefly on self-controlled studies. In these studies a single treatment is evaluated by comparing patient status before and after treatment. We consider the role of the crossover design in clinical research, using 13 crossover studies from Volumes 298 to 301 (1978 and 1979) of the *New England Journal of Medicine*[2-14] as examples. We highlight the clinically relevant issues for researchers and readers and refer them to the statistical literature[15-22] for mathematical details.

PARALLEL VERSUS CROSSOVER DESIGN

In a two-treatment crossover study, each patient's response under treatment A is compared with the same patient's response under treatment B, so that the influence of patient characteristics that determine the general level of response can be "subtracted out" of the treatment comparison.

If variation in patient characterization accounts for much variation in response, a crossover design based on a small sample of patients can provide the same statistical accuracy as a larger parallel study. Nevertheless, the decision to use a crossover design cannot be based solely on this potential saving in sample size, because powerful designs are also potential disasters. Their success balances on a narrow base of scientific and statistical assumptions. An investigator choosing between parallel and crossover designs should consider five factors that determine the effectiveness of the crossover design: (1) carry-over and period effects on treatment outcomes; (2) treatment sequencing and patient assignment; (3) crossover rules and timing of measurements; (4) dropouts, faulty data, and other data problems; and (5) statistical analysis and sample size. We will define these terms and discuss these five issues in detail after giving an example that illustrates the attractiveness of crossover designs.

Table 1 provides information relating to these factors for the 13 cross-over studies we found in the *Journal*. The studies were identified by means of a classification system developed for clinical studies. We classified a study as a crossover study only if both treatments were realistic candidates for clinical use and if either could be administered after the other. Therefore, although treatment and control readings were taken in a self-controlled study of a single treatment, we would not have classified it as a crossover study unless the use of no treatment at all was a realistic clinical alternative. Also, many medical/surgical studies do, in a sense, have cross-overs from medical to surgical therapy, but these generally compare immediate surgery with an approach calling for medical treatment followed if necessary by surgical treatment. We would classify such studies as parallel comparisons. One variety of crossover design that we did not encounter in this series matches paired organs, such as teeth or limbs. By assigning one member of each matched set of organs to each treatment the investigator may be able to compare responses within an individual. The strengths and weaknesses of the paired-organs design are similar to those of an ordinary crossover design.

EXAMPLE

The following example illustrates the power of the crossover design and the potential problems that are associated with it.

With just four diabetic patients, Raskin and Unger[6] were able to compare the effects of two insulin-infusion regimens on chemical components of blood and urine. As part of the experiment, they monitored urea nitrogen excretion for 48 hours — first while patients were receiving intravenous insulin and somatostatin and again after switching the same patients to intravenous insulin, somatostatin, and glucagon. Our Table 2 is adapted from their Table 1. The data shown there are the rate of urinary excretion of urea nitrogen in each of four patients while they were receiving insulin and somatostatin and again when they were receiving insulin, somatostatin, and glucagon.

If these eight measurements had been obtained in eight patients in a study with a parallel design, the difference in mean nitrogen excretion rate (3 g per 24 hours) would not have been statistically significant. The standard error of the difference in means would be 2.76, computed from the in-

Table 1. Characteristics of 13 Crossover Studies in the *New England Journal of Medicine*, 1978–1979.*

Article Reference No.	No. of Patients	Method of Allocation	Crossover Rule	Blind Crossover	Assessed Order Effects	Used Multivariate Methods	Disease or Subjects	Treatments	Outcome Variables
5	11	Rand	Time	No	No	Yes	Insulin-dependent diabetics, 20–29 years old	Subcutaneous insulin injection, into arm, leg, or abdomen, each with rest or exercise	Insulin absorption; plasma glucose levels
7	9	Rand	Time	Yes	No	No	Healthy non-coffee drinkers	(a) Placebo (b) Caffeine	Plasma renin; catecholamine; cardiovascular control
12	17	Rand	Time	No	No	Yes	Moderate to severe asthma	Seven different combinations of terbutaline, aminophylline, and placebo	Spirometric measurements and cardiovascular function
9	10, 21	Rand	Time	No	No	Yes	Irritable-bowel syndrome	Clidinium, placebo (normal controls untreated)	Myoelectrical and motor activity of colon
2	10	Rand	Time	No	No	Yes	Classic stable exertional angina	Passive smoking with or without ventilation; non-smoking controls	Exercise tolerance
10	12	Rand	Time	Yes	No	Yes	Men with stable angina pectoris	Five beta-adreno-receptor-blocking agents, placebo, each at 4 doses	Cardiac function

							Patients	Treatments	Outcome
4	113	Rand	State	No	No	Yes	Patients receiving cancer chemotherapy	(a) Prochlorperazine (b) Nabilone	Nausea; vomiting; side effects
13	22	Alt	Time	No	Yes	Yes	Severe burns or trauma	Three formulations for intravenous nutrition	Degree of protein catabolism
6	4	SFS	Time	No	No	Yes	Juvenile diabetes	Two diets, each with two drug treatments	Several metabolic measurements
11	20	SFS	Imp	No	No	Yes	Adults with asthma	Theophylline in various formulations	Serum concentration of theophylline
3	6	SFS	Time	No	No	Yes	Thalassemia	Drinking tea vs. water	Iron absorption
8	5	SFS	Time	Yes	No	Yes	Systemic mastocytosis	(a) Disodium cromoglycate (b) Lactose	Histamine in urine; symptoms recorded in diaries
14	15	Hap	Imp	No	No	No	Children with chronic asthma	(a) Alternate-day prednisone (b) Beclomethasone propionate (c) Combination	Hypothalamic–pituitary–adrenal function

* Rand denotes randomized, Alt denotes alternating, SFS denotes same fixed sequence, and Hap denotes haphazard. Time denotes time-dependent, State denotes state-dependent, and Imp denotes impossible to determine.

sulin and somatostatin and the insulin, somatostatin, and glucagon columns in an unpaired manner (see Table 2). Using instead the within-patient differences as the basis of the data, we obtain exactly the same mean difference (3 g per 24 hours) but with a standard error of 0.4. These differences produce a paired t-statistic of 7.5 with 3 degrees of freedom, providing strong evidence for the hypothesis that the change from insulin and somatostatin in the first period of the experiment to insulin, somatostatin, and glucagon in the second period raised the level of nitrogen excretion.

These results could be interpreted either as supporting a difference between the effects of insulin and somatostatin on the one hand, and insulin, somatostatin, and glucagon on the other, or as supporting a difference produced by the order of administration (the combination of insulin and somatostatin was always used first). The study provides no information on or control for the effects on response of treatment sequence or of changes in disease state. Although diabetes is a fairly stable disease and the investigators incorporated a washout period, a stronger study would have resulted if they had treated only half the patients first with insulin and somatostatin and the other half first with insulin, somatostatin, and glucagon. Patients could have been selected on the basis of low urea nitrogen values, and regression to the mean (the tendency of unusually high or low values to be-

Table 2. Urinary Excretion of Urea Nitrogen in Four Diabetic Patients.*

Patient No.	Treatment		Difference†
	IS	ISG	
	grams of urea nitrogen/24 hours		
1	14	17	3
2	6	8	2
3	7	11	4
4	6	9	3
Mean	8.25	11.25	3.00
S.E.M.	1.90	2.00	0.40

* Data from Raskin and Unger.[6] IS denotes intravenous insulin and somatostatin, and ISG denotes intravenous insulin, somatostatin, and glucagon.

† The S.E. of the difference between the means of ISG and IS: $2.76 = \sqrt{(1.90)^2 + (2.00)^2}$, if the groups were unpaired.

come less extreme) could have accounted for the observed difference. Also, no matter how powerful the design, studies based on data from only four patients may have relatively little clinical impact until they are confirmed with larger samples.

If the order of treatments made no difference and the variation in response remained as in Table 2, a parallel comparison would require about 14 times as many patients to achieve the same level of statistical significance (about 56 patients divided equally between treatment groups). Even if no reduction of variance were obtained, the crossover design would require only half the number of patients to produce the same precision as a parallel comparison. Each patient would contribute information on both treatments, but each would be under study longer than in a parallel design.

This reduction in sample size can be crucial in the study of treatment for an uncommon condition with a long and complex course. For example, Soter et al.[8] evaluated the use of disodium cromoglycate in the treatment of five patients with systemic mastocytosis. Graphs show the time course of symptoms in relation to the initiation and withdrawal of several administrations of drug and placebo. The high degree of patient-to-patient variability and the small number of available patients necessitated the use of the repeated crossover design to provide credible information on treatment efficacy.

KEY FACTORS

The five factors listed earlier as determining the effectiveness of the crossover design need to be considered carefully before any crossover study is mounted. We can give some general guidelines on their consideration, and in this section we continue our discussion of them as they apply to the 13 studies listed in Table 1.

CARRY-OVER AND PERIOD EFFECTS

Carry-over Effects

The therapeutic effects of the first treatment may persist (i.e., carry over) during the administration of the second. Investigators can often minimize this influence by appropriately delaying treatment administration. For example, Thadani et al.[10] compared five beta-adrenoreceptor blockers for their immediate effects on pain and other clinical signs in patients with stable angina. They administered and evaluated the drugs at weekly intervals, allowing time for the effects of one drug to dissipate before administering the other.

Period Effects

The disease may progress, regress, or fluctuate in severity during the period of investigation. For example, in the study of five patients with systemic mastocytosis, Soter et al.[8] dealt with the complex course of the disorder by switching treatment several times for each patient, with different intervals between the crossover points. The effects of any systematic trends or cycles should have balanced out in the design.

Both period and carry-over influences are called *order effects*. They complicate the task of interpretation and analysis, and weaken the scientific and statistical basis for choosing a crossover design. Carry-over effects are the most troublesome, for they suggest that a subsequent treatment's activity depends on the previous treatment. With proper design, both types of order effects can be assessed and removed from the treatment comparison. The assessment of order effects must be based on a statistical model. If the relative advantage of one treatment over another depends on which treatment the patient received first, then the difference in mean responses of treatment groups during each period will vary from period to period. This variation of treatment effect is called the *treatment by period interaction*. If such an order effect is comparable to or greater than the average treatment effect, the different treatment sequences should be considered distinct test regimens. In this situation, the design is no more powerful than that of a standard parallel comparison and can be less powerful or invalid. Brown[15] shows that estimating and adjusting for order effects requires a sample size greater than that needed for a parallel comparison. Therefore, unless order effects are known to be negligible, the crossover design loses its advantages.

Order effects may have operated in each of the 13 crossover studies in our sample. As shown in Columns 3 through 5 of Table 1, investigators tried, to various extents, to control order effects by balanced sequencing and carefully timed measurements, but only 1 of the 13 reported on these attempts (column headed "Assessed Order Effects"). The reasons for this low reporting rate are not clear. Lack of carry-over effects seems probable in most of the studies, but period effects could have been present. Readers of these reports could evaluate them more easily if the authors had included additional detail, such as an estimate and confidence interval for the treatment by period interaction.[18,21] This interaction summarizes all order effects that may be present. A descriptive summary of the interaction can be provided by a simple two-by-two table cross-classifying average treatment response and period; graphic displays are discussed in Huitson et al.[22]

Crossover designs are most appropriate in the study of treatment for a stable disease. Stable situations are studied in 12 of the 13 *Journal* articles we examined. In the one exception[13] the authors took special care to prevent bias resulting from the rapid changes in patient status.

TREATMENT SEQUENCING AND PATIENT ASSIGNMENT

The investigator must assign each patient to an initial treatment, and if there are more than two treatments or more than one administration of the treatments, he or she must specify the sequence. If all patients receive treatment according to the same fixed sequence, A followed by B, comparisons must be based on the assumption that the effects of the second treatment (B) after the first (A) do not differ from the effects B would have if it were given first. Such an experiment would provide no data by which to assess this assumption. If disease or treatment characteristics make B after A a fundamentally different treatment from B alone (e.g., chemotherapy after radiation vs. chemotherapy alone), no treatment comparison can be made. If some patients receive the sequence AB and others BA, then information is available on this issue.

If patients become available for study over time, there are four basic ways of assigning them to treatment sequences: (1) by use of the same fixed sequence for all patients (four papers in our sample); (2) by random assignment among the sequences (seven papers); (3) by deterministically balanced assignment — for example, giving the first patient AB and the second BA, then repeating as often as needed (one paper); and (4) by uncontrolled, haphazard assignment — using sequences neither fixed in advance nor governed by a randomization procedure (one paper).

Method 1 does not provide the information needed to estimate and adjust for order effects. With this method, the validity of treatment comparisons depends on the near absence of order effects, and such absence would have to be established by data or argument external to the experiment.

Random assignment protects against conscious and unconscious bias. Deterministic balancing and haphazard assignment may be as valid but are more prone to biases caused by selection of study participants and initial treatment assignments. Therefore, we recommend random assignment of treatment order, with forced balancing to get nearly equal numbers of patients assigned to each order (blocking; see Chapter 3).

CROSSOVER RULES AND TIMING OF MEASUREMENTS

A previously specified crossover rule strengthens the scientific and clinical validity of a study. Investigators commonly employ either of two types of crossover rules — one that calls for a switch in treatments after a specified length of time (*time dependent*) and one in which a crossover occurs when indicated by the clinical characteristics of the patient (*disease-state dependent*). These rules have different impacts on the magnitude and interpretation of order effects and on the general scientific strength of the study.

As with other aspects of designed experiments, the timing of measurements should be explicitly incorporated into the research protocol. The most scientifically acceptable switch points depend only on elapsed time. Initiating a treatment in response to the appearance of symptoms and withdrawing it or changing treatments when symptoms disappear makes it difficult or impossible to interpret observed treatment effects. In the 13 papers under study, 10 had time-dependent crossover rules and 1 a disease-state–dependent rule; in 2, the rules governing crossover were not clearly reported.

Whenever possible, crossover points should be concealed from patients and observers (blinded). Knowledge of a switch can influence treatment response or assessment or both, so that blinding the crossover point can reduce the influence of any subjective order effects.

DROPOUTS — FAULTY AND OUTLYING DATA

Although dropouts and implausible data points are problems for any study, their effects may be large in a study with a crossover design, because each patient contributes a large proportion of the total information and the design is sensitive to departures from the ideal plan. For example, consider again the study of diabetic patients[6] and the data in Table 2. If Patient 1 drops out during the first treatment period, only three patients are left to provide useful data. The mean difference between the two treatment regimens is still 3.00 g, but its standard error is now 0.577, and the t-statistic with 2 degrees of freedom equals 5.2, with a P value of 0.03. Although the result is still statistically significant, the P value has risen dramatically from the original 0.006. If the initial result had been less definitive, this loss of one patient would have altered the conclusions of the study. A single dropout in the comparable parallel study (with 28 patients in each group) should have had little influence on the conclusions.

Dropout rates can be high in crossover studies, since patients must receive at least two treatments to provide a complete data point. Partial information on a patient completing one treatment and then dropping out can be used in estimating the treatment effects in Period 1, provided that dropping out is not related to treatment response. A high dropout rate greatly weakens the study, and the initial sample size should be sufficiently large to compensate for this effect. Only 1 of our 13 studies reported having a subject drop out after the initial assignment of therapy. These studies were generally short and used treatments with few extreme side effects — conditions that favor the success of a crossover design.

Statistical analyses should include the identification of dropouts and deviant data points, and conclusions should not be sensitive to the latter. Protection against such sensitivity can be obtained either by setting aside deviant data or by using new techniques developed to be "resistant."[23,24]

STATISTICAL ANALYSIS AND SAMPLE SIZE

Observations made repeatedly in the same patient tend to be more similar than those made in different patients. Statistical analyses that take this relation into account are more complicated but potentially more powerful than those that are appropriate for a parallel comparison. Most important, the patient and not an individual measurement is the basic unit of statistical analysis.

Proper statistical analysis begins in effect by comparing data from a single patient over time and then combines these comparisons across patients.

When only two measurements are compared (as in the Raskin and Unger study[6]), the paired t-test provides a widely applicable procedure. In many studies, however, three or more observations are generated by each patient, and a different statistical analysis is necessary.

For example, in the study of passive smoking,[2] each patient generated six measures of the length of time to angina (a base-line and post-treatment measurement for each of three treatment days). These data were collapsed to form three values by computing the difference between post-treatment and base-line measurements for each day. A fully multivariate analysis would have used these three values as a single (though still multivariate) unit for analysis. Analysis should proceed with the use of the multivariate regression and analysis-of-variance techniques.[17,18,20,25] These techniques model the association (correlation) among the measurements for a single individual and use this association in computing standard errors of comparisons. Also, the techniques allow adjustment of P values for multiple tests on the series of measurements. In effect, they operate by linking together the results of several paired t-tests. They yield statistical tests and estimates of treatment and order effects to enable the investigator to assess the success or failure of the crossover design and then to compare response to treatment.

In our sample of 13 studies, 2 used no multivariate methods, and 11 employed them with various degrees of sophistication to assess pair-wise treatment comparisons. All authors failed to exploit the data structure fully. Billewicz[26] reported a similar failure in 9 of 20 examples from the medical literature.

Of the 13 crossover studies, 12 used a small number of patients (4 to 22), though the total number of observations of treatments was much larger. This economical use of patients both justifies and results from the powerful crossover design.

Table 1 indicates that the crossover design was used to study a wide variety of diseases and treatments, with responses generally measured by a biochemical marker. The diseases were stable, and the treatment effects transient, but this group of studies generally did not satisfy basic design requirements. For example, only 7 of the 13 employed random allocation of treatment order, and only 1 included even a basic assessment of the effect of treatment order on outcome. Although 11 of the reports included acceptable statistical analyses, each failed to report some potentially important information.

Rules calling for time-dependent, as opposed to disease-state–dependent, crossover are necessary for a valid treatment comparison and were used in 10 of the 13 studies. Only 3 of the 13 studies blinded the crossover point, but blinding was apparently impossible in 4 of the remaining 10.

The conclusions drawn in these studies could have been strengthened by empirical evidence that lasting effects of one treatment were unlikely to influence the results of subsequent treatments. The assumption that there were no residual effects can be supported only by biologic principles or previous laboratory and clinical data. The degree of credence appropriate to the reported results depends heavily on the relevance and persuasiveness of such information.

FURTHER COMPARISONS WITH PARALLEL DESIGNS

If the most favorable underlying conditions for a crossover study are met (that is, if neither time nor treatment order affects the response to the currently administered treatment), pairing each patient with himself or herself and analyzing the data properly can markedly reduce the influence of patient characteristics on the treatment comparison. This pairing represents the ultimate form of statistical adjustment for such characteristics.[1]

Even when the disease and treatments are satisfactory for a crossover trial, the choice between such a trial and a parallel comparison can be difficult. Although a crossover design has the potential advantage of economy and of providing a direct comparison of treatments in the same patient, the parallel design allows more straightforward analysis, its efficacy is less dependent on assumptions about the disease process, and it generally produces a lower dropout rate because each patient generates fewer measurements in a shorter time. In addition, the use of base-line measurements and statistical adjustments can greatly increase the precision of a parallel design.

For an example of the potential increase in precision in a parallel design, let us look again at the insulin study,[6] but suppose that the data came from a study with a parallel design using four men and four women, as shown in Table 3. Notice that the levels of urea nitrogen excretion are 5.0 g higher per

24 hours in men than in women, but that the difference between the means for the two treatment regimens is still 3 g per 24 hours. Although we do not have paired responses for four patients, we can use sex to explain some of the patient-to-patient differences in response. This explanation is accomplished by the analysis of variance, in which a level of urea nitrogen excretion is explained by the sum of a base-line value, an effect of treatment (insulin and somatostatin vs. insulin, somatostatin, and glucagon), an effect of sex (male vs. female), and a residual (the deviation between the observed response and that predicted by the model). See Table 3 for two examples.

The standard error computed from this analysis of variance is 2.09 — a 24 percent reduction from the 2.76 computed in Table 2 using the analysis that ignores the pairing. The t-statistic for the difference in treatment group means is 1.44 (3.00/2.09) with 5 degrees of freedom ($P = 0.20$). Recall that the unpaired comparison of treatment means produces a t-statistic of 1.08 (3.00/2.76) with 6 degrees of freedom ($P = 0.32$). Although the sex-adjusted analysis is still not statistically significant, paying this one degree of freedom to remove the sex effect makes the experiment more precise by reducing the estimate of statistical variability.

Table 3. Variation on the Study Shown in Table 2, with a Hypothetical Sex Variable.*

	Treatment		Mean	Sex Effect
	IS	ISG		
	grams of urea nitrogen/24 hours			
Men	14	11	12.25	2.5
	7	17		
Women	6	9	7.25	−2.5
	6	8		
Mean	8.25	11.25	9.75	
Treatment effect	−1.5	1.5		

*   Response = base-line value + (treatment effect) + (sex effect) + residual; residuals (not shown above) are the deviations between observed responses and those predicted by the model. They make the right-hand side equal the left-hand side in the general formula and numerical examples.

Examples:	Base Line		Treatment		Sex		Residual
14 =	9.75	−	1.5	+	2.5	+	3.25
8 =	9.75	+	1.5	−	2.5	−	0.75

The use of sex to help explain the data is an example of covariance adjustment. A parallel-design study with a covariance adjustment for sex would still have to be about 11 times as large as the crossover study, instead of the 14-fold size mentioned above, because some but not all of the patient variability would be accounted for by the covariate (sex). Thus, the covariance adjustment would provide some of the benefit of the crossover design. Such covariance adjustments must be interpreted with care, especially if they are based on patient characteristics discovered through an exhaustive search to produce a desired result.

In a choice between a parallel and a crossover design, we think the burden of proof should be placed on those favoring the crossover design to show that it can succeed in improving on the parallel design, though we think that such proof will often be forthcoming. Evidence of a strong likelihood of success with a crossover design in a particular study would include previous studies validating the absence of carry-over effects, low dropout rates, and a relatively stable disease process. Even in such situations a parallel design adjusted for previously specified covariates can be a strong competitor. We recommend that a design be chosen only after careful consideration of research goals, the disease process, and the trade-off between power and fragility. Such consideration requires a collaborative effort among clinical, laboratory, and statistical scientists.

PATIENTS AS THEIR OWN CONTROLS

Volumes 298 through 301 of the *Journal* contained an additional 28 reports of studies in which patients served as their own controls before, during, or after treatment. Such research designs incorporate many of the features of crossover studies, but new problems arise, as we will discuss after we give three examples of self-controlled studies. In one example, Peck et al.[27] evaluated the effect of 13-*cis*-retinoic acid on severe acne by examining changes in disease status; it appears that the acne was of such duration and intractability in each patient that substantial spontaneous improvement was unlikely to interfere with treatment evaluation. In another example,[28] patients with hypertension that had been found on screening were followed to determine whether informing them that they had elevated blood pressure was reflected in increased absenteeism for illness. Each patient's work record after diagnosis was compared with the patient's previous record. (Other parts of this study used crossover and parallel designs; we will not discuss those aspects here.)

Packer et al.[29] used a self-controlled design to investigate the clinical reaction of patients with congestive heart failure to the abrupt withdrawal of nitroprusside. Previous research had established that nitroprusside produced rapid hemodynamic and clinical improvement in patients with congestive failure, but some results had suggested that the sudden termination of treatment caused adverse clinical reactions. The self-controlled design seems ideal for the study of this problem, and it is hard for us to see how either crossover or parallel designs, usually considered more powerful, could have been employed.

Many crossover and parallel[1] studies include an element of self-controlling. We have classified such studies as crossover or parallel, and concentrate here on designs in which observations before, during, or after treatment are used as the main control on experimental variation.

Each of the five critical issues in crossover studies discussed above applies with undiminished force to self-controlled studies. We will discuss the following new problems, though at less length than the first five: lack of symmetry; the nature of the problems studied; the study of patients with refractory disease; and the absence of direct comparisons.

LACK OF SYMMETRY

A critical difference between crossover and self-controlled studies is the fundamental lack of symmetry in the latter between observations during treatment and those during control periods. These observations often differ in such matters as duration, nature, and intensity of clinical studies, and decision rules about when to modify or switch treatments. Such differences can lead to substantial bias (for example, in opportunities to observe untoward developments in the disease process) that may tilt results toward or away from the treatment under study. For example, in the Peck et al. study[27] of the effect of 13-*cis*-retinoic acid on severe acne, the scope and duration of observation before and after therapy did not match the intensity of study during the treatment period. In well-designed crossover studies, this problem does not arise, because such effects are balanced between treatments and become part of the random error.

NATURE OF THE PROBLEMS STUDIED

Another general difference between self-controlled and crossover studies is in the nature of the problems studied. In our sample, self-controlled studies were more often used at early points in the clinical development of new treatments, so that attention focused on multiple laboratory measurements rather than on one or two measures of clinical outcome. For example, Stacpoole et al.[30] used a self-controlled design in the first study in human subjects of the metabolic effects of dichloroacetate, which had been found to be effective in diabetic and starved rats. They measured blood glucose, plasma lactate, alanine, and other biochemical indexes before and after treatment. Similarly, Davis et al.[31] used this design to study the effects of captopril on arterial blood pressure, cardiac output, and plasma renin activity, and Gavras et al.[32] used it to investigate the effects of a new oral inhibitor of angiotensin-converting enzyme, SQ 14225, on cardiovascular status, liver function, and kidney function.

Although this measuring of multiple outcomes is not necessarily a weakness in self-controlled designs, it does mean that methods of analysis must be carefully tailored to exploit the data. One difficulty is the *multiple-comparisons problem*, addressed by Ingelfinger et al.[33] and others (see Chapter 12). Another is the optimal multivariate analysis of numerous dependent, as well as independent, variables. Expert help in multivariate statistical methods was rarely evident in the 28 self-controlled studies that we reviewed, though we believe it could often have led to substantially more productive use of the data.

STUDY OF PATIENTS WITH REFRACTORY DISEASE

A third common difference from crossover studies was the focus of many self-controlled studies (at least 7 of the 28 in our sample) on patients whose disease had responded inadequately to standard therapies. This may result, in part, from the use of this design at early stages of investigation in human subjects and the ethical considerations that lead to the enrollment of patients with refractory disease. We cite three examples. In a study by Bilezikian et al.[34] of mithramycin-induced hypocalcemia in patients with Paget's disease, the patients "had previously been treated with a variety of agents, including oral phosphate, calcitonin and ethane-1,hydroxy-1, 1-diphosphonate, to which they had, in general, responded poorly." In the study of 13-*cis*-retinoic acid for acne,[27] patients "uniformly had a history of minimal response to treatment with oral and topical antibiotics, oral vitamin A, topical vitamin A acid, topical benzoyl peroxide, x-rays, and other acne therapies." And a study by Tamborlane et al.[35] of portable infusion pumps for the injection of insulin focused on juvenile diabetic patients who had continued to excrete albumin while receiving intravenous insulin.

Several problems arise in studies of patients with refractory disease. One is that intense, prolonged (though perhaps incompletely effective) treatment may require a prolonged washout or recovery period, while the patient's clinical problems demand prompt relief. We found no evidence that clinical imperatives resulted in inadequate washouts in our sample, but it is not clear that any of the investigators looked for such evidence. Another problem results from regression toward the mean — the tendency of an extreme value when it is remeasured to be closer to the mean, because the original value was likely to have been unduly influenced by random variation. If patients are enrolled in a self-controlled study of some new regimen when standard treatments appear to be losing efficacy (that is, when the patients' conditions are worse than average), some general improvement may occur that has nothing to do with improved therapy. This problem can sometimes be addressed by using a first set of measurements after a washout period to establish patient eligibility for a self-controlled study and a second set, after a further stabilizing period, to establish baseline levels.

Patients whose disease fails to respond to standard treatment may include those with variant forms of the disease or those whose general conditions have deteriorated so that even a new, effective treatment is of little avail. Thus, such patients may not provide a fair test of the capabilities of a new treatment, but the value of studying such patients in the early development of a new treatment is often compelling. The main advantage is the reduced risk of compromising the well-being of patients by withholding standard treatments, which in these patients are already known to be of little effect. Although studies of nonresponders can rarely provide definitive conclusions about the efficacy of a treatment in more responsive disease, they can provide information of considerable value in planning further studies (e.g., randomized controlled trials) that are designed to give unambiguous answers to questions about broad treatment recommendations.

ABSENCE OF DIRECT COMPARISONS

Self-controlled trials provide estimates of what changes to expect when a patient receives a treatment. Investigators may need this information in the early stages of testing a new treatment to help decide whether or how to continue studying the treatment.

Self-controlled studies do not provide direct information about how a new treatment compares with standard therapies. Rather, one must combine the results of self-controlled studies with those of other studies, though this is often difficult because of inevitable differences in study design and conduct. The demographic composition of the patient groups may differ, or the disease may be more severe in one group than another. Often, one cannot even determine the direction of such differences because investigators have used subjective or partly subjective assessments of clinical status, or they have not used the same set of objective measurements to establish a diagnosis or to monitor disease progress. Different investigators may even use different outcome variables to assess the effects of treatment. One must adjust for all these matters to obtain a valid assessment of the relative merits of two treatments, though it is generally not clear how to do so, and there is rarely any way to tell whether the adjustments were appropriate and adequate. Thus, despite the clear value of self-controlled studies in the initial investigation of new treatments, one must usually use a randomized controlled trial or some other powerful research design to determine with assurance whether the new treatment should be recommended for general use.

We are indebted to the members of the Harvard study group for their comments and to Mary Schaefer for assistance in the preparation of the manuscript.

REFERENCES

1. Lavori PW, Louis TA, Bailar JC III, Polansky M. Designs for experiments — parallel comparisons of treatment. N Engl J Med 1983; 309:1291-8. [Chapter 4 of this book.]

2. Aronow WS. Effect of passive smoking on angina pectoris. N Engl J Med 1978; 299:21-4.

3. de Alarcon PA, Donovan M-E, Forbes GB, Landaw SA, Stockman JA III. Iron absorption of the thalassemia syndromes and its inhibition by tea. N Engl J Med 1979; 300:5-8.

4. Herman TS, Einhorn LH, Jones SE, et al. Superiority of nabilone over prochlorperazine as an antiemetic in patients receiving cancer chemotherapy. N Engl J Med 1979; 300:1295-7.

5. Koivisto VA, Felig P. Effects of leg exercise on insulin absorption in diabetic patients. N Engl J Med 1978; 298:79-83.

6. Raskin P, Unger RH. Hyperglucagonemia and its suppression: importance in the metabolic control of diabetes. N Engl J Med 1978; 299:433-6.

7. Robertson D, Frölich JC, Carr RK, et al. Effects of caffeine on plasma renin activity, catecholamines and blood pressure. N Engl J Med 1978; 298:181-6.

8. Soter NA, Austen KF, Wasserman SI. Oral disodium cromoglycate in the treatment of systemic mastocytosis. N Engl J Med 1979; 301:465-9.

9. Sullivan MA, Cohen S, Snape WJ Jr. Colonic myoelectrical activity in irritable-bowel syndrome: effect of eating and anticholinergics. N Engl J Med 1978; 298:878-83.

10. Thadani U, Davidson C, Singleton W, Taylor SH. Comparison of the immediate effects of five β-adrenoreceptor-blocking drugs with different ancillary properties in angina pectoris. N Engl J Med 1979; 300:750-5.

11. Weinberger M, Hendeles L, Bighley L. The relation of product formulation to absorption of oral theophylline. N Engl J Med 1978; 299:852-7.

12. Wolfe JD, Tashkin DP, Calvarese B, Simmons M. Bronchodilator effects of terbutaline and aminophylline alone and in combination in asthmatic patients. N Engl J Med 1978; 298:363-7.

13. Woolfson AMJ, Heatley RV, Allison SP. Insulin to inhibit protein catabolism after injury. N Engl J Med 1979; 300:14-7.

14. Wyatt R, Waschek J, Weinberger M, Sherman B. Effects of inhaled beclomethasone dipropionate and alternate-day prednisone on pituitary-adrenal function in children with chronic asthma. N Engl J Med 1978; 299:1387-92.

15. Brown BW Jr. The crossover experiment for clinical trials. Biometrics 1980; 36:69-79.

16. Cole JWL, Grizzle JE. Applications of multivariate analysis of variance to repeated measurements experiments. Biometrics 1966; 22:811-28.

17. Grizzle JE. The two-period change-over design and its use in clinical trials. Biometrics 1965; 21:467-80.

18. Hills M, Armitage P. The two-period cross-over clinical trial. Br J Clin Pharmacol 1979; 8:7-20.

19. Kershner RP, Federer WT. Two-treatment crossover designs for estimating a variety of effects. J Am Stat Assoc 1981; 76:612-9.

20. Layard MWJ, Arvesen JN. Analysis of Poisson data in crossover experimental designs. Biometrics 1978; 34:421-8.

21. Wallenstein S, Fisher AC. The analysis of the two-period repeated measurements crossover design with applications to clinical trials. Biometrics 1977; 33:261-9.

22. Huitson A, Poloniecki J, Hews R, Barker N. A review of cross-over trials. Statistician 1982; 31:71-80.

23. Velleman PF, Hoaglin DC. Applications, basics, and computing of exploratory data analysis. Boston: Duxbury Press, 1981.

24. Mosteller F, Tukey JW. Data analysis and regression: a second course in statistics. Reading, Mass.: Addison-Wesley, 1977.

25. Morrison DF. Multivariate statistical methods. 2nd ed. New York: McGraw-Hill, 1976.

26. Billewicz WZ. The efficiency of matched samples: an empirical investigation. Biometrics 1965; 21:623-43.

27. Peck GL, Olsen TG, Yoder FW, et al. Prolonged remissions of cystic and conglobate acne with 13-cis-retinoic acid. N Engl J Med 1979; 300:329-33.

28. Haynes RB, Sackett DL, Taylor DW, Gibson ES, Johnson AL. Absenteeism from work after detection and labeling of hypertensive patients. N Engl J Med 1978; 299:741-4.

29. Packer M, Meller J, Medina N, Gorlin R, Herman MV. Rebound hemodynamic events after the abrupt withdrawal of nitroprusside in patients with severe chronic heart failure. N Engl J Med 1979; 301:1193-7.

30. Stacpoole PW, Moore GW, Kornhauser DM. Metabolic effects of dichloroacetate in patients with diabetes mellitus and hyperlipoproteinemia. N Engl J Med 1978; 298:526-30.

31. Davis R, Ribner HS, Keung E, Sonnenblick EH, LeJemtel TH. Treatment of chronic congestive heart failure with captopril, an oral inhibitor of angiotensin-converting enzyme. N Engl J Med 1979; 301:117-21.

32. Gavras H, Brunner HR, Turini GA, et al. Antihypertensive effect of the oral angiotensin converting-enzyme inhibitor SQ 14225 in man. N Engl J Med 1978; 298:991-5.

33. Ingelfinger J, Mosteller FM, Thibodeau L, Ware J. Biostatistics in clinical medicine. 2nd ed. New York: Macmillan, 1987.

34. Bilezikian JP, Canfield RE, Jacobs TP, et al. Response of $1\alpha,25$-dihydroxyvitamin D_3 to hypocalcemia in human subjects. N Engl J Med 1978; 299:437-41.

35. Tamborlane WV, Sherwin RS, Genel M, Felig P. Reduction to normal of plasma glucose in juvenile diabetes by subcutaneous administration of insulin with a portable infusion pump. N Engl J Med 1979; 300:573-8.

6

STUDIES WITHOUT INTERNAL CONTROLS

John C. Bailar III, M.D., Ph.D., Thomas A. Louis, Ph.D.,
Philip W. Lavori, Ph.D., and Marcia Polansky, D.Sc.

ABSTRACT   Sometimes questions of clinical interest such as the efficacy and safety of a treatment can be addressed only by investigations without concurrent controls that are under the supervision of the investigator (internal controls). Such studies nearly always make use of other types of comparisons (external controls), such as historical controls.

In this chapter we examine studies of clinical treatments that have weak internal controls or lack internal controls, as illustrated by examples from the *New England Journal of Medicine*. These studies have a small but important and unique role in clinical investigation. Five interrelated features can add to the strength of such studies: (1) an intent by the investigator, expressed before the study, that the treatment will affect the outcomes reported; (2) planning of the analysis before the data are generated; (3) articulation of a plausible hypothesis before the results are observed; (4) a likelihood that the results would still have been of interest if they had been "opposite" in some sense; and (5) reasonable grounds for generalizing the results from the study subjects to a substantially broader group of patients.

Examples of successful studies of this kind include reports of dramatic or surprising results, encouraging results of innovations in desperate circumstances, studies of unique groups, and early warnings. In each successful use, we find evidence of urgency, with timeliness of the report outweighing its tentative nature.

In spite of potential pitfalls, carefully selected and reported studies without internal controls can play a substantial part in the acquisition of scientific knowledge. They can warn of dangers, point out opportunities, and help determine the focus of more powerful controlled studies, whose strength they complement but do not duplicate.

To compare the merits of treatments, some research studies, such as randomized clinical trials, make use of controls or comparison groups that are enrolled and followed simultaneously with the treated group and by the same investigators (internal controls), whereas others make use, perhaps implicitly, of controls from another period or clinical setting or the controls of other investigators (external controls). Examples of externally controlled studies are case series, comparisons with historical controls or population-based results, and explicit use of prior published reports on the same topic. In this chapter we examine clinical studies in which external controls are the primary basis for comparisons and are not just ancillary to stronger types of controls, such as those of randomized clinical trials.

Though some studies of treatment report no controls at all, we believe that the investigators nearly always have in mind some comparison group, though it may be as ill defined as "common knowledge" about what happens under standard treatment (or in the absence of treatment). Thus, studies that others might define as uncontrolled form an important subcategory of those that we designate as externally controlled.

In Chapters 4 and 5 we discussed study designs for comparing two or more treatments given concurrently to different groups of patients (parallel studies) or sequentially to the same patient (crossover studies).[1,2] These approaches have in common a control or comparison group that is internal to the study. The randomized clinical trial, which is usually a parallel study but also includes crossovers, is now widely regarded as the standard of excellence for judging other designs. We agree that, with few exceptions, one or another type of randomized clinical trial is the method of choice for establishing general clinical recommendations. Other things being equal, a study with internal controls having treatment assigned by the investigator will support substantially stronger inferences than a study without internal controls, especially if the internal controls are assigned by a random mechanism.

Externally controlled studies will continue to have widespread use in initial studies of new drugs in human beings (Phase I drug trials) to demonstrate effects that were unexpected and for which base-line data may not have been obtained. The externally controlled trial cannot replace the controlled clinical trial, but if a hypothesis cannot be tested in a randomized clinical trial, it can sometimes be examined with other kinds of data that are obtained during the usual course of therapy. New FDA rules concerning use of new drugs for dread diseases may promote wider use of externally controlled studies.

Our purposes here are to show how to make externally controlled studies stronger and more convincing and to give some advice on how to interpret the results of such work.

Consideration of the design and use of studies without internal controls has lagged behind developments regarding "stronger" research approaches (but see the report of Campbell and Stanley for a notable exception[3]). This chapter provides an empirical basis for such consideration by examining the use of this design in a series of reports that appeared in the *New England Journal of Medicine.*

Using methods described elsewhere,[1,2] we identified 20 externally controlled studies of medical treatment among the 332 Original Articles appearing in the *Journal* during 1978 and 1979 (Table 1). This set of papers provided many of our examples as well as the base of some general conclusions. We hope that this investigation will stimulate renewed consideration of this method of assessing the effects of treatment.

Studies without internal controls range from almost casual observations of possibly important relations to tightly designed exploratory investigations (such as some aspects of Phase I drug trials). Although the results of both internally and externally controlled studies may often not be confirmed, the first clues to real breakthroughs in medical science may come from comparisons with external controls.

In one of the examples of externally controlled studies[4-23] listed in Table 1, Sherins et al. reported a high incidence of gynecomastia and gonadal dysfunction in a group of 19 Ugandan boys treated with chemotherapy for Hodgkin's disease.[10] The investigators had no appropriate study group for internal controls, so they compared the findings in the Ugandan boys with the incidence of those findings as reported in North American boys. This study is a model for reporting such work. The authors were entirely fair in reporting the inferential weaknesses of their conclusions, but the findings were striking, there were no known reports of an extremely high incidence of hormonal problems in adolescent Ugandan males, and there were substantial implications regarding the long-term observation and care of other young male patients receiving the same kind of treatment. Thus, it seemed that publication was warranted before the results could be confirmed by other research approaches.

Table 1. Strength of 20 Externally Controlled Studies Reported in the *New England Journal of Medicine*, 1978–1979.

Study	Subjects	Treatment	Outcome Examined	Strength of Study				
				Direct Intent?	Planned Analysis?	Rationale Analysis?	Opposite Interest?	Discuss Generality?
Alter et al.[4]	4 chimpanzees	Frozen blood with hepatitis virus	Post-transfusion hepatitis	Yes	Yes	Yes	Some	Yes
Toft et al.[5]	100 patients with thyrotoxicosis	Thyroidectomy plus propranolol	Normal thyroid function in some	Yes	Yes	Yes	Yes	Yes
Parkman et al.[6]	2 patients with Wiskott–Aldrich syndrome	Allotransplantation of bone marrow	Hematopoiesis and graft-versus-host disease	Yes	Yes	Yes	Yes	Yes
Ahn et al.[7]	11 patients with idiopathic thrombocytopenia	Platelets loaded with vinblastine	Remission	Yes	Yes	Yes	Some	Yes
LaRossa et al.[8]	15 women with postmenopausal breast cancer	Incomplete hypophysectomy	Pain relief	Yes	Yes	Yes	Some	Yes
Shiffer et al.[9]	25 patients with leukemia in remission	Transfusion of frozen autologous platelets	Platelet count	Yes	Yes	Yes	Some	Yes

Sherins et al.[10]	19 boys with Hodgkin's disease	Chemotherapy	Gonadal dysfunction	No	Yes	Yes	Some	Some
Sundt and Whisnant[11]	310 patients with subarachnoid hemorrhage from intracranial aneurysm	Surgical management	Survival	Yes	No	Some	No	Some
Ballard et al.[12]	5 infants with hyperammonemia	Several therapies	Cure	Yes	No	No	No	Some
Graze et al.[13]	107 members of family with a history of thyroid cancer	Calcitonin test	Detection of cancer	No	Yes	Yes	No	Yes
George et al.[14]	278 patients with leukemia in remission >2.5 yr	Stopping therapy	Survival	Yes	Yes	Yes	Yes	Yes
Fefer et al.[15]	4 patients with chronic granulocytic leukemia	Chemotherapy, radiation, marrow transplant from a twin	Disappearance of Ph^1-positive cells	Yes	Yes	Yes	Yes	Yes
Dretler et al.[16]	6 patients with recurrent renal stones	Percutaneous irrigation	Stone dissolution	Yes	Yes	Yes	No	Yes

(Table continues)

Table 1. (Continued)

Study	Subjects	Treatment	Outcome Examined	Strength of Study				
				Direct Intent?	Planned Analysis?	Rationale Analysis?	Opposite Interest?	Discuss Generality?
Krikorian et al.[17]	579 patients with Hodgkin's lymphoma	Radiation and chemotherapy	Second, non-Hodgkin's cancer	No	No	Yes	Some	Yes
Stern et al.[18]	1373 patients with psoriasis	Photochemo-therapy	Skin cancer	No	Yes	Yes	Some	Yes
Fauci et al.[19]	17 patients with systemic necrotic vasculitis	Cyclophosphamide	Remission	Yes	Yes	Yes	Some	Yes
Haugen[20]	31 patients given 125 transfusions	Frozen and washed red cells	Prevention of hepatitis	Yes	No	Yes	Some	Yes
Beard et al.[21]	771 women with vaginal trichomoniasis	Metronidazole	Cancer	No	Yes	Yes	Yes	Yes
Borzy et al.[22]	3 children with immunodeficiency	Cultured thymus transplant	Fatal lymphoma	No	No	Yes	No	Yes
Koplan et al.[23]	1 million children	Pertussis vaccine	Complications, attacks	Not applicable				

In another example of strong scientific inference from observations without internal controls, LaRossa et al. investigated the relation between ablation of the pituitary gland and relief of bone pain among 15 post-menopausal patients with metastatic breast cancer.[8] They found that although pain was relieved, there was evidence of residual pituitary function in all but one of their patients. Endocrinologic function was not tested before surgery; otherwise, this study would be considered self-controlled rather than externally controlled. The authors recognized that quantitative inference would be weak, but publication was justified by the unexpected finding of good pain relief despite a degree of continued pituitary function. Although this study has some features of a report of an unexpected observation, we suspect that the authors had some prior indication that the microsurgical technique used in these patients did not result in complete hypophysectomy. If so, the targeted nature of their inquiry would make their conclusions stronger than if they had simply noticed an oddity during a review or discussion of cases in some other context.

EXTERNALLY CONTROLLED STUDIES

The studies that we classified as externally controlled ranged from a report of striking therapeutic success in two patients[6] to a large prospective risk-assessment study using population data as controls.[18] Despite their varied content and goals, each study survived an editorial review intended to ensure that (among other things) the lack of internal controls was balanced by other important features. Examples of the successful uses of external controls included reports of dramatic results of new therapy, both promising new ideas[7,9] and old ideas that had at last been made to work[15,16]; deliberate attempts at innovation (diagnostic or therapeutic) in desperate circumstances[6]; reports of groups of patients with some remarkable clinical history[14]; and early observation of previously unrecognized toxicity or other problems with a therapy.[10,17,18]

Each of these situations conveys a sense of urgency; the timeliness of the report outweighs its tentative nature. Additional study of these matters by investigations with internal controls would have substantially delayed their public availability and scientific impact; there might also be ethical problems in some circumstances (e.g., as in innovation under pressure or as in observation of new complications of therapies). Other particularly important uses of externally controlled studies, perhaps reported more often in specialty journals than in a general journal such as the *New England Journal of Medicine*, are the initial development of hypotheses that merit study by stronger methods and the development of sufficient information to permit proper design of a stronger trial, including such features as sample sizes, proper doses of drugs, and common untoward effects. Randomized clinical trials, for example, are rarely undertaken before a wealth of such information is available.

THE STRENGTH OF VARIOUS KINDS OF EXTERNAL CONTROLS

A weak (though perhaps sometimes necessary) use of external controls involves the comparison of outcomes between an internal series of cases and a differently managed series.

For example, Sundt and Whisnant compared survival in 310 patients with recent subarachnoid hemorrhage during 1969 to 1977, with survival in 63 patients from a community series treated between 1955 and 1969.[11] The authors pointed out that their own series could include only patients who survived to be admitted — a severe requirement in a condition with high early mortality. These investigators tried to reduce bias by comparing their patients who were admitted one day after hemorrhage with the patients in the community series who survived for at least one day. The temporal remoteness of the control and the arbitrariness of the subsample criterion make the comparison weak. This study stands on its merits as a careful descriptive investigation of the current treatment of subarachnoid hemorrhage by prompt surgery. It is not clear that a stronger control could have been found, but fortunately the historical comparison is not central to the main descriptive conclusions.

Sundt and Whisnant used another kind of external control. They compared long-term survival in the 280 patients who lived to operation with the survival expected in a population similar in age, race, and sex, as described in the United States life tables for 1960. Besides describing the substantial risk of early postoperative mortality, this more straightforward comparison indicated that a patient who survived six months after the operation had a nearly normal outlook for survival over the following five years. Since the authors also reported on empirical estimates of the probability of surviving for six months, this result can help physicians to advise and inform their patients. The pattern of early mortality reported by these investigators adds weight to their concern about the selection of patients according to factors affecting survival in the first hours and days; but selection according to early survival could have little impact on findings after the first six months, which are correspondingly stronger.

In another example, the external comparison carries most of the weight of the study. Stern et al. studied the incidence of cutaneous carcinoma in 1373 patients with psoriasis treated with photochemotherapy (PUVA).[18] They compared the results in their patients with the incidence of skin cancer reported in a population study cited in the article. By adjusting the expected rates for age, sex, and geographic region, the authors made the comparison more apt. Because of the well-documented nature of the comparison group (the source of the external control), the results can be easily understood by the reader. This explicit definition of the external control is one of the major strengths of the study. Nevertheless, population-based estimates of cancer incidence do not always provide reliable indications of the expected incidence in a sample of patients. A study that we have discussed[1] in another context reported that the incidence of cancer was substantially higher than expected in placebo-treated as well as in drug-treated women enrolled in a randomized trial of diethylstilbestrol (DES) to prevent spontaneous abortion.[24] The investigators used the randomized internal controls to determine the effect of DES on cancer incidence and found none. Comparison of the treated group against a population standard, rather than the randomized internal controls, would have estimated a different "effect" that confounds treatment and selection for the trial.

LIMITS ON THE GENERALIZABILITY OF RESULTS

Authors of the papers reviewed here were often careful to point out limitations on the extension of their findings to other settings, just as they were careful to present their reasons for using external controls and to discuss the resulting inferential weaknesses. Their open discussion of these matters may partly explain their success in publication. For example, the conclusions of one study were carefully limited by the authors to patients with conditions refractory to standard therapy,[7] although one might guess that patients without such conditions would have much the same pattern of response.

In other examples, the inferential leap from the study setting to other settings seems less solidly based. For example, the conclusions of Sundt and Whisnant on the treatment of subarachnoid hemorrhage[11] might be difficult to apply to a different hospital or a different patient population, and might not even be relevant to a later series of patients treated in the same hospital. This is possible in part because the authors describe the results of a loose policy of treatment ("prompt" surgery) rather than a narrowly defined therapy; thus, the reader may not know exactly how to replicate the results, even hypothetically. The main difficulty is that the specific details of the length of time that a patient had to survive in order to be included in the comparison have so much effect on the outcome that only a randomized internal control could provide the balance necessary for a solid and convincing inference.

Much can be gained by a strategy that, for such studies, limits or omits generalizing to an outside population, in favor of describing the structural features of the data in hand. For example, in their study of psoriasis treated with PUVA, Stern et al. first provided a descriptive analysis of the relative risk of skin cancer in subgroups of patients defined by skin type and exposure to ionizing radiation.[18] They then provided a more speculative and interpretive analysis of the number of PUVA treatments delivered to the 27 patients in whom basal-cell or squamous cancer developed. This shift of focus from description to inference and exploration requires appreciation of the study group in its own right, rather than as a sample from some larger universe. Exploratory procedures can sometimes only suggest possibilities, but these will be the prior hypotheses for more narrowly focused (and perhaps internally controlled) future studies. Such exploration should be presented as such, without apology, but we believe that authors should usually go further and explicitly warn readers against drawing inferences that the authors themselves have eschewed.

NATURE AND SCOPE OF DATA ANALYSIS AND SUMMARIZATION

In externally controlled studies, particularly those with small numbers of subjects, the issues of data analysis and summarization must be considered with special care. When there are few patients, investigators may collect a great deal of information on each one and seek to compensate for small numbers with exhaustive individual information. Serial measurements of some clinical variable are a common example; Graze et al. measured serum calcitonin at several times after thyroidectomy in patients with C-cell hyperplasia.[13] A thorough exploration of these serial data might include an attempt to model the temporal course of the measurements, thus putting the multiple observations to greater use. Methods of data analysis appropriate to such repeated, correlated measures are sophisticated and often require expert statistical help, but serial data that are considered worth collecting may be worth the extra effort in analysis.

Important internal relations observed in externally controlled studies may require detailed presentation, or even representation in full as case reports or tabulations of individual data. Authors and editors must adapt their material to meet requirements for space and readability, but by the nature of these studies and their role in the literature the reader often has a substantial stake in having more detail available than would be appropriate in more structured work. For example, the reader interested in designing an investigation that builds on the results of an externally controlled study may use information on the variability of outcomes to calculate the necessary sample sizes for an internally controlled study. Observing the apparent duration of effect of treatments may help an investigator guess the feasibility of a crossover study instead of a parallel comparison. Estimates of cost of measurements and their reliability can affect the design of future studies.

IMPORTANCE OF PRIOR HYPOTHESES

We referred above to the importance of distinguishing between prior and post hoc hypotheses in drawing scientific inferences. It is now widely recognized that statistical tests of hypotheses (P values) are strictly applicable only to hypotheses fully and explicitly formed before the data are examined. Otherwise a P value has a distinctly different meaning and carries a much reduced weight of inference. The usual methods and formulas for calculation of tests of hypotheses may be applied to post hoc hypotheses, but the usual interpretation of these tests may be profoundly weakened. A good way for an investigator to protect the strength of inferences and to convince others of the validity of statistical conclusions is to describe carefully the scientific rationale that led to the particular limited set of hypotheses that the study was designed to confront. This advice applies to all statistical analyses.

More specifically, if a hypothesis is to be supported by a set of data, there must in concept be some chance that the data will turn out to be contrary to prediction, if the hypothesis is in fact not correct. Unfortunately, such specificity about hypotheses generated before the data are inspected in any manner or degree is often lacking in research reports of all types. The lack is more acute in externally controlled studies than in randomized trials and other internally controlled studies because the structure of those investigations may yield clear implications about specific prior hypotheses.

This problem is illustrated by the study of Toft et al., who reported on thyroid function in 100 patients with thyrotoxicosis treated with propranolol before and immediately after subtotal thyroidectomy.[5] As part of their report, they analyzed postoperative thyroid function and related it to the estimated weight of the thyroid remnant after surgery. They asserted, on the basis of evidence that we have reproduced in Table 2, that postoperative thyroid function was closely related to remnant weight. We see at least three possible prior hypotheses. Without knowing which (if any) hypothesis motivated the collection of data, one cannot judge the support that the data give to any of the three.

(1) A possible prior hypothesis was that euthyroid subjects with normal serum thyrotropin (the authors' thyroid-status Group 1) would have higher remnant weights than the other subjects. Then the P value at the extreme right of Table 2 is the only P value in the table that could be given the standard interpretation. It tests the differences between Group 1 and the pooled Groups 2, 3, and 4. The other P values reflect post hoc comparisons that may be suggestive but not definitive regarding subgroups of the main comparison group.

(2) If the four thyroid-status groups were deliberately ordered post hoc according to the decrease in thyroid activity, then the prior hypothesis may have been that the groups with more function had higher average remnant weights. This hypothesis is not examined or supported by the summary data in Table 2, nor are any of the t-tests relevant. There may be such a relation, but it should be summarized and tested by a more powerful method

Table 2.   Thyroid Status and Thyroid-Remnant Weight after Thyroidectomy in 78 Patients of Toft and Colleagues.[5]

Group	Status	No. of Patients	Remnant Weight (g)*	P Value†	
1	Euthyroid, with normal serum thyrotropin	18	7.1 ± 0.4	—	
2	Euthyroid, with raised serum thyrotropin	29	5.3 ± 0.4	P<0.005	
3	Temporary hypothyroidism	17	6.0 ± 0.5	0.05<P<0.10	P<0.001
4	Permanent hypothyroidism	14	4.8 ± 0.4	P<0.001	

* Means ± S.E.

† Mean weight of Group 1 was compared with means of Groups 2, 3, and 4, both individually and collectively, by Student's t-test (two-tailed P values).

for analyzing ordered categorical data, such as the method described by Moses, Emerson, and Hosseini[25] or its extensions.

(3) If the important hypothesis was that the groups did not all come from a population with the same mean remnant weight, an analysis of variance or a distribution-free alternative should have been applied first to determine whether there was any evidence of a difference among the group means.[26,27]

Reports of new therapies may require a preliminary field test to demonstrate sufficient promise to make a major clinical comparison worthwhile, but the preliminary studies themselves are worth reporting. For example, Ahn et al. loaded platelets with vinblastine and infused them in 11 patients with idiopathic thrombocytopenic purpura that was refractory to standard treatment.[7] The authors' hypothesis was that macrophages were responsible for platelet destruction and that as a result of engulfing and destroying the loaded platelets the macrophages would be destroyed as well. This well-defined scientific hypothesis led to the experiment and therefore to the observations; the observations did not lead to the hypothesis.

In another example, George et al. reported on 278 patients with acute lymphocytic leukemia in whom maintenance chemotherapy was stopped after 2½ years of continuous remission.[14] Though no concurrent internal controls were continued on treatment indefinitely, the relatively low recurrence rate after cessation of treatment and the well-known problems of long-continued treatment argue for the cessation of therapy under conditions similar to those of the reported series.

These comments would apply regardless of the research design. However, such ambiguity in the precise nature of the prior hypotheses is characteristic of reports on externally controlled studies. By contrast, in an internally controlled study, the most important comparisons and analyses (with their associated hypotheses) are usually determined in advance. Often these plans develop in the process of designing the study — e.g., what control group should be chosen, or what is the best way to measure the relative merits of two therapies. In an externally controlled study, there may be no natural point in the design at which these issues are automatically considered. The investigators may have to set aside time and effort to set down the details of the planned analyses clearly. The result of this effort will be a convincing rationale that can form the basis for a logical description of the experimental design and thus back up the author's assertion that the hypotheses described and tested were chosen before the results were arrived at.

Explicit prior hypotheses that could — at least in concept — be refuted by observation add considerable strength to externally controlled studies, as they do to studies with more rigorous designs. It is in general crucial that authors tell readers, honestly and explicitly, which hypotheses were made before the experiment and which were made afterward, as well as what the authors consider to be the scientific justification for these hypotheses. This rule has great force in all sorts of designs, but it seems to be easily overlooked when the study has no internal controls. There is often room for post hoc debate over which hypotheses were truly in mind before some study, and investigators should write down a list of such hypotheses before any data are examined. Also, since scientific inferences are dependent on the intent of the investigator or analyst, persons who come to a single set of data with different prior beliefs and expectations may well reach incompatible conclusions. In the absence of such a list, authors should do their best to be objective in reporting their prior hypotheses, and readers should retain a degree of skepticism about reported findings and inferences. Authors should tell readers how and when a hypothesis was formed, why a particular research approach was used, what kinds of analyses were undertaken, and why some results of the analyses were presented rather than others. This is the only way that readers can judge which of the findings are most likely to be replicated and which are interesting but only indicative of what a repeat experiment might show.

THE USES OF EXTERNALLY CONTROLLED STUDIES

Studies with poor or incomplete controls have often led to serious errors, thus introducing worthless treatments. Chalmers offers case histories of the impact of well-designed controlled studies of therapeutic measures that were initially promoted on less solid grounds.[28] The examples include gastric freezing for peptic ulcer and a low-protein diet for hepatic failure. All the same, we have argued above that some kinds of reports that rely entirely on external controls can be valuable and convincing. Are these ideas incompatible?

We think not, because the uses of internally controlled and externally controlled studies are different. Randomized trials and other strong research designs are useful and even necessary for developing broad clinical recommendations about how to prevent disease or treat patients in the ordinary (nonresearch) course of events, whereas externally controlled studies — normally inappropriate for this purpose — have other very important uses for which the stronger designs may not even be applicable. Of the 20 externally controlled studies in our series, 4 dealt with the late effects of treatment for potentially lethal or intractable and disabling diseases, such that a low incidence of late complications would not necessarily contraindicate the treatment.[10,17,18,21] Four other studies dealt with technical feasibility (infecting chimpanzees with hepatitis B from frozen human blood,[4] removing the platelets of patients with leukemia and freezing the cells for later reinjection,[9] dissolution of kidney stones in vivo,[16] and detecting thyroid hyperplasia and carcinoma in relatives of patients with medullary carcinoma[13]). Six other studies implied therapeutic recommendations but did not seem to be intended to be definitive; interim use of results might be justified by desperate and otherwise insoluble clinical problems.[6,7,12,15,19,22] One study was purely descriptive of interesting results.[8]

Only the remaining five seemed to be offered as definitive solutions to clinical problems.[5,11,14,20,23] These studies require critical evaluation to avoid the sort of mistakes that Chalmers and others have documented. With the absence of a rigid framework for design comes increased difficulty

— for reader, writer, and journal editor — in evaluating the persuasiveness of the arguments and conclusions of the study.

A SCHEMA FOR ASSESSING INFERENCES FROM EXTERNALLY CONTROLLED TRIALS

We propose the following list of five closely related (and sometimes overlapping) questions, whose answers provide a preliminary indication of the strength of evidence provided by an externally controlled clinical study. (1) Does it appear that the intervention was applied with the primary intent of affecting the outcomes reported, such as cure, survival, or the incidence of complications? (2) Is it clear that the authors' intent to analyze and report their findings preceded the generation of the data (though the data may have been gathered for a different primary purpose)? (3) Have the authors shown that they had a plausible rationale for their interpretation of the data before the data were inspected or the analysis was undertaken? (4) Would the results have been interesting (i.e., publishable) if they had been different in some important sense from those actually obtained? Would negative findings have had a chance of being reported? (5) Do the authors present reasonable grounds for generalizing their results?

Each of these questions is designed to elicit some information on whether standard interpretations of statistical confidence limits and P values may be more or less justified. A large number of favorable answers will tend to support statistical inferences. If the answers to all these questions are yes, the data should still be regarded with some skepticism, but if all answers are no, even the most extreme results would scarcely be worth marginal attention.

We present our own answers to these questions for the 20 studies in our sample. Table 1 shows that most of the papers made a strong showing. If one interprets "some" as midway between yes and no, the proportion of favorable answers (yes in each case) varies from 92 percent in the third and fifth columns regarding the strength of the study (prior rationale and generalizability) to a low of 50 percent in the fourth column (opposite interest). Four of the studies rated a 5.0 (yes on all five questions), and only one study was rated as low as 1.5 (one yes, one some).

CONCLUSIONS

The design and interpretation of externally controlled studies must be responsive to the peculiar goals and pitfalls of this part of the research effort. In the absence of a well-defined sampling plan or other support for inference, a finding of *statistical significance* may be misleading or impossible to interpret, whereas the finding *not statistically significant* may be meaningless. Explicit recognition — by authors, editors, and readers — that these studies have a valid though limited role that differs from that of more structured investigations would help to reduce confusion over the substance and interpretation of reports of uncontrolled studies.

Only a few research studies that rely solely on external controls will meet stringent tests for the validity of scientific inferences and hence for publication in a journal with a broad, general readership. However, we believe that any research program or journal that will take no risks at all with externally controlled (or even uncontrolled) studies of potentially important advances runs some risk of becoming irrelevant to the most important current developments in science and medicine. Careful attention to problems of scientific inference will help investigators — and, at a later stage, journal editors and readers — to tell which externally controlled studies are most worth performing, publishing, and accepting as valid contributions.

REFERENCES

1. Lavori PW, Louis TA, Bailar JC III, Polansky M. Designs for experiments — parallel comparisons of treatment. N Engl J Med 1983; 309:1291-8. [Chapter 4 of this book.]

2. Louis TA, Lavori PW, Bailar JC III, Polansky M. Crossover and self-controlled designs in clinical research. N Engl J Med 1984; 310:24-31. [Chapter 5 of this book.]

3. Campbell DT, Stanley JC. Experimental and quasi-experimental designs for research. Boston: Houghton Mifflin, 1966.

4. Alter HJ, Tabor E, Meryman HT, et al. Transmission of hepatitis B virus infection by transfusion of frozen-deglycerolized red blood cells. N Engl J Med 1978; 298:637-42.

5. Toft AD, Irvine WJ, Sinclair I, McIntosh D, Seth J, Cameron EHD. Thyroid function after surgical treatment of thyrotoxicosis: a report of 100 cases treated with propranolol before operation. N Engl J Med 1978; 298:643-7.

6. Parkman R, Rappeport J, Geha R, et al. Complete correction of the Wiskott–Aldrich syndrome by allogeneic bone-marrow transplantation. N Engl J Med 1978; 298:921-7.

7. Ahn YS, Byrnes JJ, Harrington WJ, et al. The treatment of idiopathic thrombocytopenia with vinblastine-loaded platelets. N Engl J Med 1978; 298:1101-7.

8. LaRossa JT, Strong MS, Melby JC. Endocrinologically incomplete transethmoidal transsphenoidal hypophysectomy with relief of bone pain in breast cancer. N Engl J Med 1978; 298:1332-5.

9. Schiffer CA, Aisner J, Wiernik PH. Frozen autologous platelet transfusion for patients with leukemia. N Engl J Med 1978; 299:7-12.

10. Sherins RJ, Olweny CLM, Ziegler JL. Gynecomastia and gonadal dysfunction in adolescent boys treated with combination chemotherapy for Hodgkin's disease. N Engl J Med 1978; 299:12-6.

11. Sundt JM Jr, Whisnant JP. Subarachnoid hemorrhage from intracranial aneurysms: surgical management and natural history of disease. N Engl J Med 1978; 299:116-22.

12. Ballard RA, Vinocur B, Reynolds JW, et al. Transient hyperammonemia of the preterm infant. N Engl J Med 1978; 299:920-5.

13. Graze K, Spiler IJ, Tashjian AH Jr, et al. Natural history of familial medullary thyroid carcinoma: effect of a program for early diagnosis. N Engl J Med 1978; 299:980-5.

14. George SL, Aur RJA, Mauer AM, Simone JV. A reappraisal of the results of stopping therapy in childhood leukemia. N Engl J Med 1979; 300:269-73.

15. Fefer A, Cheever MA, Thomas ED, et al. Disappearance of Ph^1-positive cells in four patients with chronic granulocytic leukemia after chemotherapy, irradiation and marrow transplantation from an identical twin. N Engl J Med 1979; 300:333-7.

16. Dretler SP, Pfister RC, Newhouse JH. Renal-stone dissolution via percutaneous nephrostomy. N Engl J Med 1979; 300:341-3.

17. Krikorian JG, Burke JS, Rosenberg SA, Kaplan HS. Occurrence of non-Hodgkin's lymphoma after therapy for Hodgkin's disease. N Engl J Med 1979; 300:452-8.

18. Stern RS, Thibodeau LA, Kleinerman RA, et al. Risk of cutaneous carcinoma in patients with oral methoxsalen photochemotherapy for psoriasis. N Engl J Med 1979; 300:809-13.

19. Fauci AS, Katz P, Haynes BF, Wolff SM. Cyclophosphamide therapy of severe systemic necrotizing vasculitis. N Engl J Med 1979; 301:235-8.

20. Haugen RK. Hepatitis after the transfusion of frozen red cells and washed red cells. N Engl J Med 1979; 301:393-5.

21. Beard CM, Noller KL, O'Fallon WM, Kurland LT, Dockerty MB. Lack of evidence for cancer due to use of metronidazole. N Engl J Med 1979; 301:519-22.

22. Borzy MS, Hong R, Horowitz SD, et al. Fatal lymphoma after transplantation of cultured thymus in children with combined immunodeficiency disease. N Engl J Med 1979; 301:565-8.

23. Koplan JP, Schoenbaum SC, Weinstein MC, Fraser DW. Pertussis vaccine — an analysis of benefits, risks and costs. N Engl J Med 1979; 301:906-11.

24. Bibbo M, Haenszel WM, Wied GL, Hubby M, Herbst AL. A twenty-five-year follow-up study of women exposed to diethylstilbestrol during pregnancy. N Engl J Med 1978; 298:763-7.

25. Moses LE, Emerson JD, Hosseini H. Analyzing data from ordered categories. N Engl J Med 1984; 311:442-8. [Chapter 13 of this book.]

26. Snedecor GW, Cochran WG. Statistical methods. 8th ed. Ames, Iowa: Iowa State University Press, 1989.

27. Colton T. Statistics in medicine. Boston: Little, Brown, 1974.

28. Chalmers TC. A challenge to clinical investigators. Gastroenterology 1969; 57:631-5.

7

THE SERIES OF CONSECUTIVE CASES AS A DEVICE FOR ASSESSING OUTCOMES OF INTERVENTIONS

Lincoln E. Moses, Ph.D.

ABSTRACT   An important part of the medical literature consists of reports of series of cases. Typically, a series has been accumulated over time and consists of all patients meeting certain criteria over a specified interval. Ordinarily, if assessing an intervention, the report entails comparison — either explicit, between subclasses in the series, or implicit, between the whole series and ordinary expectations and experience. Interpretation of the series report depends on the author's clarity about the definitions actually applied, the integrity of counting (who was excluded and why), and the consistency of diagnosis, outcome measurement, and other factors. In addition, such variables as age, parity, and stage of disease may influence outcomes; thus, information concerning these factors may be needed for confident interpretation of the series. Acceptance of a reported series result at face value is, therefore, unlikely to be justified, and interpretation typically calls for analysis. If the necessary ancillary information is not reported, the implications of the results may remain obscure.

An air of serving the common good clings to the process of reporting as general information the results of one's own extensive experience. Medicine enjoys a long tradition of such literature, and valuable results have sometimes ensued. Moreover, a large share of medical knowledge has been accumulated in just this way, through the publication of reports on series of cases. The purpose of this chapter is to examine the usefulness and limitations of series for assessing the safety and efficacy of medical interventions. Two historical examples help clarify the issues involved: the first demonstrates results with a new technique; the second compares outcomes between two differently treated subsets of a single series of patients.

In 1847 John Snow published an epochal work, *On the Inhalation of the Vapour of Ether in Surgical Operations.*[1] Snow described the equipment he had devised, the procedure he had used, and 75 operations (52 at St. George's Hospital and 23 at University College Hospital) performed before September 16, 1847, in which he had delivered ether anesthesia.

These two series (with four and two deaths, respectively) were doubtless, in the eyes of the author and his readers, harbingers of the future. For a modern reader they also serve as a window on the past; in each of the 75 operations, neither the thorax nor the abdomen was entered. The two series showed that Snow's apparatus was effective in vaporizing ether for patient inhalation and that, with the new apparatus and procedure, anesthesia was induced in all patients, all patients revived from the anesthesia, and the surgery was performed more easily than without it. These results helped to dispel the mistrust of ether that had been engendered by inept applications in England in 1846.

In 1836, drawing from his practice over the years, Pierre Louis published an account of 77 patients with pneumonia uncomplicated by other disease.[2] He classified them by whether or not they had survived the disease and by when in the course of their illness he had begun bleeding them. Early bleeding turned out to be associated with reduced survival. The series of observations was an important part of his attack on bleeding as a panacea.

These accounts, though much abbreviated, show some of the characteristics of the series as a source of information in medicine. First, a series typically contains information acquired over time. Second, the patients in a series are all similar in some essential way: all the patients in Snow's series received ether, though with various operations; all those in Louis' series had the same disease (and physician) but varied in the way they were treated. Third, all the patients in a defined class are discussed: Snow described all administrations of ether on or before September 16, 1847; Louis described all patients with pneumonia for whom he had records indicating no other disease and whom he had bled. Fourth, a comparison is involved; it may be direct, as in the study by Louis, or indirect, as in Snow's study. Thus, fairness of comparison becomes a crucial issue. (Louis assured his readers that the two groups, survivors and decedents, were as alike in the initial severity of their disease as he could make them by including and excluding cases from his files.) Finally, the reported series, whatever its value as evidence, may or may not be influential. Louis was a member of the faculty at Paris, which lent weight to his report. (The influence of his report contrasts notably with the small impact of James Lind's beautifully controlled experiment demonstrating the curative power of lemons in scurvy; Lind was a naval surgeon without high standing, and his study did not affect the policy of the Royal Navy until some 40 years later.)

Just as a series can advance correct understanding, so it can promote the pursuit of bad leads. Presumably, most once-popular but later-discarded therapies were initially supported by series of favorable cases. Even in recent times there was support for portacaval shunt for the treatment of esophageal varices and gastric freezing for the treatment of ulcers.

In this chapter, the aim is to analyze the strengths and weaknesses of series as sources of information for appraising the efficacy and safety of diagnostic and therapeutic interventions. Perhaps there are straightforward ways to distinguish the trustworthy from the spurious, misleading information that is sometimes conveyed in a series.

WHAT IS A SERIES?

I shall apply the term "series" to a study of the results of an intervention that has certain characteristics. First, it is longitudinal, not cross-sectional; postintervention outcomes are reported for a group of subjects known to the investigator before the intervention. Second, all eligible patients in some stated setting over a stated period are described. These eligible patients are seen as alike; they may have a common disease, they may have received the same treatments, or they may share some other essential characteristic. Series may have other important design factors; comparison groups may be present or absent, for example, and the research may have been planned either before the data were acquired or afterward.

Not every published series reports outcomes of interventions; descriptive features of patients with a certain disease might be reported in a series, for example. But here I shall deal with those series that do report the outcomes of an intervention applied to all eligible subjects; these subjects are chosen according to criteria that depend only on pretreatment status. The actual data collection may go forward in time according to a research plan, or it may be undertaken after all cases are complete. Intermediate or mixed

cases can also occur. The data are regarded as if the subjects were first identified on the basis of eligibility, then treated, and then observed with respect
to outcome.

Much of the medical literature consists of articles that meet this description. Feinstein[3] reviewed all issues of the *Lancet* and the *New England Journal of Medicine* appearing between October 1, 1977, and March 31, 1978.
Of the 324 "structured research papers" that he identified, 47 (15 percent)
contained reports on series (with "transition cohorts" and "outcome
cohorts") as I use the term. This proportion of articles was approximately
equaled by the 53 papers (16 percent) that reported clinical trials. Bailar et
al.[4] reviewed all Original Articles published in the *Journal* during 1978 and
1979. Of the 332 articles, 80 (24 percent) apparently met the definition of a
series used here.

INTERPRETING A SERIES

A number of factors bear on the interpretation of a report of a series. Several of the most important are discussed below.

THE INTEGRITY OF COUNTING

My definition of a series includes the word "all," and the word is essential. Conclusions based on "selected cases" are notoriously treacherous because selection can grossly affect the data; at the extreme, only the successes
or only the failures might be reported. Presumably, the limitations of selected cases underlie the skepticism sometimes voiced about the usefulness
of voluntary disease registries.

At a minimum, the reader needs to know what criteria were used to determine the inclusion and exclusion of subjects, how many subjects were
included, and what happened to each of them. Here is where the concept
"all" enters as essential to the study's value. Two kinds of problems lurk
here. The first is operational: it may be difficult or impossible to learn some
of the essential information in retrospect. The outcomes for some patients
who belong in the series may be unknown. Patients who cannot be followed often differ on average from those who can be followed; more of the
former may be dead or cured. Without complete follow-up, the available
figures lose much of their meaning.

The second type of counting problem is definitional. The series report, to avoid being a recital of selected cases, needs to describe what was done to all eligible patients and the outcome for each of them. Who is (or was) eligible may depend on diagnostic criteria that require judgment calls. What was done to the patients may involve judgment as well; if the intervention is a new surgical procedure that has changed somewhat with time, then determination of which patients did and which did not receive "the" new operation requires a decision by the author. Even identification of the outcome for a patient may demand a judgment call. If the patient dies on the operating table, for instance, there may be a question (and a decision) about whether death was due to anesthesia or to treatment failure.

The definitional and operational problems of counting are likely to loom larger if the series study is planned after the data have been collected.

ADJUSTMENT FOR INTERFERING VARIABLES

In the United Kingdom, a series report on 5174 births at home and 11,156 births in hospitals stated that perinatal mortality was 5.4 per 1000 home births and 27.8 per 1000 hospital births.[5] What use can be made of these numbers? A moment's thought fills the mind with questions about the comparability of the two series of mothers. How did they differ in age, parity, prenatal care, prenatal complications, home circumstances, general health, and disease status? Without answers to these questions, we must withhold any firm interpretation of the data. With information on all these variables — and doubtless some others — we are better off. But even with such information, we still face the hard question of how to adjust the raw results for differences in these other variables. They are likely to be relevant to perinatal mortality, but we do not know "the right way" to adjust numerically for these factors, even if we have the information.

The complexities of adjustment are nicely exemplified in a study of more than 15,000 consecutive (eligible) deliveries at Beth Israel Hospital in Boston, about half involving electronic fetal monitoring, which was the intervention being studied.[6] The authors identified many variables as risk factors; among them were gestational age, hydramnios, placental and cord abnormalities, multiple birth, breech delivery, and prolonged rupture of membranes. Their primary analysis used 18 variables in a multiple-regression–derived risk index, on which each delivery was scored. Each case was then assigned to one of five (ordered) strata, depending on the risk score. In addition to the primary analysis, the authors applied risk stratification in two other ways and also analyzed the data independently in terms of log-linear models. Clearly, the answer to the question of how to adjust the data is not always straightforward. The authors qualified their results with this observation: "Since we are applying our risk score to the set of data from which the weights for the score are computed, we may be overstating the concentration of benefit in the high-risk categories." This candid caveat further attests to the intrinsic difficulty of adjusting for relevant variables in the effort to interpret the results in a series.

The message here is that interpretation of even an apparently sharp difference in a series-based study may require additional information about the data in the series; moreover, even with such additional information the results may remain ambiguous.

CONTROL OF INTERFERING VARIABLES BY MATCHING

Another strategy for assessing treatment outcome after adjusting for interfering variables like patient mix, referral patterns and so forth, is to use matching. The idea is to construct a second series consisting of patients "like" the ones who get the intervention under study, and then compare the outcomes. Cameron and Pauling[7] used this approach in a series-based assessment of supplemental ascorbate on extending survival in terminal cancer patients. The investigators selected from the files 100 patients who were judged to be no longer able to benefit from any conventional treatment and who were given ascorbate. (This is not readily perceivable as "all" of any well-defined group; the investigators regarded them as "represent[ing] a random selection of all the terminal patients in [the] hospital, even though no formal randomization process was used.") Then, for each of these patients, 10 comparison subjects were chosen from the files of the same hospital; matching variables were sex, age (within five years), cancer of the same primary organ, and histological tumor type. The response variable was the survival time past initiation of ascorbate for the treated subjects, and the response variable for the controls was length of survival past the judged date when disease became untreatable. In all there were 100 cases and 1000 controls. The authors reported a marked prolongation of survival with ascorbate. Later Moertel et al.[8] reported a randomized clinical trial of the same question. The results of that trial slightly favored the placebo; the authors' analysis concluded, "There is no reasonable likelihood that vitamin C produces even a 25 percent increase in (terminal) survival times over placebo ($P = 0.017$)." The conflicting results suggest that effective matching — especially in retrospect — can be hard to accomplish. In particular here, the determination from records of just when a patient's disease became untreatable would be hard to equate with a determination of that same matter made face-to-face with the patient (i.e., when vitamin C treatment began).

· The generic limitations of matching include: (1) some important variables may have gone unmeasured, and so cannot be used for matching; (2) though it may be possible to match on one or two variables, the difficulty of finding subjects who match on them all grows as the number of matching variables is increased; (3) the two groups of subjects may differ in ascertainment (of eligibility or outcome, as would seem to be the case in the ascorbate study). A full treatment of matching is given by Cochran.[9]

NECESSARY INFORMATION

At a minimum, to interpret the findings of a series-based study accurately, we need to know answers to the cub reporter's legendary questions: *Who* were the subjects (i.e., what were their relevant characteristics)? *What* was done? (This calls for definitions of treatment, diagnosis, staging, adjuvant care, follow-up, and so forth.) By *whom* was it done? (By world-class experts? By teaching-hospital staff? By community-hospital staff?) *When* was it done? (Over a time span long enough to permit the appearance of large trends of various sorts?) *Why* was a certain treatment used? (Because other treatments had already failed? Because the patients were not strong enough to tolerate other treatments? For palliation? For cure?)

The randomized clinical trial is especially effective in addressing these questions for two reasons. First, since a protocol is prepared in advance, the key design questions have explicit answers that are unlikely to change. Second, randomization provides escape from the reality that differently constituted groups are hard to compare; this escape is effected by constructing a *single* group of patients eligible for and willing to receive either intervention and by then applying the different interventions to random subsets of this one group.

EFFECTS OF THE ABSENCE OF A PROTOCOL

The absence of a protocol prepared before data are acquired may give rise to certain kinds of defects in the study. Exactly what interventions were performed on what kinds of patients for what indications may be unclear in hindsight. Decisions about which patients to count as eligible, and therefore to include, and which to omit may have been based on subjective judgment. Withdrawals may be poorly documented or even undocumented, with possibly major effects on results. The reader may wonder whether the results reported have been selected from among many possible end points and are therefore less likely to be reproducible than significance tests indicate. Increasing numbers of studies have been performed in which the research was planned after the data had been collected. Fletcher and Fletcher[10] studied articles in the *Lancet* and the *Journal of the American Medical Association* and found that 24 percent of those published in 1946 fit this description; for articles published in 1976, the corresponding figure was 56 percent. Of course, difficulties of this type can be mitigated by careful reporting, but they can be eliminated only if the data have been gathered so systematically as to conform to an implicit protocol in all important respects.

EFFECTS OF NONRANDOMIZATION

The series-based study is vulnerable to the many dangers that randomization forestalls. The key considerations are the comparability of patients receiving different interventions and the equivalence of outcome evaluations. In the absence of randomization, doubts about interpretation can, and should, nag the reader. The effects of crossing over illustrate the difficulties.

Suppose that a serious disease can be treated effectively by surgery, but operative mortality and postsurgical sequelae are drawbacks. A medical therapy is therefore an attractive alternative. If there is a class of patients for whom the two treatments appear to be equally reasonable, then a suitably designed and executed randomized controlled trial should indicate which treatment is actually superior in that class of patients. Now, it may happen that some medically treated patients do not respond, and the gravity of the

disease may require that they receive the surgical treatment after initial assignment to the medically treated group. This is called *crossing over.*

In a randomized controlled trial, the effect of crossing over is simply to
change the research question from its original form to this one: For patients of the class originally defined, is the superior policy to perform surgery immediately or to give medical treatment and avoid surgery unless it
becomes indicated? This is very likely to be a better — i.e., a more realistic
and practical — question than the original one, so no harm is done.

In nonrandomized studies, crossing over is more likely to cause serious
problems of interpretation. The two policies — surgery immediately and a
medical therapy until surgery may be necessary — can be difficult to compare, because in retrospect it may not be clear which policy has been applied to a patient treated surgically.

ASSESSING INDIVIDUAL STUDIES

A summary of the discussion to this point would suggest that series-
based studies are liable to grave difficulties, although of course not every
study comes to a false conclusion. Determining the evidential value of the
individual study under consideration amounts to assessing whether the selection of patients has biased the results, whether the evaluation of treatment outcomes has differed according to the treatment, and whether the
withdrawal of patients from the study has biased the results. The reader
may recognize the questions but be powerless to answer them with the information published. The investigator may be unable to answer them on
the basis of the records. These difficulties are more formidable when the research plan is established after the data have already been recorded.

The timing of observations also affects a series in another way. The cases
can be defined as all those present at one of several points in time. Thus,
the series study may (1) take up each patient with a certain disease as he or
she presents in a given setting during a stated time interval, or (2) it may
first take cognizance of a patient at the time of initiating treatment A or
treatment B, or (3) the study may look backward and capture all subjects
who began treatment A or B during that stated interval. Difficulties of interpretation grow as the series originates at later stages in the sequence
above. Starting at (2) rather than (1) blurs information about inclusion

and exclusion criteria that were applied. Starting instead at (3) further adds problems of data availability and quality.

THE "CLEAR-CUT" SERIES

The reader may think that the picture of series has been presented too negatively and may reason, "If a small series is studied, clear differences may be observed at once, and the complexities referred to may not need to be unraveled. If a new approach is so good that it has an explosive impact, then an acceptable study can be devised readily enough." This objection raises fair questions. Isn't a large fraction of medical practice based on the results of series-based studies? Aren't there many instances in which such studies — e.g., those of penicillin and ether anesthesia — have unambiguously demonstrated the truth? Certainly, medical practice has evolved largely from series-based information. But we also know that much of today's accepted doctrine will be discarded when more and more careful evaluations are performed. The problem is to ascertain which series (i.e., which uncontrolled studies) point to the right answers and which do not.

What about penicillin for syphilis, sulfa drugs against pneumococcus, and similar examples? These have been called "slam-bang" effects. When they occur, they are dramatic. The very fact that they are so dramatic should remind us that they are also rare. An effort to enumerate brings to mind the use of vitamin B_{12} against pernicious anemia, penicillin for subacute bacterial endocarditis, x-ray films for setting fractures, cortisone for adrenal insufficiency, insulin for severe diabetes, propranolol for hypertrophic aortic stenosis, methotrexate for choriocarcinoma, indomethacin for patent ductus arteriosus, and perhaps two or three times as many. But have there been more than one or two such cases per year in the last half century? Perhaps not.

Slam-bang effects are uncommon. They account for only a small number of the thousands of studies published each year. Furthermore, they are not always open-and-shut cases. In 1847, the year that Snow published his report on the ether series, J.Y. Simpson reported his results lauding chloroform anesthesia. Controversy about the comparative merits of the two anesthetics persisted at least until the Lancet Commission examined the matter in 1893. Between 1848 and 1893, a total of 64,693 administrations of chloroform and only 9380 administrations of ether had been identified. The commission recommended ether as safer than chloroform in general surgery "in temperate climes."[11] Similarly, x-ray films for diagnosing fractures clearly work, but how long did it take to discard x-ray therapy for the treatment of acne? Prefrontal lobotomy for schizophrenia stands as a reminder that a treatment may come to be widely used on the basis of inadequate evidence — only to be discarded later. The occasional slam-bang effect, clearly detectable from an uncontrolled study, is at the favorable end of a spectrum; at the other end lie series-based studies that defy interpretation.

The Office of Health and Technology Assessment (OHTA)[12] reviewed the nine published series that report at least 10 cases of transsexual surgery; its assessment report states:

> These studies represent the major clinical reports thus far published on the outcome of transsexual surgery. None of these studies meets the ideal criteria of a valid scientific assessment of a clinical procedure, and they share many of the following deficiencies:
>
> a. There is often a lack of clearly specified goals and objectives of the intervention, making it difficult to evaluate the outcomes;
> b. The patients represent heterogeneous groups because diagnostic criteria have varied from center to center and over time;
> c. The therapeutic techniques are not standardized, with varying surgical techniques being combined with various other therapies;
> d. None has had adequate (if any) control groups (perhaps this is impossible);
> e. There is no blinding, with the observers usually being part of the therapeutic team;
> f. Systematically collected base-line data are usually missing, making comparison of pre- and post-surgery status difficult;
> g. There is a lack of valid and reliable instruments for assessing pre- and post-surgery status and the selection of scoring of outcome criteria usually involve arbitrary value judgments;
> h. A large number of patients are lost to follow-up, apparently due in great part to the desire of transsexuals to leave their past behind; and,
> i. None of the studies are presented in sufficient detail to permit replication.

Although the procedure under consideration is quite unusual, most of the difficulties listed are threats to the majority of assessments using series of patients receiving a new treatment or procedure. They also amount to a list of most of the problems that the protocol of a randomized clinical trial is intended to forestall.

ADDITIONAL ISSUES IN INTERPRETATION

A number of other issues also affect the usefulness of a series in providing information on the value of an intervention or procedure.

SUBGROUPS

The difficulties associated with direct reliance on data in a whole series are exacerbated if one attempts to pick out subclasses marked by strikingly good or bad results. The idea seems reasonable enough, but it ignores a somewhat subtle, inescapable fact: one can always expect to find some good-looking and some bad-looking subsets in any body of data, even when no bias has influenced any part of it. Furthermore, such differences among subsets can easily be large enough to look quite convincing to the unsophisticated analyst.

Suppose that n subjects, a random sample from some population with standard deviation σ, are further divided at random into k equal-sized subgroups. Then the standard error for the mean of the undivided sample is S.E. = $\sigma/\sqrt{n}$. Now, of course, the largest of the subgroup means must exceed the mean for the whole group, but it can be surprising how large this excess must be. If $k = 4$, the average excess of the largest subgroup mean over the whole-group mean to be expected from random division is 2.06 S.E.; if $k = 7$, it is 3.56 S.E.; and if $k = 10$, it is 4.87 S.E. Comparison of the best and the worst of subgroups can produce differences that appear even more vivid but are meaningless. The following rule of thumb demonstrates this point: If a group of n subjects is divided at random into k equal-sized subgroups, then the difference between the largest and the smallest subgroup means has an expected value that is approximately k times the standard error of the mean of the whole group (when k is not more than 15).

With such large subgroup differences to be expected by random division, we must temper our enthusiasm when we identify a series subset that differs strikingly from the whole group; we should face the question, Is this difference large enough to accept as meaningful, considering what chance alone would produce? There are methods for answering the question, which also arises with randomized controlled trials.[13]

TEMPORAL DRIFT

If a series has been accumulated over a long time (as happened with Louis' study, but not with Snow's), additional problems are likely. Over a long period, shifts can occur — indeed, they are to be expected. The patient population may change as referral patterns alter; thus, the demographic composition of the sample may drift. Supportive care, diagnostic criteria, and exposure to pathogenic agents may change over time; even the treatments themselves may change. It follows that information about the sequence and timing of the cases in the series may be essential to a realistic analysis. The issue here is not hypothetical; Schneiderman[14] gives examples of clinical trials in which a second control group was initiated because treatments were modified during the trial. Analysis then showed that the two successive control groups, although seen in the same setting and meeting the same criteria, differed importantly with respect to survival.

GRAB SAMPLES

Statistical inference is a powerful tool for learning from experience. It is at its best when data are obtained in ways that permit the correct application of probability theory — e.g., with random sampling from a population or with data from a randomized clinical trial. When the probabilistic structure is chaotic or unknown, the data constitute what is often called a "grab sample." Inference from such samples, whether by application of formal statistical methods or by other means, is treacherous. The experience of a single physician is in some sense a grab sample; so are the case study and the series of successive cases. The fact that we can learn from experience shows that it is not impossible to reach valid conclusions from grab samples, but the process is fraught with difficulty, uncertainty, and error.

Data from a grab sample may be especially useful when two conditions obtain: first, when the data come from a well-defined setting that is relatively stable over time and, second, when the data are taken from this setup at regular intervals over a protracted period.

Two examples help explain this point. First, statistical reports from the Metropolitan Life Insurance Company have long given useful indications of trends in longevity and disease attack rates despite the fact that, because of selective factors that apply to insurance policyholders, its statistics could not wisely be used to estimate the average longevity and the attack rates of diseases for the U.S. population. Similarly, cross-sectional information from the populations covered by particular health care facilities such as the Kaiser Health Plan, the Mayo Clinic, and Professional Activities Study (PAS) hospitals would not be expected to apply directly to larger or different groups, although changes in such series over time might so apply.

Second, air pollutants are monitored at stations situated in particular locations; the relation of pollutant levels at such stations to the levels experienced by people in the schools, homes, and factories or on the roads near the monitoring stations is in general poorly known. Nonetheless, when those monitored levels rise or fall, we feel justified in thinking that the pollution levels experienced by the nearby population rise or fall as well. The monitored pollutant levels would serve less well or even not at all as measures of the absolute dose to the population nearby.

Even such restricted use, to indicate trends, depends strongly on the assumption of stability in the system. For example, a change in membership, fees, or reporting methods could affect the interpretation of trends observed in the statistics of a health plan. Similarly, a seasonal change in the prevailing wind direction might cause some areas to receive higher levels of pollution, even though every monitoring station showed lower levels. Consider, for instance, a location near a major pollution source that is upwind of that source during the region's high-pollution season but is downwind from it during the region's low-pollution season.

So the message here, again, is that the meaning and reliability of the data in a series are generally obscure until clarified by detailed study. It is possible that detailed study will reveal essential flaws that bar trustworthy interpretation of the kind that one might initially hope would be feasible.

CONCLUSIONS

A series study is a record of experience; as such, it has prima facie value. It may provide useful information about how to apply a new technique and about what kinds of difficulties and complications are encountered in its application. This is the case with Snow's book. Postmarketing surveillance produces what might be called partial series, in which the total number of cases under observation can only be estimated. It is a method of study that has its just role in medical investigation. The series is most liable to be inadequate as an indicator of the effectiveness of treatments. The two principal threats to its validity in such cases are vagueness and bias.

Advance planning and full, careful reporting can do much to mitigate problems of interpretation. The planning should be done while the investigators contemplate the way in which they would study the problem if they could use a randomized controlled trial; they can identify probable disturbing variables and can measure and report them. It is impossible to make two differently constituted groups reliably comparable by means of statistical adjustments, so doubts about selection and assessment bias cannot ordinarily be entirely removed. Nevertheless, fuller information helps with the difficult task of interpretation.

The description of a series of successive cases provides readers with vicarious experience, acquired with little outlay of effort. Often, the investigator has also had to expend relatively little effort to collate the experience and report it. Thus, in terms of effort, the series may be regarded as an efficient information source.

Useful interpretation of this vicarious information, however, is likely to involve considerable difficulty. A thorough knowledge of surrounding circumstances is ordinarily necessary, but the series study may not adequately report these facts. Even if the necessary supplementary information is reported, quantitative methods for taking accurate account of it may be hard, or even impossible, to devise.

Acceptance of the results of a series study at face value is almost never justified. Any statistic is simply the reported outcome of some process; until the details of the process are known, one cannot know what the statistic means, however it may be labeled. Thus, the interpretation and use of results from a series-based study typically call for analysis, which will prove to be feasible in some instances but infeasible in others.

REFERENCES

1. Snow J. On the inhalation of the vapour of ether in surgical operations. London: John Churchill, 1847.

2. Louis PCA. Researches on the effects of bloodletting in some inflammatory diseases, and on the influence of tartarized antimony and vesication in pneumonia. Boston: Hilliard Gray, 1836.

3. Feinstein AR. Clinical biostatistics. XLIV. A survey of the research architecture used for publications in general medical journals. Clin Pharmacol Ther 1978; 24:117-25.

4. Bailar JC III, Louis TA, Lavori PW, Polansky M. A classification for biomedical research reports. N Engl J Med 1984; 311:1482-7. [Chapter 8 of this book.]

5. Is home a safer place? Health Soc Serv J 1980; 90:702-5.

6. Neutra RR, Fienberg SE, Greenland S, Friedman EA. Effect of fetal monitoring on neonatal death rates. N Engl J Med 1978; 299:324-6.

7. Cameron E, Pauling L. Supplemental ascorbate in the supportive treatment of cancer: prolongation of survival times in terminal human cancer. Proc Natl Acad Sci U S A 1976; 73:3685-9.

8. Moertel CG, Fleming TR, Creagan ET, et al. High-dose vitamin C versus placebo in the treatment of patients with advanced cancer who have had no prior chemotherapy: a randomized double-blind comparison. N Engl J Med 1985; 312:137-41.

9. Cochran WG. Planning and analysis of observational studies. New York: John Wiley, 1983.

10. Fletcher RH, Fletcher SW. Clinical research in general medical journals: a 30-year perspective. N Engl J Med 1979; 301:180-3.

11. Report of the Lancet Commission appointed to investigate the subject of the administration of chloroform and other anaesthetics from a clinical standpoint. Lancet 1893; 1:629-38, 693-708, 761-76, 899-914, 971-8, 1111-8, 1236-40, 1479-98.

12. Office of Health Research Statistics and Technology. Transsexual surgery. Washington, D.C.: Department of Health and Human Services, 1981 (Assessment report series. Vol. 1. No. 4).

13. Ingelfinger JA, Mosteller F, Thibodeau LA, Ware JH. Biostatistics in clinical medicine. 2nd ed. New York: Macmillan, 1987:280-2.

14. Schneiderman MA. Looking backward: is it worth the crick in the neck? or: pitfalls in using retrospective data. Am J Roentgenol Radium Ther Nucl Med 1966; 96:230-5.

8

~

A CLASSIFICATION FOR BIOMEDICAL RESEARCH REPORTS

John C. Bailar III, M.D., Ph.D., Thomas A. Louis, Ph.D.,
Philip W. Lavori, Ph.D., and Marcia Polansky, D.Sc.

ABSTRACT   Biomedical research uses a wide range of designs applied to problems in laboratory, clinical, and population settings. Whatever the nature of the study, a few key features — such as the admission rule, the method of allocating subjects to treatments, and the use of controls — largely determine the strength of scientific inferences.

We used these and other features to classify the 332 Original Articles published in the *New England Journal of Medicine* during 1978–1979. This classification directs attention to critical aspects of study design and performance and can help in the choice of suitable research approaches and protocols. It emphasizes the critical role of the investigators' intent in performing and analyzing a study, and it alerts readers to important aspects of interpretation. We recommend that authors always report enough detail about their work for readers to apply this or a similar classification. Omission of such detail may limit the interpretation of a research study, because a study that cannot be classified has probably been incompletely reported.

Research classification schemes have been developed by Feinstein,[1] Fletcher and Fletcher,[2] and Makrides and Richman,[3] among others. Our scheme differs from these in three important ways: we do not use the same sets of descriptive categories as other authors; we focus specifically on the research report rather than the research activity itself; and our classification is based on empirical study of a well-defined collection of such reports.

A classification of research reports according to issues of scientific inference has several kinds of uses. First of all, it can alert even experienced readers to critical issues in parts of the literature that are unfamiliar to them. For example, an epidemiologist understands well the implications of using a cohort rather than a case–control approach, but an expert on metabolic diseases who needs to use epidemiologic results may not. Secondly, it provides a didactic tool for persons not yet familiar with the scope, power, and potential failings of various scientific methods. Thirdly, classification can promote more complete reporting of major components of the study design. A research report that cannot be classified within some reasonable scheme is likely to have failed in some aspect of communication. Furthermore, a classification can bring out unsuspected features of studies or data or both that illuminate common but vaguely stated problems. For example, our classification in Table 1 has already been used to determine the prevalence of the use of various statistical procedures,[4] to isolate and study prospective investigations of deliberate interventions,[5-8] and to select a subset of these for a study of reporting quality.[9] Our identification of pseudo-longitudinal studies emerged in this way, and such studies seem not to have been previously discussed as a general class. Finally, a classification can be used to track the relative frequencies of various designs over time or across medical disciplines. Such information may be useful in identifying the underuse of powerful research methods, assessing and improving the state of the art in various medical specialties or geographic areas, guiding libraries and other major purchasers in the selection of journals, and conducting bibliometric and cliometric studies of how new medical science develops.

METHODS

The classification presented here was developed from a study of the 332 research reports published as Original Articles in the *New England Journal of Medicine* during 1978 and 1979. Each Original Article was reviewed and assigned to a broad category according to an initial, simple structure. Additional readers then repeatedly studied, sorted, restudied, and resorted the articles, using an evolving classification scheme designed to group articles with similar components of scientific inference, regardless of the disease studied, the medical specialty, or the research context.

At least three of us independently classified each paper at each iteration. After each round of sorting we examined the sorted sets for several purposes: to assess and possibly redefine category boundaries; to identify subgroups of papers that might be used in the next, more detailed level of classification; and to identify problems in classification whose resolution might clarify important concepts or issues. This process was repeated many times over a period of months until we believed that the process had stabilized and that there would be few further important changes in either the classification scheme itself or our categorization of individual papers. We undertook this entire process with the conscious purpose of letting the study materials determine the classification and avoiding preconceptions based on other work or on common knowledge about the structure of research.

PRINCIPLES FOR CLASSIFYING RESEARCH REPORTS

Several principles and observations emerged from our initial attempts to develop a classification.

First of all, each paper in a defined universe should have one and only one place in the structure. Consequently, no paper should be ambiguous in its position or left over at the end. We did not find the latter to be a problem in our sample of 332 Original Articles, except in 2 articles[10,11] that dealt primarily with methods of data analysis rather than with data themselves. We arbitrarily kept them in our sample, and they are classified with the subject matter of the papers they cite as references. This principle also means that papers with features of two or more categories must be classified in only one category. Reports with two or more sets of features of nearly equal weight were uncommon in our sample, but when they occurred, we arbitrarily gave preference to classification of a study as longitudinal rather than cross-sectional, and within each of these broad groups to the class or subclass that occurs first (highest) in Table 1. This rule prevented the need for a small category of mixed or undetermined research reports.

Secondly, the classification should be based on information that is, or ought to be, reported. Authors usually provide the most important facts, and these were often prominent in the Abstract, though information in the Abstract had to be checked against the text. For reasons we assume are related to editorial convention in the *Journal*, the penultimate sentence of the Abstract was often the most revealing, and a preliminary classification based on it, the title, and any subtitles often agreed well with a more considered determination.

Thirdly, although our classes contain research structures that are homogeneous in certain important ways, it was difficult to find even two reports that were similar in every important aspect of research strategy and analytic method. For example, one of the most homogeneous small groups consisted of clinical trials with crossover, but such papers in our sample differed in such critical aspects as whether the patients were randomly assigned to treatment, whether the change in treatment was at a fixed time after entry or triggered by changes in patient status, and whether the physician responsible for direct patient care, the physician responsible for assessing treatment effects, or both, were blinded to the assigned treatment.[6] Thus, even the terminal branches of our classification contain papers that differ in important ways.

Furthermore, we classified published reports, not the research investigations themselves, so that two reports based on the same study and data may well have been classified in different categories. Thus, data generated in the course of a randomized clinical trial (longitudinal) may also have been useful in developing a new classification of disease pathology (cross-sectional).

Finally, classifying papers in this way must not in itself be interpreted as judging the studies or the conclusions reported by the authors. For that, one would need a deep knowledge of each subject area covered and an understanding of the constraints imposed on each investigator (time, cost, staff, and number of patients available), the initial purpose or purposes of the study, the extent of knowledge at the time about the research question, the size of any effects or differences that the investigator attempted to detect, and many other matters. Furthermore, we suspect that some papers did not report all the relevant strong points of design. This lack indicates a failure of communication rather than a weak research strategy. In a study assessing reporting in randomized controlled trials, DerSimonian et al.[9] have documented such weaknesses with respect to many facts that the authors must surely have known but did not report, such as the method of randomization.

THE CLASSIFICATION

LONGITUDINAL VERSUS CROSS-SECTIONAL STUDIES

Table 1 shows the classification we used, outlined to include several levels of detail. It also gives the numbers of Original Articles classified in each classification branch. The first, most fundamental split separates research reports according to whether the work was longitudinal (category I) or cross-sectional (category II). Longitudinal reports study a process over time to investigate changes; an example is a randomized clinical trial. The responding systems may be such things as individual human beings, cells in culture, microorganisms, health maintenance organizations, or entire human populations. Cross-sectional reports focus on describing a state or phenomenon at a fixed or indefinite time; an example is a descriptive study of the metabolic defects in diabetes.

Classification of a study as cross-sectional or longitudinal does not depend directly on the time period covered by the observations. A study of the cardiovascular effects of a new general anesthetic would be considered longitudinal even though the observations in a given patient covered less than an hour, whereas a study of the prevalence of the use of nasal decongestants would be considered cross-sectional even though questions might be based on the frequency of use over a period of months or years. (The latter would become longitudinal if we examined the change in use over that period, perhaps as a result of some educational effort or other intervention.) Most of the laboratory studies of biologic mechanisms in our sample were cross-sectional, whereas most of the studies of the effects of external factors on human beings (such as the treatment of disease) were longitudinal.

We recognize that very few research reports are devoid of elements of change during the observation period and that deciding between the longitudinal and cross-sectional categories can sometimes be difficult if the phenomenon under study has an important time dimension. The critical concept here is the difference between change used solely as a tool in the study (cross-sectional) and change as an object of study (longitudinal). When this concept is understood, most problems in deciding between the longitudinal and cross-sectional categories diminish or disappear. For example, a glucose-tolerance test involves change over a period of several hours, but if glucose tolerance is used to determine the prevalence of diabetes among persons working with a toxic chemical, the study will be cross-sectional in our scheme; the real object of study is disease prevalence, not the rise and fall of the blood glucose level over time. A single set of data can sometimes be used for both purposes, though we found few examples of this duality in the 332 articles we examined.

The longitudinal–cross-sectional dichotomy isolates key considerations of study design, conduct, and analysis. Generally, longitudinal designs require patient follow-up, documentation of intervening events, and analysis

Table 1. Classification of 332 Original Articles in the *New England Journal of Medicine*, 1978–1979.*

Classification	Number of Articles (%)		
I. Longitudinal studies			202
A. Prospective studies		181 (55)	
1. Studies of deliberate interventions		110	
a. Sequential	41		
i. Crossover	13		
ii. Self-controlled	28		
b. Parallel	49		
c. Externally controlled	20		
2. Observational studies		64	
a. Causes and incidence of disease	21		
b. Deliberate but uncontrolled interventions	13		
c. Natural history; prognosis	30		
3. Pseudoprospective		7	
B. Retrospective studies		21 (6)	
1. Deliberate intervention		2	
2. Observational		13	
3. Pseudoretrospective		6	
II. Cross-sectional studies			130 (39)
A. Disease description		26	
B. Diagnosis and staging		34	
1. Normal ranges		6	
2. Disease severity		28	
C. Disease processes		70	
1. Exploratory		35	
2. Observational		19	
3. Case reports		16	
Total			332

* A few papers with mixed content are classified according to the dominant feature. If two criteria seemed to be equally applicable, we assigned the article to the category that is listed first in the table. Figures above and to the right are totals for those below and to the left.

of a series of measurements. These features are less important in cross-sectional studies, so that other kinds of problems — such as errors of measurement, the interpretation of transient effects, and definitions of disease states — become more prominent.

COHORT VERSUS CASE–CONTROL STUDIES

The second level of classification in Table 1 subdivides longitudinal reports according to whether the objects of study were selected on the basis of input or output variables. *Input variables* refers to presumed determinants of outcome, and *output variables* to the status of subjects after treatment or follow-up. Sampling on the basis of input variables produces studies called cohort (or prospective) studies, whereas sampling on the basis of output variables leads to case–control (or retrospective) studies.[12] For example, a randomized clinical trial comparing survival in patients receiving different treatments is a cohort study (the primary classification is based on treatment, an input variable), whereas identifying patients with bladder cancer and matching them with cancer-free controls to compare the two groups with respect to saccharin use produces a case–control study (the classification of patients is based on whether they have cancer — an output variable).

Scientific inferences and statistical techniques depend critically on whether the prospective or retrospective approach is used. For example, a case–control study can often identify an association between the occurrence of a disease or condition and various outcomes, but it is rarely able to pin down causation. This matter has been extensively discussed in the epidemiologic literature.[13]

Determining whether a study is primarily a cohort or case–control study is sometimes difficult. Although these categories often relate to the sequence in which states or events are observed as having occurred, time relations are secondary to the mode of selection of objects for study, and the selection process is not always clear from a published report. Some authors still use these words to refer to time relations rather than to the sampling plan; thus, their usage may be inconsistent with ours.

Examples of sampling on the basis of input variables include classic follow-up studies of disease treatment (with output variables then ascertained over a period that ranges from minutes to many years); short-term studies of population mortality rates (as in national life tables, in which persons are individually observed for only a short period but their aggregated experience is used to create long-term syntheses in such areas as life expectancy); and the use of old records to identify (today) a complete cohort of persons who can be traced from some past time. Thus, a 20-year cohort study can be conceived, implemented, and completed within months if the right sorts of records are available. For example, determination of the cancer incidence rates from 1955 to 1985 of army personnel known to have been exposed to ionizing radiation from atomic blasts in 1952 can be conducted as a cohort study even though the study is initiated in 1992. Of course, there may be serious problems with the quality of data available for cohort studies based on information gathered years ago and for another purpose.

The distinction between *cohort* and *case–control*, as we and others use the terms, is sometimes confusing because matched controls are on rare occasions used in cohort studies, though the matching is not based on an outcome variable. We may today identify all the patients with, say, cholelithiasis who were seen at a specific hospital since 1950, select controls matched according to the time of diagnosis of that condition (and perhaps other variables), and compare the numbers of myocardial infarctions between that time and the present. Although such a study would compare cases with matched controls, it would be classified as a cohort study because selection was based on the presence (cases) or absence (controls) of cholelithiasis, not on outcome (the incidence of myocardial infarctions).

Another variation of the cohort study has been called the *nested case–control* design. In this approach, complete cohorts of (say) exposed and nonexposed persons are identified, but some or all of the additional information needed for analysis is collected only after the outcome of interest has been observed. A hypothetical example is a possible study of the relation between exposure to dioxin (TCDD) and the later appearance of cancer. Measurement of dioxin is very expensive, so a study might collect and store blood samples from persons likely to have a range of exposures, but not determine dioxin levels until after the follow-up period. Dioxin could then be measured for all subjects who developed cancer but for only a matched random sample of those who did not. Because cancer is likely to be relatively uncommon, this approach could substantially reduce the number and cost of dioxin measurements. This approach has been developed largely since the time of publication of the papers classified in Table 1, and does not appear in the table.

A disease state or other feature or item that defines eligibility for the study of both cases and controls should not be confused with input and output variables. For example, in a comparison of treatments for diabetes, "diabetes" is neither an input nor an output variable; rather, it defines the universe of study. Authors sometimes seemed to have trouble with this point, especially in prospective observational studies with comparisons among patient subgroups (parts of category IA2), in which having a single, specific treatment was a requirement of admission but the investigators attempted to use the data in inappropriate ways for the evaluation of that treatment.

CONTROL OF INTERVENTIONS

We have further divided reports of prospective studies according to whether they focused on the outcome of deliberate interventions that were under the control of the current investigators or were solely observational. Observational studies were more often directed toward the natural history of disease than to therapy. Classification of a study as involving "deliberate intervention" does not imply that the observed results were intended. We classified reports of drug toxicity and other untoward results of treatment as studies of deliberate interventions because the treatment was intentional, though the outcome was not.

FURTHER SUBDIVISIONS OF LONGITUDINAL STUDIES

Longitudinal categories were again subdivided. For example, category IA1 (longitudinal, prospective, deliberate intervention) contains parallel, sequential, and externally controlled treatment comparisons. All these studies were controlled in one way or another, but since control was not the critical feature, it was not itself the basis of the classification scheme. Parallel studies compare treatments given to different patients in a single clinical trial (49 studies in our series), sequential studies compare different treatments given to the same patient in a single clinical trial (13 in our series with crossover and 28 more with other kinds of self-controls), and externally controlled studies compare treatments given to different patients at different times or places (20 in our series). These reports can be classified further on the basis of such features as methods of patient allocation and of statistical analysis. The table shows additional levels of classification of category IA1, but these are not discussed further here because they have been analyzed in detail elsewhere.[5-7]

Longitudinal observational studies were divided into studies of (1) the causes and incidence of disease, (2) uncontrolled interventions, and (3) natural history. Although this partition is still based on issues of scientific inference, it largely reflects the kinds of data that are likely to be of the greatest immediate interest: antecedents of the disease, attempts to treat it, and results or outcome. Thus, the sources of bias differ, as do the methods of analysis. For example, the biases in an "unselected" case series at a tertiary referral center may be tolerable for studies of natural history, but they are likely to be disastrous for studies of disease antecedents. Conversely, incomplete follow-up is often of limited concern in a study of antecedents, but it can destroy a study of natural history.

The first subcategory, "causes and incidence of disease," contains 21 reports. Only 10 of the 21 were reports of disease incidence in defined groups of persons; we had expected the number to be larger. This finding may reflect the editorial policies of the *Journal*, the preferences of investigators to publish such material elsewhere, or the generally attenuated state of research on the causes as opposed to the treatment of disease. The remaining 11 papers in this set dealt with the sequelae of diseases or their treatments; 4 included control groups of normal persons, and 7 did not.

In the 13 studies classified as involving "deliberate but uncontrolled interventions," someone acted to produce an effect, but that intervention was not under the control of the investigator, whose role was to observe and report what happened. For example, Foster et al.[14] and Greenwald et al.[15] studied women with newly diagnosed breast cancer to determine the relation between an intervention not under the investigator's control (breast self-examination) and the extent of the cancer. Of the papers in this category, one dealt with populations rather than individuals as the main object of study (cities with fluoridation as compared with those without it). Two of the others had before-and-after observations about patients, and the remaining 10 had only concurrent but nonrandomized controls.

We found 30 longitudinal observational studies of the natural course of illness. Of these, seven used normal controls, and five used patients with other, related diseases as controls. Such controls lend some strength to inferences about outcomes, especially when the outcomes are common or long delayed. Examples in our series included growth and cognitive development, the recurrence of cancer, and measures of cardiac function. The remaining 18 papers in this set had no external controls, and for most of them such controls would have been infeasible or inappropriate. This last group included five reports about biologic predictors of the response to cancer chemotherapy. These were not classified as studies of deliberate interventions because the focus was on subgroups of patients and not on the treatments, which had in general already been established as being effective.

PSEUDOLONGITUDINAL STUDIES

We use the term *pseudolongitudinal* to refer to research that was in fact cross-sectional, though the investigators treated the data as if they were longitudinal. We call such studies pseudoprospective or pseudoretrospective because, though all the data were gathered at one time, the underlying concepts of analysis and inference were essentially the same as those in ordinary cohort and case–control studies. An example of a pseudoprospective study is a report on the relation between lead toxicity in early childhood and behavioral problems in school,[16] in which the investigators used lead levels in deciduous teeth (lost at about the time the data on behavior were collected) to infer lead exposures in early childhood (when the teeth were formed).

The defining feature of this and other pseudolongitudinal studies is that the data collection occurred over such a brief time that few or no relevant changes occurred, and from these data the authors inferred the values of one or more input variables at a prior time.

No question arises about such imputations for variables like sex and race (which do not change) and age (which changes in a known way), and we do not classify studies with such solidly based inferences as pseudolongitudinal. When investigators make similar assumptions about other kinds of variables, however, the prior status may be less certain. We have considered studies based on one-time questionnaire data referring to both past and present status as pseudolongitudinal.

In some respects, pseudolongitudinal studies may be considered cross-sectional or perhaps given a category of their own. We considered this matter carefully and decided to group them with longitudinal studies because in all our examples the intent of the investigator, the method of analysis, and therefore the scientific issues were those of a longitudinal study. The implied mode of sampling still determined whether a study was pseudo-prospective or pseudoretrospective. In the former, persons were admitted to the study on the basis of (imputed) input variables, whereas entry in the latter was determined from (known) output variables.

CROSS-SECTIONAL REPORTS

Cross-sectional studies were divided into three categories: those with a primary focus on the description of new disease entities, those assessing diagnostic and staging techniques, and those studying disease processes and mechanisms. Again, the issues of scientific inference seemed to run parallel to broad aspects of subject matter.

We found 26 Original Articles that described new diseases or disease variants, and we were surprised to find that 15 of the 26 described what appeared to be familial diseases. This high frequency of research reports on familial diseases in a general medical journal suggests a need for a survey and review of both present practice and desirable criteria for investigating and reporting familial diseases that would be similar to recent survey papers on treatment.[5-8]

The collection of 34 papers reporting on the presence and severity of disease ("diagnosis and staging") included 6 that established normal ranges for various physical and laboratory measurements; the remaining 28 had widely scattered contexts.

The final set of 70 papers included 19 reporting observations about specific disease processes, another 35 that attempted to investigate and explain such processes, and 16 more detailed investigations of from one to several patients with conditions of special interest ("case reports").

INVESTIGATIVE INTENT

Our classification is based in concept on objective criteria, but it frequently requires interpretation of the intent of research investigators. Intent is not, however, used in isolation. For example, Gugler et al.[17] reported on work that was designed to be a crossover trial, but we classified the study as a parallel comparison because all the patients continued with their initial therapy without crossover.

Intent has a major role in the interpretation of P values and other statistical measures that are valid only for hypotheses developed independently of the data at hand. The P value for the principal treatment comparison in a prospective clinical trial carries substantially more weight than a P value reported in an observational study, in which the treatment comparison may have been made specifically because odd-looking data caught an author's attention and led to an unplanned calculation of a P value. This issue is discussed by Ware et al.[18]

Because of the need to determine intent, a research report may be classified differently by two reviewers, depending on their understanding of the investigator's motives or other aspects of the analysis. Although we were initially reluctant to include this subjective element, we found that discussion of intent almost always helped to reduce or dispel uncertainty or disagreement about other aspects of the classification or about the interpretation of results.

Two examples illustrate the importance of investigators' intent even at the first level of classification — longitudinal as opposed to cross-sectional studies. The first was a report that some feature of a disease was related to prognosis.[19] Such a study would be considered longitudinal if the author's main objective was to improve prediction of the course of specific cases. However, if the same data and the same analytic methods were viewed primarily as defining previously unrecognized but relevant disease categories, the study would be cross-sectional.

The second example was a report about techniques for screening treated cancer patients to detect recurrences before their usual clinical appearance.[20] Such a report would be viewed as longitudinal if the emphasis was on the relation between early detection of recurrences and outcome (does it do the patient any good?) or as cross-sectional if the emphasis was on follow-up assessments of the diagnostic accuracy of the procedure. If viewed as longitudinal, the study must be assessed for its handling of certain biases in assessing the impact of earlier diagnosis (these are often called the *length bias* and *lead-time bias*[21-23]), issues of relatively less import for the cross-sectional study.

Classification of a few of the pseudoprospective and pseudoretrospective studies was difficult because of some overlap in concept between those papers and ordinary cross-sectional papers. Individual judgment was involved, and we have done our best to interpret the intent of the authors. Also, some papers were clearly "pseudo-X," but it is not clear what "X" is; the investigators reported on whatever case material they could get, regardless of input or output variables. We have somewhat arbitrarily put them into one classification or another on the basis of what seems to be the dominant intent of the authors regarding interpretation, not sampling.

Our system of classification was developed on the basis of a single collection of research reports — those published as Original Articles in the *Journal* during a two-year period. We believe that it should be broadly applicable to laboratory, clinical, and epidemiologic studies in human subjects, though the relative frequencies of various kinds of studies are likely to vary widely from those reported here. We believe that the system will also have relevance to research reports in, for example, education, sociology, and economics. It is clearly inappropriate for many important articles, such as state-of-the-art reviews, reports on new methods of medical education, or analyses of current medical policy issues.

We have not undertaken a formal study of the reproducibility of either our development of the structure itself (which would have required independent sets of investigators to develop classifications de novo) or the classification of individual papers in the structure, but we think that similar schemes would tend to bring out the same aspects of scientific inference. Because judgment was sometimes required, others might not have applied the classification in just the way we did, and might not have placed all the 332 papers in the same terminal branches of the classification tree that we did. This is true in part because critical information was sometimes missing from the published record, and different readers may interpret incomplete information in different ways. It is also partly due to the fact that the classification system was developed for application as a whole rather than step by step, and untrained users may not recall all the nuances of classification at the time they are needed to classify a specific paper. We do not believe that these unresolved questions of reproducibility are serious.

Within its field of application, the classification system requires certain items of information that readers will need anyway if they are to understand the nature and implications of a research study. A study that cannot be classified according to this or some similar scheme has probably been incompletely reported. We recommend that authors review each manuscript before submission for publication to be sure that critical items have not been omitted.

REFERENCES

1. Feinstein AR. Clinical biostatistics. XLIV. A survey of the research architecture used for publications in general medical journals. Clin Pharmacol Ther 1978; 24:117-25.

2. Fletcher RH, Fletcher SW. Clinical research in general medical journals: a 30-year perspective. N Engl J Med 1979; 301:180-3.

3. Makrides L, Richman J. Writing the research proposal. In: Research methodology and applied statistics: a seven-part series. Physiother Canada 1981; 33:163-8.

4. Emerson JD, Colditz GA. Use of statistical analysis in the *New England Journal of Medicine*. N Engl J Med 1983; 309:709-13. [Chapter 3 of this book.]

5. Lavori PW, Louis TA, Bailar JC III, Polansky M. Designs for experiments — parallel comparisons of treatment. N Engl J Med 1983; 309:1291-8. [Chapter 4 of this book.]

6. Louis TA, Lavori PW, Bailar JC III, Polansky M. Crossover and self-controlled designs in clinical research. N Engl J Med 1984; 310:24-31. [Chapter 5 of this book.]

7. Bailar JC III, Louis TA, Lavori PW, Polansky M. Studies without internal controls. N Engl J Med 1984; 311:156-62. [Chapter 6 of this book.]

8. Louis TA, Bailar JC, Lavori P. Experimental designs for clinical investigations. In: Lemberger L, Reidenberg MM, eds. Proceedings of the Second World Conference on Clinical Pharmacology and Therapeutics, Washington, D.C., July 31-August 5, 1983. Bethesda, Md.: American Society for Pharmacology and Experimental Therapeutics, 1984:19-30.

9. DerSimonian R, Charette LJ, McPeek B, Mosteller F. Reporting on methods in clinical trials. N Engl J Med 1982; 306:1332-7. [Chapter 17 of this book.]

10. Diamond GA, Forrester JS. Analysis of probability as an aid in the clinical diagnosis of coronary-artery disease. N Engl J Med 1979; 300:1350-8.

11. Horwitz RI, Feinstein AR. Alternative analytic methods for case-control studies of estrogens and endometrial cancer. N Engl J Med 1978; 299:1089-94.

12. White C, Bailar JC III. Retrospective and prospective methods of studying association in medicine. Am J Public Health 1956; 46:35-44.

13. Feinstein AR, Wells CK. Randomized trials vs. historical controls: the scientific plagues of both houses. Trans Assoc Am Physicians 1977; 90:239-44.

14. Foster RS Jr, Lang SP, Costanza MC, Worden JK, Haines CR, Yates JW. Breast self-examination practices and breast-cancer stage. N Engl J Med 1978; 299:265-70.

15. Greenwald P, Nasca PC, Lawrence CE, et al. Estimated effect of breast self-examination and routine physician examinations on breast-cancer mortality. N Engl J Med 1978; 299:271-3.

16. Needleman HL, Gunnoe C, Leviton A, et al. Deficits in psychologic and classroom performance of children with elevated dentine lead levels. N Engl J Med 1979; 300:689-95.

17. Gugler R, Lindstaedt H, Miederer S, et al. Cimetidine for anastomotic ulcers after partial gastrectomy: a randomized controlled trial. N Engl J Med 1979; 301:1077-80.

18. Ware JH, Mosteller F, Delgado F, Donnelly C, Ingelfinger JA. P values. [Chapter 10 of this book.]

19. Bloomfield CD, Gajl-Peczalska KJ, Fizzera G, Kersey JH, Goldman AI. Clinical utility of lymphocyte surface markers combined with the Lukes–Collins histologic classification in adult lymphoma. N Engl J Med 1979; 301:512-8.

20. Goldenberg DM, DeLand F, Kim E, et al. Use of radiolabeled antibodies to carcinoembryonic antigen for the detection and localization of diverse cancers by external photoscanning. N Engl J Med 1978; 298:1384-8.

21. Shapiro S, Goldberg JD, Hutchison GB. Lead time in breast cancer detection and implications for periodicity of screening. Am J Epidemiol 1974; 100:357-66.

22. Albert A, Gertman PM, Louis TA. Screening for the early detection of cancer. I. The temporal natural history of a progressive disease state. Math Biosci 1978; 40:1-59.

23. Feinleib M, Zelen M. Some pitfalls in the evaluation of screening programs. Arch Environ Health 1969; 19:412-5.

●

SECTION III

Analysis

9

DECISION ANALYSIS

Stephen G. Pauker, M.D., and Jerome P. Kassirer, M.D.

Excellent clinical judgment requires optimal decision making. Many of the decisions that physicians make in their practices involve little uncertainty and little risk: these rote or routine choices need no special contemplation because they are "tried and true" practices. But for each routine problem there are several for which no easy solution is at hand. To deal with these, the tough problems, a physician can search for a properly designed, double-blind controlled study that examined patients of the same age, sex, and race and with the same conditions in the same stage; use an algorithm developed for such patients; use the problem-oriented approach to data gathering and hope that the solution to the problem will emerge; or ask for the help of one or more consultants. The frustrations encountered with each of these approaches are familiar.

For 15 years we and others have been developing decision analysis and applying it to difficult clinical problems,[1-6] and after experience with several hundred such analyses tailored to individual patients,[7] we are convinced that this quantitative approach warrants careful consideration as a tool for making decisions not only for individual patients but also for classes of clinical problems. This actively evolving field provides insights not available from clinical studies or expert opinion.

In this chapter we provide a few examples of some advances in the methods and the application of decision analysis. We consider both the advantages of the method and its limitations and offer our thoughts about the extent of the dissemination of decision analysis in medicine.

BAYES' RULE

The modern physician is inundated by data — both clinical information that has been obtained intentionally and unanticipated results of screening tests and imaging procedures. In most circumstances, clinical information does not establish diagnoses with certainty; instead, each finding allows the physician to revise the probability of various diagnostic alternatives. In this sequential, iterative process, three sets of probabilities are defined: (1) the probabilities of the diagnoses *before* the presence of a new finding is revealed (prior probabilities); (2) the probabilities that a given finding is observed in each disorder diagnosed (conditional probabilities); and (3) the probabilities of the diagnoses *after* the presence of a new finding is revealed (posterior or revised probabilities). The terms "prior" and "posterior" are defined with respect to a given diagnostic finding. In the sequential diagnostic process, the posterior probabilities for one finding become the prior probabilities for the next. Thus, the diagnostic implications of a given test result vary from patient to patient, depending on the presence of other findings.

A mathematical combination of prior and conditional probabilities produces posterior probabilities. The relation among the three sets of probabilities — Bayes' rule — has been understood for two centuries, but this formulation has been applied to clinical reasoning only in the past several decades. Initial applications of Bayes' rule presented the physician with an equation* or with a computer program[8-12] that had the characteristics of a "black box." Other popular alternatives to the formal equation have been introduced, including nomograms[13-15] and tables.[16] Unfortunately, these latter approaches are practical only when a single disease is considered to be either present or absent and when the test result can be considered to be binary — i.e., either positive or negative. Recently, two different tabular formulations of Bayes' rule have appeared: a two-by-two table,[5,17] best used for calculating measures of test performance in a given study, and a

$$*P_{dis|find} = \frac{P_{dis} \times P_{find|dis}}{\sum_{i=1}^{n} P_{dis\ i} \times P_{find|dis\ i}},$$

*where $P_{dis\ i}$ denotes the prior probability of disease i, $P_{find|dis\ i}$ the conditional probability of the finding in patients with disease i, and $P_{dis\ i|find}$ the posterior probability of disease i, given the presence of the finding. The particular disease, among the i possible diseases, is denoted as dis.

Table 1.   Spreadsheet Template Using Bayes' Rule to Interpret a Negative Thallium Test.*

A — Diagnosis: Coronary Artery Disease	B — Prior Probability	C — Conditional Probability of Negative Scan	D — Product (B × C)	E — Posterior Probability (100 × D/Sum)
	percent	percent		percent
Critical	10	5	50	1.6
Noncritical	70	20	1400	43.6
Negligible	20	88	1760	54.8
			3210	

*   This template can be easily built on any standard spreadsheet program.

posterior-probability calculator, designed to provide the interpretation of test information in a given clinical setting.[18,19] The latter technique can be used when several alternative diagnoses are possible and when test results lie along a continuum — e.g., serum enzyme levels. The calculation is performed easily with pencil and paper or with a hand-held calculator. With the almost ubiquitous availability of personal computers and spreadsheet programs, templates for this calculation are easily created. Table 1 demonstrates the use of a spreadsheet to interpret the results of preoperative dipyridamole–thallium perfusion scanning in a 67-year-old man with peripheral vascular disease. This formulation of Bayes' rule uses a table with five columns: Column A, a list of mutually exclusive and exhaustive diagnoses; Column B, the prior probability of each diagnosis; Column C, the conditional probability of the observed finding, given each diagnosis; Column D, the product of Columns B and C and the total of all of these products; and Column E, the posterior probability, which is calculated by dividing each product in Column D by the sum of the products. A negative dipyridamole–thallium test diminishes the likelihood of critical coronary disease in this patient from 10 percent to less than 2 percent (Table 1), but almost half of comparable patients with such a test result will nonetheless have clinically important disease.

   This simple technique can help the physician avoid the common reasoning error of neglecting the base rate.[20-22] In fact, because of the availability of probabilistic data in the medical literature and because clinicians are taught to quote and rely on the literature, this type of reasoning error is quite prevalent. Such errors are most likely to arise when the diagnostic test provides an unexpected result; the unwary clinician may rely too heavily on a highly "accurate" diagnostic test, neglecting the critical influence of disease prevalence.

DECISION TREES

Decision analysis, a derivative of operations research and game theory, involves identifying all available choices and the potential outcomes of each and structuring a model of the decision, usually in the form of a decision tree. Such a tree consists of nodes, which describe choices, chances, and outcomes. The tree is used to represent the strategies available to the physician and to calculate the likelihood that each outcome will occur if a particular strategy is employed. The relative worth of each outcome is also described numerically, as a utility, on an explicitly defined scale — e.g., a life expectancy of 17 years, or a score of 50 on a scale on which immediate death is defined as 0 and normal life expectancy in good health is defined as 100. The utility of a chance node is calculated as the weighted average of the utilities of its possible outcomes, where the weights are the probabilities that each outcome will occur. For example, a chance node describing a 5 percent chance of immediate death (with a life expectancy of 0), a 20 percent chance of survival with disabling angina (with a life expectancy of 7 years), and a 75 percent chance of survival free of angina (with a life expectancy of 15 years) would represent a life expectancy of 12.65 years, i.e., $(5\% \times 0) + (20\% \times 7) + (75\% \times 15)$. The utility of a decision node is the maximum of the utilities of its component strategies, since the rational decision maker should choose the alternative that, on average, provides the highest value. Even when objective data are not available from the literature or from local experience, probabilities and utilities nevertheless must be quantified to preserve the logic of the decision process and to make optimal use of whatever data are available.

STRUCTURING PROBLEMS WITH SUBTREES

A decision-analysis model is used to provide insight about real-world problems. Because the real world of clinical medicine is complex, such models often must be rather complex. The insights and conclusions that a decision model provides can be helpful only if the model represents the clinical problem with sufficient fidelity. To create realistic models, the analyst needs a notation that is compact and that helps avoid certain mistakes. These seemingly contradictory demands can be resolved because most decision trees contain many repetitive structures. Even when management plans are vastly different, the prognosis is often described by the same series of chance events but with different frequencies of occurrence. These homologous structures can often be represented by a common tree fragment, called a subtree, that can be shared among different strategies and events. Figure 1 shows a rather complex decision tree representing alternative strategies for treating a patient whose thyroid was irradiated in childhood. Figure 2 shows the same model with subtree notation. Not only is the representation in Figure 2 readily understood, but it emphasizes analogies among events. When a common subtree appears in different places within a decision model, the likelihoods and the values of the outcomes may differ. For example, the probability of a benign lesion with a defect on a scan is somewhat lower than the probability of a benign lesion in a palpable nodule. On the other hand, the chance of recurrence of a cancer in a palpable nodule is 10 times higher than the chance of recurrence of a cancer found on scanning. Thus, the features of a subtree are often represented as variables, which assume different values when the same subtree is used in different contexts.

Subtree notation emphasizes relations among factors in a decision model. For example, when considering whether or not to perform surgery in a patient with unstable coronary disease, the analyst might be tempted to consider as separate variables the likelihoods of survival with and without such therapy, estimating likelihood from either a single report in the literature or several reports. In fact, survival in both circumstances often reflects the underlying state of the patient: patients with more severe coronary disease or poorer ventricular function survive less well under either plan. Thus, these two factors are linked, either to each other or to a common underlying factor. For example, if the efficacy of bypass surgery in lowering the annual mortality rate from coronary artery disease is denoted by e, and the mortality rate among patients with coronary artery disease (CAD) by μ_{CAD}, then the annual mortality rate among patients in whom sur-

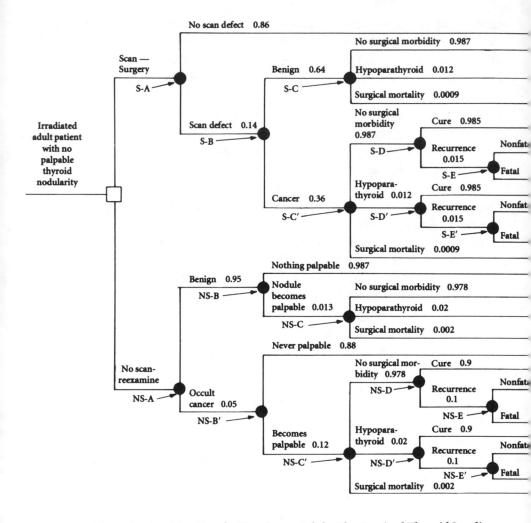

Figure 1.   Decision Tree for Treating an Adult Who Received Thyroid Irradiation during Childhood.*
The decision, represented by the node (open square) at left, is between obtaining
a thyroid scan, with surgery if a defect is found, and not obtaining a scan, with a
plan to reexamine the patient for palpable thyroid nodules. Each chance event is
represented by a node shown as a solid circle. A nodule or a defect may represent
either benign disease or cancer, surgery may be complicated by hypoparathyroid-
ism or death, and a cancer may or may not recur. This tree contains 17 chance
nodes and 23 outcomes. S denotes scan, and NS no scan.
The numbers on each branch are probabilities.
* Reproduced from Stockwell et al.,[23] with the permission of the publisher.

gery is successful can be expressed as $\mu_{CAD} \times (1 - e)$. Because subtree notation encourages the analyst to look for symmetries in a decision problem and to press probabilities and values symbolically, it provides a new language and technique for expressing and examining such relations.

Occasionally, using subtree notation can suggest additional strategies. For example, certain treatments are traditionally given only after a diagnosis is considered to have been confirmed. As an alternative, the decision analyst might consider whether such therapy might be given "empirically" — before a diagnosis has been established definitively. In our experience, empirical therapy is a viable alternative when one is considering issues such as steroids for idiopathic nephrotic syndrome,[24,25] amphotericin for unexplained fever in a patient receiving immunosuppressants,[26] or radiation for a new pulmonary nodule in an octogenarian with anorexia.[27]

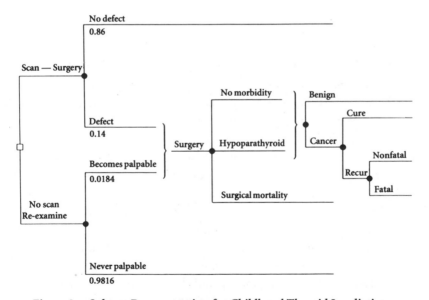

Figure 2.   Subtree Representation for Childhood Thyroid Irradiation.
This tree is equivalent to the one shown in Figure 1, but contains only six chance nodes and seven outcomes.

PRESENTING SENSITIVITY ANALYSES

The full benefit of the effort required to design and implement a decision tree is not obtained if the model is used simply to determine the optimal management strategy. One of the principal benefits of a decision model is the capacity to ask, What if? — What if the disease is really more likely? What if the test is actually less accurate? What if the risk of surgery is greater? Such questions are answered by performing sensitivity analyses — varying the values assigned to one or several variables in a systematic fashion and repeating the calculations to determine whether the optimal decision changes.

The simplest (one-way) sensitivity analysis involves changing the value of a single variable and recalculating the expected utility of each strategy. It can be presented as a table of values or, often more informatively, as a graph, such as Figure 3. The figure reveals an interesting and frequent phenomenon — namely, that strategy lines may intersect. Higher on the graph means greater expected utility, so crossing lines define regions where different decisions are best. These intersections are called decision thresholds[29,30]: if a given variable (in this case, the probability of herpes simplex encephalitis) has a value less than a threshold value, then one action is optimal (in this case, brain biopsy); if the variable has a value greater than a threshold, then another action is optimal (in this case, empirical drug therapy). In fact, the threshold values summarize the results of the one-way sensitivity analysis. Threshold values can tell the analyst whether a change in a given variable would change the optimal decision, but they do not indicate how much would be gained or lost by choosing a given strategy. That insight requires knowledge of the expected utilities of each strategy and the differences among them. Such differences are readily identified by examining Figure 3.

Of course, one-way sensitivity analyses provide only limited insight because they examine only changes in a single variable; the other variables are held to base-line values. The clinician, on the other hand, must sometimes explore the best strategy for a combination of factors — e.g., what if both the risk associated with lung biopsy and the probability of pneumocystis pneumonia are increased in a particular patient? Such complex yet important questions can be addressed by performing two-way sensitivity analysis — varying the values of two variables independently over broad ranges and

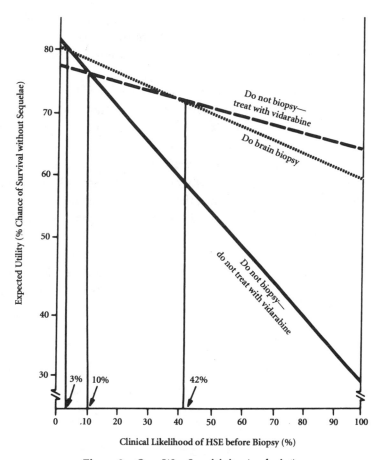

Figure 3.   One-Way Sensitivity Analysis.*

   This analysis examined whether or not to administer vidarabine or perform a
brain biopsy in a patient with suspected herpes simplex encephalitis (HSE). The
variable being examined is the probability that the patient has the disease (horizontal axis); the expected utilities are shown on the vertical axis. Each strategy
under analysis corresponds to a single line. The vertical lines at 3 percent, 10 percent, and 42 percent indicate diagnostic and therapeutic thresholds at which the
optimal strategy changes. Below 3 percent the best policy is "Do not biopsy, do
not treat," but beyond 10 percent it is much the worst policy.
   * Reproduced from Barza and Pauker,[28] with the permission of the publisher.

determining the best strategy for all combinations. The calculational demands of such an analysis are only the first hurdle; once performed, the analysis must be presented in a format that provides the physician with clinical insight.

Two somewhat different formats have been developed for summarizing such analyses. The one shown in Figure 4 demonstrates how thresholds and expected utilities vary, and is useful in displaying the differences between options. Such formats, however, can only compare two strategies. A more compact and understandable representation is shown in Figure 5, in which every combination of the two variables corresponds to a unique point on the graph. In the left panel, the graph is divided into two regions that specify which strategy is optimal for each combination of values. If the point lies within the shaded area, administering amphotericin empirically is optimal; if the point lies within the unshaded area, avoiding amphotericin administration is optimal. In contrast to the representation in Figures 3 and 4, however, the graphs in Figure 5 do not indicate how strongly one strategy should be preferred over another.

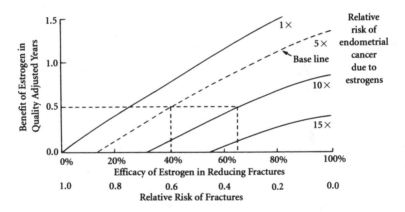

Figure 4.   Two-Way Sensitivity Analysis.*
This analysis examined whether or not to administer estrogens to postmenopausal women to prevent osteoporosis. The efficacy of postmenopausal estrogens in reducing fractures is shown on the horizontal axis, and the benefit of estrogen therapy (i.e., the difference between the calculated expected utilities of two strategies) is shown on the vertical axis. Each line corresponds to a different relative risk of endometrial cancer induced by estrogen therapy.
*Reproduced from Hillner et al.,[31] with the permission of the publisher.

A format similar to the one in the left panel of Figure 5 can be used to summarize the results of a so-called three-way sensitivity analysis, in which three clinically relevant factors are varied simultaneously and independently, shown in the middle and right panels of Figure 5. In both these panels, a family of curves depicts how the optimal strategy for each combination of the first two variables might be altered by changes in a third, independent factor. For example, in the middle panel the region bounded by the curves P = 0.1 and P = 0.3 represents the circumstances in which

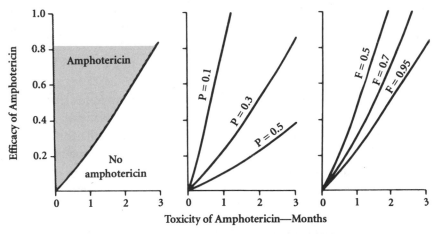

Figure 5. Two-Way and Three-Way Sensitivity Analyses.*
This analysis examined whether or not to administer amphotericin
to an immunosuppressed patient with persistent fever. The toxicity
of amphotericin is shown on the horizontal axis, and its efficacy on the
vertical axis.
The left panel contains a two-way sensitivity analysis, and the middle and right
panels contain three-way sensitivity analyses in which each line corresponds to
a different value of a third variable. In the middle panel, the probability of fungal infection (P) has been varied from its base-line value of 30 percent: the
lower the probability of fungal infection, the smaller the set of circumstances
in which empirical amphotericin is indicated — i.e., the smaller the region
above the curve. In the right panel, the mortality rate for fungal infection (F)
is varied from the base-line assumption of 95 percent. Again, the lower the
risk of fungal infection, the smaller the set of circumstances in which empirical
therapy is indicated.
*Reproduced from Gottlieb and Pauker,[26] with the permission of the publisher.

the optimal strategy would be to administer amphotericin if the probability of fungal infection was greater than 0.3, but to withhold such therapy if the probability was less than 0.1.

AUTOMATION

As should be evident from the discussions of tree representation and sensitivity analysis, clinical applications of decision analysis impose substantial calculational burdens. If even a moderately complex problem is examined manually, even with the help of a calculator, the analyst must devote many hours to multiplications and additions. In fact, the questions that an analyst wishes to ask of a decision-tree model are severely limited by the time required to calculate the answers. If a sensitivity analysis requires 20 or 30 hours of computation, even the most ardent analyst may turn to other tasks.

Over the past several years, many microcomputer programs, developed in the medical arena[32-35] and elsewhere,[36] have become available, allowing the experienced analyst to explore decision-tree models efficiently. These programs are cumbersome to use, however; they require substantial time to learn; and they cannot teach the inexperienced physician how to design and interpret a decision tree or identify and avoid errors in a model.

ASSIGNING VALUES TO OUTCOMES

Some major criticisms of decision analysis have focused on the assignment of utilities to various outcome states. Early models used arbitrary scales[3,4]; it was difficult to understand the meaning of the scales and to determine whether small differences had any clinical importance. Recently, clinical decision analysts have begun to use meaningful utility scales — e.g., quality-adjusted life expectancy.[37] Several techniques have also been developed for helping patients assess their attitudes toward alternative health outcomes and to express these attitudes quantitatively so that their preferences can be explicitly incorporated into decision analyses.

LIFE EXPECTANCY

The literature typically summarizes the prognosis of patients with a single disease process — i.e., a select and fairly "clean" population or, at best, average patients. The clinician, however, cares for individual patients, who often differ in important ways from patients described in the literature — e.g., in age or coexisting diseases. Thus, one of the clinician's central tasks is to identify the experience reported in the literature and tailor it to the individual patient. Such "massaging" of data is always somewhat arbitrary and empirical, but it is an essential part of traditional, implicit decision making. For quantitative decision making, an approximation of life expectancy has been developed to provide a mechanism for calculating and understanding how independent mortality risks operate.[38] According to this method (the declining exponential approximation of life expectancy), the mortality rate is viewed as the exponent in a declining exponential curve, similar to a drug half-life in the familiar single-compartment model of pharmacokinetics. Independent "forces of mortality" can be combined by addition, yielding an overall mortality rate.

When an average rate is known, life expectancy or average survival can be approximated by the reciprocal of the rate. For example, the life expectancy of a 70-year-old man with three-vessel coronary artery disease and Dukes' stage B carcinoma of the colon can be approximated by adding together the average mortality rate among patients of his age, sex, and race (μ_{ASR}) and the independent excess mortality rates associated with these diseases (μ_{CAD} and $\mu_{ColonCa}$). As shown in Table 2, this technique can approximate life expectancy under various management strategies that might be specified in a decision tree. One must be careful, however, to notice that the disease-specific mortality rates are excess rates and not crude rates.

QUALITY OF LIFE

Increased participation by patients in making decisions about their medical care requires not only that they be informed about alternatives and given the opportunity to express their wishes but also that they be guided in assessing their attitudes; these results need to be incorporated into the decision-making process.[39] When using decision analysis, the physician's role as *a* decision maker (as opposed to being *the* decision maker) is not abdicated to the patient; instead, patient and physician work as a team. The physician's expertise clearly consists of knowledge of the medical facts; the patient's expertise often consists of conceptualizing the effects of a potential outcome on him and his family.

Table 2. Calculation of Life Expectancy.

Step	Action	Result
1	Life expectancy of 70-year-old man	8 yr
2	Reciprocal of Result 1 yields μ_{ASR}*	0.125/yr
3	Excess mortality† for coronary disease	0.080/yr
4	Excess mortality for colon cancer	0.090/yr
5	Total average mortality (sum of Results 2 through 4)	0.295/yr
6	Reciprocal of Result 5 yields life expectancy	3.39 yr
	If surgery has 50% efficacy, then	
7	Excess mortality of coronary disease	0.040/yr
8	Total average mortality (sum of Results 2, 4, and 7)	0.255/yr
9	Reciprocal of Result 8 yields life expectancy	3.92 yr

* μ_{ASR} denotes average mortality among patients of the same age, sex, and race as the patient under evaluation.

† If the point survival at a given time is known, then the average mortality rate for a period can be calculated with the equation $\mu_{Crude} = -(1/n) \ln S_n$, where S_n is the probability of surviving n years. For example, if a series reports that the five-year survival among 56-year-old men with three-vessel coronary artery disease is 55 percent, then the crude mortality rate would be $-(1/5) \ln 0.55$, or 0.12 per year. The excess mortality rate is defined as the force of mortality over and above the rate among average patients of the same age, sex, and race. In this case, the series described 56-year-old men, whose life expectancy is approximately 25 years; thus, μ_{ASR} is 1/25 year, or 0.04 per year. The excess mortality rate for three-vessel coronary artery disease treated medically is then 0.12 per year minus 0.04 per year, or 0.08 per year.

Several approaches are available to help patients understand the dynamics of a decision and to help them formulate ideas about the relative merits of its outcomes. With most techniques, the patient is presented with a limited set of scenarios and is asked to choose between pairs of alternatives. For example, a prospective parent, informed in advance of the medical terms, might be asked[40] whether a pregnancy in which there is a 20 percent risk that the fetus has trisomy 21 should be aborted or carried to term. By presenting a sequence of such choices in which the probability of an outcome in one scenario is varied — e.g., from 20 percent to 50 percent to 80 percent — the physician can help assess in what circumstances the patient would be indifferent to choosing between two scenarios — i.e., when both scenarios would be perceived as equally bad. Such points of indifference can then be used to create a utility scale. Presenting a sequence of scenarios involving chance events is called the lottery technique.

Another common technique for assessing attitudes is the time trade-off approach,[41] in which the patient is asked to consider two scenarios that differ not in the probabilities of their outcomes but in their duration. For example, a patient with carcinoma of the larynx might be asked,[42] "Would you rather live for 8 years with normal speech or live for 10 years after a laryngectomy?" The duration of life with normal speech is varied in a sequence of questions until the patient recognizes his or her indifference point, which can then be used to create a utility scale.

THE MARKOV PROCESS

Prognosis can often be described as a series of chance events for which the patient is at risk. For example, a patient with silent gallstones may have an episode of acute cholecystitis in any given year.[43] If an episode occurs and the patient requires cholecystectomy, the risks of surgery depend on how old the patient is at the time. If this sequence of events were modeled as a simple decision tree with a set of chance nodes describing the events that might occur each year (i.e., acute cholecystitis, death due to surgery, or death due to other causes), the tree would double in breadth each year. After a mere 30 years (a reasonable time frame for clinical events in such a patient), the tree would contain more than 1 billion branches and would be impossible to evaluate by hand and even cumbersome to assess by computer.

Fortunately, a vast segment of the tree structure is repetitive, describing the same events year after year. In recent years it has become increasingly popular and convenient to use state-transition, or Markov-process, models in such decision problems.[44] These models define a small set of "health states" and specify the allowed transitions between the states. For example, a patient in the "silent gallstones present" health state may, in any given year, move to the state "cholecystectomy" and then to the state "post-cholecystectomy" or to the state "dead." The likelihood that a patient will move from one health state to another in any given period is called the transition probability for such a change. Each state of health is also assigned an incremental value when a patient remains in the state for a given period. For example, a patient who has silent gallstones for a year might be credited with

Table 3.   Markov Simulation of Prognosis of Cholelithiasis.

Year*	Patient Age	Patients with Silent Gallstones	Patients Surviving Cholecystectomy This Year	Patients Post-Cholecystec-tomy	Patients Dead
0	30	100,000	0	0	0
1	31	98,838	994	0	168
2	32	97,676	982	992	349
3	33	96,516	970	1,971	544
4	34	95,356	958	2,934	752
5	35	94,198	945	3,883	974
10	40	90,698	452	6,600	2,251
20	50	83,936	206	8,582	7,277
30	60	71,403	171	9,246	19,180
40	70	50,509	114	7,906	41,472
50	80	27,785	53	5,061	67,100
60	90	867	1	178	98,954
70	100	0	0	0	100,000
Total		3,788,787	13,571	380,022	
Quality adjustment		1.0	0.9	1.0	
Quality-adjusted total		3,788,787	12,213	380,022	
Total quality-adjusted years for cohort				4,181,025 yr	
Average quality-adjusted survival for member of cohort				41.8 yr	

* Although the actual calculation was done for each year, to conserve space the table shows only the results for years 0 through 5 and for each 10th year after year 10.

1 quality-adjusted year, whereas a patient who has an episode of cholecystitis and undergoes cholecystectomy in a given year might be credited with only 0.9 quality-adjusted year.

The model is used by placing a hypothetical cohort of patients in one or more states at the beginning of the horizon of analysis (e.g., placing 100,000 patients in the silent-gallstones state) and following their course year by year. As shown in the simplified example in Table 3, after a sufficiently long time horizon, all patients will have died. The number of quality-adjusted years of life in the cohort is added up and then divided by the size of the original cohort, yielding the expected quality-adjusted survival for a member of the cohort. Such decision models, quite feasible with computer support, provide important insights into clinical disorders that evolve over time.

SOME EXPECTATIONS

Over the decades during which decision analysis has been evolving in medicine, many of the principal concerns of its critics[45-51] have been satisfied: the time-consuming calculational burden has been eliminated by automation[32-36]; arbitrary utilities have been replaced by meaningful scales; and the threat to physicians of a mathematical approach to medical decision making simply has not materialized. Despite critically important and substantive advances in the application of decision analysis to clinical problems, most physicians faced with a difficult clinical problem do not immediately reach for a sketch pad or a microcomputer to create a decision tree. Why not? First, expertise in using the method is limited. Second, formal decision analysis takes time — time to construct a model (the tree) properly, time to gather and tailor the data, and time to interpret the model and decide which assumptions to test with sensitivity analysis. In short, it is often impractical in the hectic arena of clinical practice. Beyond these serious problems are several that aficionados worry about: the necessarily simplified models do not always reflect the real problems of the patient; the results are often distorted because the data available are stretched to the extreme to fit the problem; the utilities used to reflect patients' feelings about the quality of life associated with various outcomes are "soft" and inconstant over time; and the methods available to examine the effect of the data used in the analysis are still in need of considerable refinement.

Are these problems so immense that decision analysis will be relegated to a footnote in medical history? Not from our vantage point. Any assessment of decision analysis must be made against the "usual" approach to medical decision making; in that traditional mode, when we encounter a difficult problem for which no controlled study has provided a solution, we either contemplate the decision ourselves, implicitly, or we gather the opinions of consultants, hoping to build a consensus. Whether or not consensus is reached, the decision is usually made implicitly, a tacit approach that may or may not consider all reasonable alternatives or weigh the outcomes of competing choices appropriately. Implicit decision making almost never identifies situations in which the choice simply does not matter, despite the clear existence of such situations.[52] Alternative strategies of solving problems, such as the problem-oriented record, algorithms, flow charts, and the new discipline of clinimetrics, have not yet proved advantageous over the traditional approach. In contrast to the traditional approach, decision analysis is explicit; it forces us to consider all pertinent outcomes, it lays open in stark fashion all our assumptions about a clinical problem, including numerical representations of the chances and values of outcomes; it forces us to consider how patients feel about the quality of outcomes; and it allows us to come to grips precisely with the reasons why colleagues differ about actions to be taken.

We believe that applying decision analysis to individual and generic clinical problems is worth the effort but that it will require further investment in methodologic research and expanded effort to teach quantitative problem solving to a generation of students and house officers. As we delve deeper and deeper into the molecular nature of the diseases we battle, so should we dissect the day-to-day medical decisions that so critically influence the quality of the care we deliver.

REFERENCES

1. Ledley RS, Lusted LB. Reasoning foundations of medical diagnosis. Science 1959; 130:9-21.

2. Lusted LB. Decision-making studies in patient management. N Engl J Med 1971; 284:416-24.

3. Schwartz WB, Gorry GA, Kassirer JP, Essig A. Decision analysis and clinical judgment. Am J Med 1973; 55:459-72.

4. Kassirer JP. The principles of clinical decision making: an introduction to decision analysis. Yale J Biol Med 1976; 49:149-64.

5. Weinstein MC, Fineberg HV, Elstein AS, et al. Clinical decision analysis. Philadelphia: WB Saunders, 1980.

6. Kassirer JP, Moskowitz AJ, Lau J, Pauker SG. Decision analysis: a progress report. Ann Intern Med 1987; 106:275-91.

7. Plante DA, Kassirer JP, Zarin DA, Pauker SG. Clinical decision consultation service. Am J Med 1986; 80:1169-76.

8. Warner HR, Toronto AF, Veasey LG, Stephenson R. A mathematical approach to medical diagnosis: application to congenital heart disease. JAMA 1961; 177:177-83.

9. Warner HR, Toronto AF, Veasy LG. Experience with Bayes' theorem for computer diagnosis of congenital heart disease. Ann NY Acad Sci 1964; 115:558-67.

10. de Dombal FT, Leaper DJ, Staniland JR, McCann AP, Horrocks JC. Computer-aided diagnosis of acute abdominal pain. Br Med J 1972; 2:9-13.

11. Gorry GA, Barnett GO. Sequential diagnosis by computer. JAMA 1968; 205:849-54.

12. Gorry GA, Kassirer JP, Essig A, Schwartz WB. Decision analysis as the basis for computer-aided management of acute renal failure. Am J Med 1973; 55:473-84.

13. Vecchio TJ. Predictive value of a single diagnostic test in unselected populations. N Engl J Med 1966; 274:1171-3.

14. Fagan TJ. Nomogram for Bayes' theorem. N Engl J Med 1975; 293:257.

15. Sackett DL, Haynes RB, Tugwell P. Clinical epidemiology: a basic science for clinical medicine. Boston: Little, Brown, 1985.

16. Galen RS, Gambino SR. Beyond normality: the predictive value and efficiency of medical diagnoses. New York: John Wiley, 1975.

17. Griner PF, Mayewski RJ, Mushlin AI, Greenland P. Selection and interpretation of diagnostic tests and procedures: principles and applications. Ann Intern Med 1981; 94:553-600.

18. Gorry GA, Pauker SG, Schwartz WB. The diagnostic importance of the normal finding. N Engl J Med 1978; 298:486-9.

19. Schwartz WB, Wolfe HJ, Pauker SG. Pathology and probabilities: a new approach to interpreting and reporting biopsies. N Engl J Med 1981; 305:917-23.

20. Tversky A, Kahneman D. Judgment under uncertainty: heuristics and biases. Science 1974; 185:1124-31.

21. Eraker SA, Politser P. How decisions are reached: physician and patient. Ann Intern Med 1982; 97:262-8.

22. Eddy DM. Probabilistic reasoning in clinical medicine: problems and opportunities. In: Kahneman D, Tversky A, eds. Judgment under uncertainty: heuristics and biases. Cambridge: Cambridge University Press, 1982:249-67.

23. Stockwell RM, Barry M, Davidoff F. Managing thyroid abnormalities in adults exposed to upper body irradiation in childhood: a decision analysis: should patients without palpable nodules be scanned and those with scan defects be subjected to subtotal thyroidectomy? J Clin Endocrinol Metab 1984; 58:804-12.

24. Lau J, Levey AS, Kassirer JP, Pauker SG. Idiopathic nephrotic syndrome in a 53-year-old woman: is a kidney biopsy necessary? Med Decis Making 1982; 2:497-519.

25. Kassirer JP. Is renal biopsy necessary for optimal management of the idiopathic nephrotic syndrome? Kidney Int 1983; 24:561-75.

26. Gottlieb JE, Pauker SG. Whether or not to administer amphotericin to an immunosuppressed patient with hematologic malignancy and undiagnosed fever. Med Decis Making 1981; 1:75-93.

27. Moroff SV, Pauker SG. What to do when the patient outlives the literature, or DEALE-ing with a full deck. Med Decis Making 1983; 3:313-38.

28. Barza M, Pauker SG. The decision to biopsy, treat, or wait in suspected herpes enceph-
    alitis. Ann Intern Med 1980; 92:641-9.

29. Pauker SG, Kassirer JP. Therapeutic decision making: a cost-benefit analysis. N Engl J
    Med 1975; 293:229-34.

30. *Idem.* The threshold approach to clinical decision making. N Engl J Med 1980; 302:1109-
    17.

31. Hillner BE, Hollenberg JP, Pauker SG. Postmenopausal estrogens in prevention of os-
    teoporosis: benefit virtually without risk if cardiovascular effects are considered. Am J
    Med 1986; 80:1115-27.

32. Pauker SG, Kassirer JP. Clinical decision analysis by personal computer. Arch Intern
    Med 1981; 141:1831-7.

33. Silverstein MD. A clinical decision analysis program for the Apple computer. Med Decis
    Making 1983; 3:29-37.

34. Pass TM, Goldstein LG. CE Tree: a computerized aid for cost-effectiveness analysis. In:
    Hefferman HG, ed. Proceedings of 5th annual Symposium on Computer Applications
    in Medical Care. Washington, D.C.: IEEE Computer Society, 1981:219-21.

35. Hollenberg J. SMLTREE: the all-purpose decision tree builder. Boston: Pratt Medical
    Group, 1985.

36. Franke DW, Hall CR. Arborist: decision tree software. Dallas: Texas Instruments, 1984.

37. Pliskin JS, Shepard DS, Weinstein MC. Utility functions for life years and health status.
    Oper Res 1980; 28:206-24.

38. Beck JR, Pauker SG, Gottlieb JE, Klein K, Kassirer JP. A convenient approximation of
    life expectancy (the "DEALE"). II. Use in medical decision-making. Am J Med 1982;
    73:889-97.

39. Kassirer JP. Adding insult to injury: usurping patients' prerogatives. N Engl J Med 1983;
    308:898-901.

40. Pauker SP, Pauker SG. The amniocentesis decision: an explicit guide for parents. In:
    Epstein CJ, Curry CJR, Packman S, Sherman S, Hall BD, eds. Risk, communication,
    and decision making in genetic counseling: part C of Annual Review of Birth Defects,
    1978. New York: Alan R Liss, 1979:289-324.

41. Sackett DL, Torrance GW. The utility of different health states as perceived by the
    general public. J Chronic Dis 1978; 31:697-704.

42. McNeil BJ, Weichselbaum R, Pauker SG. Speech and survival: tradeoffs between quality
    and quantity of life in laryngeal cancer. N Engl J Med 1981; 305:982-7.

43. Ransohoff DF, Gracie WA, Wolfenson LB, Neuhauser D. Prophylactic cholecystectomy
    or expectant management for silent gallstones: a decision analysis to assess survival.
    Ann Intern Med 1983; 99:199-204.

44. Beck JR, Pauker SG. The Markov process in medical prognosis. Med Decis Making
    1983; 3:419-58.

45. Feinstein AR. Clinical biostatistics. XXXIX. The haze of Bayes, the aerial palaces of
    decision analysis, and the computerized Ouija board. Clin Pharmacol Ther 1977;
    21:482-96.

46. Ingelfinger FJ. Decision in medicine. N Engl J Med 1975; 293:254-5.

47. Schwartz WB. Decision analysis: a look at the chief complaints. N Engl J Med 1979; 300:556-9.

48. Brett AS. Hidden ethical issues in clinical decision analysis. N Engl J Med 1981; 305:1150-2.

49. Feinstein AR. The "chagrin factor" and qualitative decision analysis. Arch Intern Med 1985; 145:1257-9.

50. Politser P. Decision analysis and clinical judgment: a re-evaluation. Med Decis Making 1981; 1:361-89.

51. Cebul RD. "A look at the chief complaints" revisited: current obstacles and opportunities for decision analysis. Med Decis Making 1984; 4:271-83.

52. Kassirer JP, Pauker SG. The toss-up. N Engl J Med 1981; 305:1467-9.

10

P VALUES

James H. Ware, Ph.D., Frederick Mosteller, Ph.D.,
Fernando Delgado, M.S., Christl Donnelly, M.S.,
and Joseph A. Ingelfinger, M.D.

ABSTRACT   *P* values measure the strength of statistical evidence in many scientific studies. They indicate the probability that a result at least as extreme as that observed would occur by chance. *P* values are a way of reporting the results of statistical tests, but they do not define the practical importance of the results. They depend upon a test statistic, a null hypothesis, and an alternative hypothesis. Multiple tests and selection of subgroups, outcomes, or variables for analysis can yield misleading *P* values. Full reporting and statistical adjustment can help avoid these misleading values. Negative studies with low statistical power can lead to unjustified conclusions about the lack of effectiveness of medical interventions.

We discuss the role and use of *P* values in scientific reporting and review the use of *P* values in a sample of 25 articles from Volume 316 of the *New England Journal of Medicine*. We recommend that investigators report (1) summary statistics for the data, (2) the actual *P* value rather than a range, (3) whether a test is one-sided or two-sided, (4) confidence intervals, (5) the effects of selection or multiplicity, and (6) the power of tests.

R eaders of the medical literature encounter *P* values and associated tests of significance more often than any other statistical technique. In their review of Volumes 298 through 301 of the *New England Journal of Medicine*, Emerson and Colditz[1] found that 147 (44 percent) of the 332 articles classified as Original Articles used the *t*-test and 91 (27 percent) used Fisher's exact test or a chi-square test (see Chapter 3). In other articles *P* values were used in association with nonparametric tests, in life-table analyses, and in the study of regression and correlation coefficients.

Because *P* values play a central role in medical reporting, medical investigators and clinicians need to understand their origins, the pitfalls they present, and controversies about their use. *P* values depend on assumptions about the data, and therefore analyses involving the calculation of many *P* values can be misleading. Although *P* values can be useful as an aid to reporting, they provide the most information when they are reported with descriptive information about study results.

Investigators compute *P* values from the measurements provided by the sample of participants included in a scientific study and use these values to draw conclusions about the population from which the observations were drawn. For example, in a study of angina following myocardial infarction, Schuster and Bulkley[2] reported that mortality was 72 percent (31 of 43) among patients with ischemia at a distance from the zone of infarction and 33 percent (9 of 27) among those with ischemia in the infarct zone ($P<0.005$). The authors concluded from the small *P* value that patients with ischemia at a distance from the infarction were at increased risk, as compared with patients with ischemia in the infarct zone. This example illustrates a major use of *P* values: to determine whether or not an observed effect can be explained by chance, i.e., random variation in patient outcomes. The specific meaning of the statement "$P<0.005$" is that the observed difference *or a more extreme difference* in mortality rates would occur with probability less than 0.005 if the true mortality rates were identical for patients with ischemia in the infarct zone and patients with ischemia at a distance. We return to these concepts below.

The frequent reporting of *P* values attests to a wide belief in their usefulness in communicating scientific results. Moreover, the *P* value associated with the primary results of a scientific study can be a factor in the editorial decision to publish those results.[3] Thus, clear and consistent practices in the calculation and citation of *P* values are important elements of good scientific reporting. The interpretation of a *P* value can depend substantially on the design of the study, the method of collecting data, and the analytic practices used. Scientific reports frequently underemphasize or omit information that readers need to assess an author's conclusions. As a result, readers sometimes misunderstand the importance of either a highly significant or a nonsignificant *P* value.

This chapter describes the basic ideas in a *P* value calculation. The actual calculation of *P* values is discussed in many textbooks.[4-7] We also discuss the value of confidence intervals as an alternative to *P* values, review current controversies regarding the use of *P* values in medical reporting, and recommend six specific practices for the reporting of *P* values.

To sharpen our understanding of the current use of *P* values in medical reporting, we reviewed the use of *P* values and related statistical information in 25 Original Articles selected from Volume 316 (January–June, 1987) of the *New England Journal of Medicine*. In discussions of controversies and recommendations regarding use of *P* values, we report how the issue was managed in these 25 articles.

WHAT ARE *P* VALUES?

P values are used to assess the degree of dissimilarity between two or more sets of measurements or between one set of measurements and a standard. A *P* value is a probability, usually the probability of obtaining a result as extreme as or more extreme than the one observed if the dissimilarity is entirely due to random variation in measurements or in subject response — that is, if it is the result of chance alone. For example, Davis et al.,[8] in reporting a study of 10 patients with congestive heart failure, state that, at single daily doses of captopril of 25 to 150 mg, "the cardiac index rose from 1.75 ± 0.18 to 2.27 ± 0.39 liters per minute per square meter ($P<0.001$)." Here the *P* value indicates that a difference at least as great as that observed would occur with a probability of less than 0.001, or in less than 1 in 1000 trials, if captopril had no effect on the cardiac index in the population of patients with congestive heart failure and if the assumed probability model was correct. The *P* value depends implicitly on three elements: the test statistic, the null hypothesis, and the alternative hypothesis.

THE TEST STATISTIC

To summarize the dissimilarity between two sets of data, we choose a statistic that reflects the differences likely to be caused by the treatment under study, such as the difference in means, which may use a *t* statistic, or a difference in death rates, which may use a chi-squared statistic. Davis et al.[8] determined the cardiac index before and after treatment, patient by patient, and used the *t* statistic for the differences to assess the changes associated with treatment. In the study of angina cited earlier, Schuster and

Bulkley used a different statistic, the chi-square test for association between death and location of the ischemia, which they computed from the numbers of subjects and deaths in each of the two groups.

When only random variation acts, the *P* value properly reflects the relative frequency of getting values as extreme as the one observed. When systematic effects are present, such as treatment increasing length of life or increasing probability of survival in a surgical operation, the statistic should tend to reflect this by being more likely to produce extreme, usually small *P* values. When small *P* values occur, they hint either that a rare random event has occurred or that a systematic effect is present. In a well-designed study, the investigator usually finds the systematic effect more plausible.

THE NULL HYPOTHESIS

The information needed for the calculation of the *P* value comes from expressing a scientific hypothesis in probabilistic terms. In the study by Davis et al., the scientific hypothesis to be tested statistically (though not necessarily what the authors expected or hoped to find) was that cardiac output would be unchanged by treatment with captopril. Such a hypothesis is called a null hypothesis, the "null" implying that no effect beyond random variation is present. If this null hypothesis were true, the mean change in cardiac index in a sample of patients with congestive heart failure would vary around zero in repeated sampling.

Although the null hypothesis is central to the calculation of a *P* value, only 4 of the 25 articles we reviewed from Volume 316 of the *New England Journal of Medicine* stated the null hypothesis explicitly. In most articles, the null hypothesis was defined implicitly by statements such as "calorie-adjusted potassium intake [was] significantly lower in the men and women who subsequently had a stroke-associated death, as compared with all other subjects ($P = 0.06$)." [9] Although such implicit definitions could, in principle, be ambiguous, we found no instances of such ambiguities in the 25 articles we reviewed.

One might ask why small or large *P* values give us any feeling about the truth of the null hypothesis. After all, if chance alone were at work, a small *P* value would merely tell us that a rare event has occurred, just as a large *P* value goes with a likely event. This argument calls attention to another concept needed to understand *P* values: the alternative hypothesis.

THE ALTERNATIVE HYPOTHESIS

When investigators report P values, they have an underlying but usually unstated notion that the null hypothesis may be false and that some other situation may be the true one. For example, Davis et al. had an alternative hypothesis in mind — that captopril would increase the cardiac output of patients with congestive heart failure. (Because this alternative hypothesis includes many different sizes of effect on cardiac output of placebo and captopril, statisticians sometimes speak of alternative *hypotheses.*) Statistical tests are designed so that small P values are more likely to occur when the null hypothesis is false.

Thus, we reject the null hypothesis not only because the probability of such an event is low when there is no effect, but also because the probability is greater when there is an effect. These two ideas together encourage us to believe that an alternative hypothesis is true when the test of the null hypothesis yields a small P value for the observed effect.

Notice the similarity between this way of thinking and proof by contradiction. Logicians state as a premise whatever they intend to disprove. If a valid argument from that point leads to a contradiction, the premise in question must be false. In statistics, we follow this approach, but instead of reaching an absolute contradiction, we may observe an improbable outcome.[10] We must then conclude that either the null hypothesis is correct and the improbable has happened or the null hypothesis is false. Deciding between these two possibilities can be difficult. One question that arises is how small the probability should be before we can conclude that the null hypothesis is mistaken.

Authors rarely state the alternative hypothesis explicitly. In our review of 25 articles from Volume 316 of the *New England Journal of Medicine*, we found no instances in which authors did so. There is a greater potential for ambiguity with implicit specification of alternative hypotheses than with null hypotheses. Nevertheless, we could infer the alternative hypothesis in each of the 25 articles we reviewed.

THE 0.05 AND 0.01 SIGNIFICANCE LEVELS

The P value measures surprise. The smaller the P value, the more surprising the result if the null hypothesis is true. Sometimes, a very rough idea of the degree of surprise suffices, and various simplifying conventions have come into use. One popular approach is to indicate only that the P value is smaller than 0.05 (P<0.05) or smaller than 0.01 (P<0.01). When the P value is between 0.05 and 0.01, the result is usually called "statistically significant"; when it is less than 0.01, the result is often called "highly statistically significant." This standardization of wording has both advantages and disadvantages. Its main advantage is that it gives investigators a specific, objectively chosen level to keep in mind. Sometimes, it is easier to determine whether a P value is smaller than or larger than 0.05 than it is to compute the exact probability. The main disadvantage of this wording is that it suggests a rather mindless cut-off point, which has nothing to do with the importance of the decision to be made or with the costs and losses associated with the outcomes.

The 0.05 level was popularized in part through its use in quality-control work, where the emphasis is on the performance of a decision rule in repeated testing. This viewpoint carries over reasonably well to the relatively small and frequently repeated studies of process and mechanism that represent the building blocks of scientific understanding. In the large and expensive clinical trials and descriptive studies that are increasingly common in modern science, however, the protection provided by repetition is rarely available.

CONFIDENCE INTERVALS

Some methodologists argue that medical reports rely too heavily on P values, especially when the P value is the only statistical information reported.[11-13] Because no single study determines scientific opinion on a subject, it is incumbent upon the investigator to provide a more complete analysis of study results. In particular, these observers recommend that investigators routinely report confidence intervals rather than P values.[11]

We create confidence intervals for a given degree of confidences, such as 95 percent. The 95 percent tells how often such intervals include the true value of the estimated statistic, such as the mean or a correlation coefficient. If we assert that the true value is in the interval, we will be right in 95 percent of the assertions, barring breakdowns in the assumptions. In the study by Davis et al., the two-sided 95 percent confidence interval for the mean increase in cardiac index achieved by treatment with captopril is 0.32

to 0.71 liter per minute per square meter of body-surface area. The confidence interval gives information not given by the P value; it reports information in terms of the measurements actually used and the width of the confidence interval gives an indication of the informativeness of the study. The width is usually closely related to the standard error of the statistic being assessed, often about four times as wide. For example, in large samples the 95 percent confidence interval for population means reaches 1.96 standard errors above the observed mean and 1.96 standard errors below, and therefore is about 4 standard errors wide. Confidence intervals can also be calculated for confidence levels other than 0.95, but this is done infrequently.

We agree that sole reliance on the P value is incomplete reporting, especially when the P value is reported only as a range. The P value does, however, provide information not provided by the confidence interval. The statement "$P = 0.002$" gives an indication of degree of extremity that may not be readily apparent from noting that a null value is outside the 95 percent confidence interval. Citing both the actual P value and the confidence interval provides more information than either report can provide alone. Reporting basic summary statistics of the data, such as the mean and the standard error, will usually aid readers in understanding the study and its conclusions. This practice is well established in the *New England Journal of Medicine*. All but 3 of the 25 articles we reviewed used either confidence intervals or standard errors to describe the precision of an estimate.

OTHER CRITICISMS OF P VALUES

Some statisticians criticize the P value as a measure of strength of evidence. Diamond and Forrester[14] pointed out that, in the language of diagnostic testing, the P value corresponds to the false-positive rate, the probability of the observed result or one more extreme if the null hypothesis is true. It is *not* the probability that the alternative hypothesis is true given the data. They argued that the popularity of the P value as a measure of the strength of evidence results in part from the mistaken belief among many physicians and other scientists that it does have this interpretation.

Diamond and Forrester noted that the probability that an alternative hypothesis is true depends not only on the P value but also on the prior probability that the hypothesis is true, which may not be known. This issue is identical to that encountered in the interpretation of diagnostic test results. For example, if practically no one in the population has the disease, most diagnoses based on fallible tests will be false positives, whereas if the disease is common, fewer diagnoses will be false positives. Similarly, if effects in the real world almost never or almost always happen, then positive findings are usually mistaken in the first case and almost certain in the second. Many statisticians believe that this is not a limitation of the *P* value, preferring that data should be described using statistics that depend only on the data, not on subjective information such as prior probabilities. In this latter view, readers of scientific reports should provide their own prior probabilities of various hypotheses.

P VALUES AND SIGNIFICANCE TESTS

P values are an integral part of the statistical technique known as hypothesis testing or significance testing. Hypothesis testing is a method for choosing between the null hypothesis and alternative hypotheses. To explain the connection between *P* values and hypothesis testing, we can describe formal hypothesis testing as consisting of the following steps: (1) choose a test statistic; (2) choose the significance level, α, of the test; (3) compute the *P* value; and (4) if the *P* value is smaller than α, reject the null hypothesis in favor of the alternative hypothesis or hypotheses; otherwise, accept the null hypothesis. For example, when the significance level, α, is 0.05, hypothesis testing leads to rejecting the null hypothesis if the *P* value is less than 0.05.

Given the test used to compute the *P* value, the significance level, and the number of observations made, one can determine in advance the set of values of the statistic that would result in rejection of the null hypothesis. This set of values is called the *critical region*. Thus, another description of the fourth step in hypothesis testing is this: (4) if the statistic falls into the critical region, reject the null hypothesis; otherwise, do not reject the null hypothesis. (Statisticians say that we "fail to reject" the null hypothesis to emphasize that the data cannot prove the null hypothesis to be true. Future studies may show that it is highly improbable.)

Schuster and Bulkley[2] used the chi-square test for association to compare the mortality rates for patients with ischemia in the infarct zone and for patients with ischemia at a distance from the infarction. In this instance, the critical region consisted of all mortality outcomes that led to a value for chi-square of more than 3.84, the upper 0.05 critical value of the chi-square distribution with 1 degree of freedom.

Note that the words "accept" and "reject" are merely formal labels like "success" and "failure"; they do not correspond to any particular action. The implication of the word "reject" is simply that substantial evidence against the null hypothesis has been supplied. The use of the words "accept" and "reject" is an unfortunate convention, possibly a holdover from sampling methods for quality control, in which the sampling was used to determine what to do about a batch of manufactured materials, where options might have included selling the batch at usual prices, selling it at reduced prices, scrapping it altogether, and so on.

ISSUES IN REPORTING AND INTERPRETING *P* VALUES

When one considers the frequent use of *P* values in the medical literature and the emphasis sometimes placed on the statistical significance of a study's results, it is not surprising that controversies have arisen about the methods used to compute and interpret *P* values. The next sections of this chapter discuss some of these controversies.

ONE- AND TWO-SIDED TESTS

In many calculations of *P* values, the investigator must choose between the one-sided and the two-sided approach. The one-sided *P* value, used mainly in comparisons between two treatment groups, is the probability of observing a difference *favorable to the innovation* as extreme as or more extreme than the one observed, if the null hypothesis is true. The two-sided *P* value is the probability of observing a difference *in either direction* as extreme as or more extreme than the one observed. The two-sided *P* value is often approximately and sometimes exactly twice as large as the one-sided *P* value.

In some situations, a good argument can be made for reporting the one-sided *P* value. In studies comparing an innovation with a standard regimen, for example, the alternative hypothesis of greatest interest is that the innovation is superior. When the innovation is not used outside investigational settings, the distinction between an ineffective and a detrimental innovation may be unimportant for medical practice. When data on patient outcomes are accumulated gradually, it may even be unethical to continue a trial for the purpose of making this distinction.

Despite this argument, some statisticians and journal editors believe that one-sided *P* values should never be used. First, they argue that readers will benefit from uniform practices in reporting of *P* values, so that a *P* value of a given size has the same meaning across all articles. Second, they believe that the situations justifying a one-sided test are extremely rare, and that most uses of one-sided *P* values are more properly viewed as attempts to exaggerate the strength of a finding. Thus, investigators planning to cite one-sided *P* values must be prepared to defend them. This is one of the issues that investigators should address while planning a study, and they should defend their decision from the outset. The early decision to use a one-sided test for the specific comparison presented should be documented. Even when this is done, investigators may encounter editorial resistance if they report one-sided *P* values in scientific articles.

A third concern with one-sided *P* values arises when a study is important to a scientific or regulatory decision about the efficacy of a drug. To approve the introduction of a new drug, a regulatory group should be convinced that the expected benefits of the drug outweigh its costs and risks. The *P* value from a scientific study is only one factor in this decision, and the criterion that the *P* value be less than 0.05 may sometimes appear to be insufficiently stringent for establishing efficacy. Insisting on two-sided tests is one way to be more conservative. When issues like this are behind a debate about "sidedness," the ideal resolution would be for investigators to discuss more directly the criteria for sufficient evidence. Usually, the actual average value of the treatment effect and a measure of its variability will be more helpful than the *P* value to those who must weigh costs and benefits.

We believe that the distinction between one- and two-sided *P* values is not of fundamental importance in interpreting the results of a study, provided that the investigator clearly states which type of statistic was used. If readers are told what the test was about, what measure was used, and whether one-sided or two-sided results are being reported, they can usually recalculate the *P* value, at least approximately, in the form they believe most appropriate. Investigators should be aware, however, that this position is not uniformly held, and that any use of one-sided *P* values may be regarded skeptically by some editors and referees.

In our review of 25 articles from Volume 316 of the *New England Journal of Medicine*, we found one use of one-sided P values[15] to test a null hypothesis. In 11 other articles, the information provided was not sufficient to determine whether P values were one- or two-sided. The other 13 used two-sided values.

ONE-SIDED, TWO-SIDED, OR THREE-SIDED?

In fact, the usual two-sided formulation of a test for assessing the difference in performance of two treatments offers only two decisions; either the treatments are alike (there is no evidence of a difference) or they differ. From a statistical point of view, the report based on a two-sided P value should be "alike" (not significantly different) or "different," but not that "the innovation is more effective than standard therapy." Both investigators and readers might be surprised to learn that a study has shown only that two treatments are different, without commenting on directionality, but this is the proper interpretation of the two-sided test.

If we want to know which treatment is preferable, we are in at least a three-decision problem, with one tail representing "innovation preferred," one representing "standard preferred," and the center for "insufficient evidence to declare a difference." It may be useful to keep track of the total probability of declaring a difference, the usual two-sided total, but what matters in the decision is the probability associated with a particular extreme outcome.

MULTIPLICITY

Often investigators report many P values in the same study, while still other P value calculations are unreported. Because P values smaller than 0.05 occur by chance in 5 percent of tests of true null hypotheses, the probability that at least one P value will be smaller than 0.05 increases with the number of comparisons, even when the null hypothesis is correct for each comparison. This increase is known as the *multiple-comparisons* problem. When several comparisons are performed simultaneously, both the calculation and the interpretation of P values must be reexamined.

The rate at which statistically significant results occur in tests of null hypotheses is sometimes called the false-positive rate. For tests at the 0.05

level, the false-positive rate is 1 in 20 tests. Consider the investigator who studies several treatments in a single clinical trial. If all pairs of treatments are compared with respect to a single outcome (such as recovery), and if a standard significance test for two groups is used for each comparison, then the probability of finding at least one significant difference increases rapidly with the number of treatments, even when the null hypothesis — that the treatments are of equal efficacy — is true. To ensure that the probability of one or more false-positive results is no greater than a specified value, say 0.05, the investigator must use special statistical methods that take into account the number and type of comparisons made. These methods are called multiple-comparisons procedures. Glantz[16] recommends that these procedures be used routinely in the analysis of studies comparing three or more groups, and Godfrey[17] (see Chapter 12) reviews some of the multiple-comparisons procedures for comparing several group means.

When an investigation is complex and involves many different kinds of comparisons and tests of interrelated hypotheses, an attempt to adjust for multiplicity may be inconsistent with the objectives of the study. Indeed, in attempting to decide when to adjust for multiplicity, one encounters a conceptual difficulty with the idea of multiple-comparisons tests. If it includes multiple tests of different hypotheses, the study or scientific paper may not be the natural unit for controlling the overall probability of a false-positive result, yet no consensus has been reached about the appropriate use of multiple-comparisons procedures in the typical multifaceted scientific study. This difficulty has led some statisticians to reject multiple-comparisons procedures entirely in favor of simply recognizing the effects of multiple testing. In that view, 1 of every 20 tests will produce a *P* value smaller than 0.05 merely by chance, and this likelihood must simply be kept in mind in the interpretation of any collection of hypothesis tests and their *P* values.

Although many of the articles reviewed from Volume 316 of the *New England Journal of Medicine* reported numerous *P* values, only one[18] made an adjustment for multiplicity. This paper used Bonferroni's method to adjust for multiple comparisons in repeated analyses of a group evaluated on several occasions.

The problem of multiplicity can arise in a variety of ways. Three important practices — repeated analyses of accumulating data, the use of a variety of end points, and the selection of subgroups of interest by the results of the analysis — can substantially increase the probability of finding incorrectly at least one comparison that is statistically significant. In a notable example, the Food and Drug Administration refused to approve a claim based on the Anturane Reinfarction Trial[19] that sulfinpyrazone (Anturane) was effective in the prevention of sudden death in the six months after myocardial infarction. One of the reasons given by the FDA officials for this decision was that a reported P value of 0.003 for comparing the frequency of sudden death in the first six months after infarction exaggerated the strength of the evidence against the null hypothesis because the end point, the time interval after infarction, and the subset of enrolled patients included in the final analysis had all been identified through repeated analysis of accumulating data rather than by criteria specified before the study began.[20]

Selection of subgroups, end points, or analytic methods leads to multiple-comparisons problems because each selection creates opportunities for choice and thus for significance. Selection effects may be difficult to recognize unless the investigators report fully on their methods. For example, if 10 end points are studied and only the 1 that changed significantly is reported, the reader cannot recognize the selection problem. The reader may therefore be wise to take a jaundiced view of unusual end points offered without justification. A more subtle form of the multiple-comparisons problem, in which investigators inspect the data and actually perform the test only where they think it might yield "positive" results, is discussed below under post hoc hypotheses.

When investigators mount large, multicenter clinical trials or observational studies that are likely to be unique, they may make special efforts to protect against problems of multiplicity in interpreting study results. If several end points are to be considered, a subset of these are defined as primary end points, and these determine the outcome of the study regarding, for example, the efficacy of a new treatment. The overall error rate for the study is then allocated to these end points by some method that ensures that the probability of a false positive will not exceed the desired level. Similarly, when patient safety or concerns about the ethics of the study require multiple analyses of the data as they accumulate over time, special procedures called sequential methods are used to ensure that the probability of a positive result at any one of these multiple analyses does not exceed the desired false-positive rate.

PRIMARY, SECONDARY, AND POST HOC HYPOTHESES

In some studies, investigators specify a small set of primary hypotheses to be tested by the study. The methods for testing these hypotheses may be described in detail in the protocol, and statistical procedures are specified that ensure that the probability of any false-positive result does not exceed a specified level, say 0.05, for the set of tests of the primary hypotheses. Additional hypotheses of interest to investigators may also be identified in advance and described as secondary hypotheses. Typically, each test of a secondary hypothesis or of post hoc hypotheses (those that emerge in the data analysis) is reported with the nominal P value arising when that test is considered alone.

Tukey[21] formalized this strategy by introducing the terms *confirmatory* and *exploratory* testing. Confirmatory tests are those based on primary hypotheses, and should be performed so as to justify the use of the results in decision making — for example, regarding the acceptance of a new drug. Exploratory hypotheses are those arising from a thorough exploration of study data; they may be more or less anticipated when the study is designed, but are interpreted as "post hoc"; that is, they are generated from the data analysis. Post hoc findings must be interpreted cautiously, in light of the regular occurrence of false-positive findings in multiple comparisons. This distinction is useful in both study design and data analysis. It offers a practical strategy for resolving some aspects of the multiple-comparisons problem.

BIAS

Although small P values may make chance an unlikely explanation for differences between two groups, they do not necessarily imply that the differences are due to the therapy or exposure. Instead, the differences may result from other characteristics of the groups. The tendency of groups within a study to respond differently because of such noncomparability is called a nonrandom error or a nonsampling error or bias. It is assumed to be absent in the usual interpretation of both P values and confidence intervals. Methods for reducing bias are discussed elsewhere.[22,23]

PRECISE *P* VALUES

In any situation, later investigators will find the exact *P* value more helpful than a range, such as $P \leqslant 0.05$, when they combine the results of one study with those of other investigations. As a practical matter, however, the investigator who determines *P* values by reference to statistical tables will find that for many problems, tables are conveniently available for only a few significance levels.

Although practice varied somewhat in the *Journal* articles we reviewed, many authors quoted *P* values less than 0.05 as a range, for example, $P<0.05$, $P<0.02$, or $P<0.01$, and values above 0.05 as NS (nonsignificant). Each of these practices, but especially reporting a *P* value simply as NS, is a form of incomplete reporting. This is easily repaired by giving the actual number found.

STATISTICAL POWER AND SAMPLE SIZE

Investigators often ask statisticians to calculate an appropriate sample size for a particular study. Sample size calculations of this kind depend on the end point to be analyzed, the statistical method to be used, and the magnitude of the difference that the investigators want to detect. Then once a significance level has been chosen, one can calculate the probability that the null hypothesis will be rejected when there is an effect of a given size. This probability is the *power* of the test for that size of effect. There is a close connection between sample size and the power of the statistical test.

Statistical power affects both the design and the interpretation of scientific studies. Studies with low power are likely to be negative, and published results may discourage further research in that direction if the limitations of the study are not appreciated. Reviews of papers reporting the results of therapeutic trials have shown that studies frequently have low power for important alternative hypotheses. For example, Freiman et al.[24] found that of 71 randomized trials for which the authors reported "no effect," 67 (94 percent) had a power smaller than 0.9 for an alternative hypothesis of a 25 percent therapeutic improvement, and 50 (70 percent) of the trials had a power smaller than 0.9 for a 50 percent improvement. This means that in 94 percent of the studies there was a greater than 10 percent chance of missing a 25 percent therapeutic improvement, and in 70 percent of the studies there was a similar chance of missing a 50 percent improvement. (Freiman et al. used a one-tailed test with a significance level of 0.05.) An update of that work is reported in Chapter 19 of this book.

The reluctance of authors and editors to publish the results of studies that describe nonsignificant comparisons can produce the opposite problem. A proliferation of small trials can lead to overrepresentation of the 1 in 20 trials that achieves a *P* value of less than 0.05 by chance. Zelen[3] estimated that 6000 to 10,000 therapeutic investigations of patients with cancer were in progress during 1982 and that these trials could generate as many as 300 to 500 false-positive reports.

The problem of false-positive findings can be addressed in part by reporting confidence intervals as well as significance levels. The confidence interval identifies the range of values, such as differences in treatment effects, that is compatible with the study results; wide confidence intervals help to identify studies with low reliability. Discussions of power also have a place in reports of study results. For example, the statement that the study had a power of 15 percent to detect a 50 percent reduction in mortality conveys information different from the corresponding confidence interval because it says, "We scarcely had a chance," and it facilitates proper weighing of negative studies. Only 1 of the 25 articles from Volume 316 of the *New England Journal of Medicine* provided quantitative information about the power of a negative result,[25] perhaps in part because only 4 articles reported negative results for hypotheses of primary interest.

STATISTICAL AND MEDICAL SIGNIFICANCE

Statisticians have adopted the word significance to describe the results of statistical tests of hypotheses, but confusion between the everyday meaning of the word and its technical meaning muddies the waters. Although *statistical significance* is a technical term, *medical significance* is more vague because it usually means "importance." Four possibilities exist in describing the implications of study results. Some findings are both statistically significant and medically significant; other findings are neither. The more troublesome results are significant in only one of these senses. When samples are very large, small differences may be statistically significant even

Table 1.   Two-Sided *P* Values for Increases in Mean Cardiac Index Observed in Studies with Several Sample Sizes.

Mean Increase*	Sample Size		
	10	25	100
0.10	$P = 0.29$	$P = 0.09$	$P < 0.001$
0.25	$P = 0.01$	$P < 0.001$	$P < 0.001$
0.50	$P < 0.001$	$P < 0.001$	$P < 0.001$

* Liters per minute per square meter.

though they have no importance in clinical practice or possibly even in public health. At the other extreme, small samples may produce large differences so imprecisely determined that they are not statistically significant.

To illustrate this point, we return to the study of captopril and cardiac index.[8] Suppose that the standard deviation of change in cardiac index is 0.3 liter per minute per square meter, a value close to the estimated standard deviation of 0.27 reported by Davis et al. Whereas the medical significance of a treatment effect depends on the size of the average increase in cardiac index, the *P* value depends on both the observed mean increase and the sample size. Although a mean increase of 0.50 liter per minute per square meter would be statistically significant for any sample size between 10 and 100 patients, an increase of 0.10 liter per minute per square meter would generate a *P* value between 0.29 (defined as nonsignificant) and less than 0.001 (highly significant), depending on the size of the sample (Table 1). The point is that *P* values are useful for assessing the role of chance in producing an observed effect of treatment, but the medical importance of that effect depends on its magnitude. This example shows once again the value of citing confidence intervals in scientific reports. For instance, a mean increase of 0.10 liter per minute per square meter in a study with sample size of 100 yields the 95 percent confidence interval (0.04–0.16), while a mean increase of 0.50 in a study with 10 patients yields the 95 percent confidence interval (0.29–0.71).

Although we have recommended that *P* values be reported as actual values rather than as a range, we report only $P < 0.001$ in Table 1. We see little value in reporting the exact *P* value when it is smaller than 0.001 because the precise P value associated with very extreme results is sensitive to small biases or departures from the assumed probability model.

RECOMMENDATIONS

One should not equate P values or the results of hypothesis testing with decisions. P values are a way of reporting the results of statistical analyses. Similarly, the result of a hypothesis test might be best described as a conclusion, rather than a decision, to emphasize that the results of hypothesis tests are another way of reporting data. Decisions depend on conclusions but also on such factors as costs, risks, size of effect, consequences, and policy considerations. Issues of institutional decision making and many other factors can also affect the decision. Hypothesis tests and P values give us a form of reporting that has value because it can be standardized. The practical decision is a separate matter that is based on P values and related information but not on the P values alone.

Our review of the issues that arise in reporting and interpreting P values suggests several practices that we think should routinely be followed in scientific reporting (see also the Uniform Requirements for Manuscripts Submitted to Medical Journals[26] and Chapter 16 of this book[27]):

(1) Report basic summary statistics, such as the mean and the standard deviation, not just the results of a statistical test.

(2) When it is computationally feasible, report the actual P value (such as $P = 0.032$) rather than an inequality (such as $P<0.05$). This practice provides additional information, usually at the cost of little added effort.

(3) State clearly, whenever the distinction is relevant, whether a P value is for a one-sided or a two-sided test.

(4) Report confidence intervals or standard errors to communicate the range of values consistent with the study data. Although confidence intervals may not always be necessary, they should be included more frequently in reports of major findings.

(5) Discuss the effect of multiplicity on the reported P values. Although formal multiple-comparisons procedures may not be available for a particular kind of analysis, the interpretation of P values often de-

pends on the extent to which the final results arose from comparisons of several groups or measurements made on several occasions, as well as on multiplicities in data analysis, such as the use of several statistical tests, the exclusion of "atypical" patients, and data exploration.

(6) Discuss the statistical power of negative studies. Small studies are likely to be interpreted as unduly discouraging about the value of further investigation if the limited power of the study is not reported.

These proposals have implications for the design and conduct of scientific studies. They imply that investigators should consider power in designing studies. They also indicate the importance of specifying the hypotheses to be studied and the associated analyses before the study begins, in order to avoid ambiguities that may arise from analyses that were suggested only by the data. The steps necessary to ensure the validity of P values are not technically difficult, yet they can substantially improve the prospects of achieving the goals of a scientific study.

REFERENCES

1. Emerson JD, Colditz GA. Use of statistical analysis in the *New England Journal of Medicine*. N Engl J Med 1983; 309:709-13. [Chapter 3 of this book.]

2. Schuster EH, Bulkley BH. Early post-infarction angina: ischemia at a distance and ischemia at the infarct zone. N Engl J Med 1981; 305:1101-5.

3. Zelen M. Strategy and alternate randomized designs in cancer clinical trials. Cancer Treat Rep 1982; 66:1095-100.

4. Ingelfinger JA, Mosteller F, Thibodeau LA, Ware JH. Biostatistics in clinical medicine. 2nd ed. New York: Macmillan, 1987.

5. Snedecor GW, Cochran WG. Statistical methods. 8th ed. Ames, Iowa: Iowa State University Press, 1989.

6. Colton T. Statistics in medicine. Boston: Little, Brown, 1974.

7. Armitage P. Statistical methods in medical research. New York: John Wiley, 1971.

8. Davis R, Ribner HS, Keung E, Sonnenblick EH, LeJemtel TH. Treatment of chronic congestive heart failure with captopril, an oral inhibitor of angiotensin-converting enzyme. N Engl J Med 1979; 301:117-21.

9. Khaw K-T, Barrett-Connor E. Dietary potassium and stroke-associated mortality: a 12-year prospective population study. N Engl J Med 1987; 316:235-40.

10. Gore SM. Assessing methods — art of significance testing. Br Med J 1981; 283:600-2.

11. Rothman KJ. A show of confidence. N Engl J Med 1978; 299:1362-3.

12. Rothman KJ. The role of significance testing in epidemiologic research. Chapter 9 in Modern epidemiology. New York: Little, Brown, 1985.

13. Walker AM. Reporting the results of epidemiologic studies. Am J Pub Hlth 1986; 76:556-8.

14. Diamond GA, Forrester JS. Clinical trials and statistical verdicts: probable grounds for appeal. Ann Intern Med 1983; 98:385-94.

15. McPherson DD, Hiratzka LF, Lamberth DC, et al. Delineation of the extent of coronary atherosclerosis by high-frequency epicardial echocardiography. N Engl J Med 1987; 316:304-9.

16. Glantz SA. Biostatistics: how to detect, correct and prevent errors in the medical literature. Circulation 1980; 61:1-7.

17. Godfrey K. Comparing the means of several groups. N Engl J Med 1985; 313:1450-6. [Chapter 12 of this book.]

18. Totterman TH, Karlsson FA, Bengtsson M, Mendel-Hartvig I. Induction of circulating activated suppressor-like T cells by methimazole therapy for Graves disease. N Engl J Med 1987; 316:15-22.

19. Anturane Reinfarction Trial Research Group. Sulfinpyrazone in the prevention of sudden death after myocardial infarction. N Engl J Med 1980; 302:250-6.

20. Temple R, Pledger GW. The FDA's critique of the Anturane Reinfarction Trial. N Engl J Med 1980; 303:1488-92.

21. Tukey TW. Analyzing data: sanctification or detective work? Am Psychologist 1969; 24:83-91.

22. Cochran WG. Variation, control, and bias. In: Moses L, Mosteller F, eds. Planning and analyzing observational studies. New York: John Wiley, 1983:1-14.

23. Anderson S, Auquier A, Hauck WW, Oakes D, Vandaele W, Weisberg HI. Statistical methods for comparative studies. New York: John Wiley, 1980.

24. Freiman JA, Chalmers TC, Smith H Jr, Kuebler RR. The importance of beta, the type II error, and sample size in the design and interpretation of the randomized controlled trial: survey of 71 "negative" trials. N Engl J Med 1978; 299:690-4. [Chapter 19 of this book.]

25. Riis BJ, Thomsen K, Christiansen C. Does calcium supplementation prevent postmenopausal bone loss? A double-blind, controlled study. N Engl J Med 1987; 316:173-7.

26. Uniform Requirements for Manuscripts Submitted to Medical Journals. N Engl J Med 1991; 324:424-8.

27. Bailar JC, Mosteller F. Guidelines for statistical reporting in articles for medical journals. Ann Intern Med 1988; 108:266-73. [Chapter 16 of this book.]

11

Simple Linear Regression in Medical Research

Katherine Godfrey, Ph.D. *

ABSTRACT   This chapter discusses the method of fitting a straight line to data by least squares and focuses on examples from 36 Original Articles published in the *New England Journal of Medicine* in 1978 and 1979. Medical authors generally use linear regression to summarize the data (as in 12 of 36 articles in my survey) or to calculate the correlation between two variables (21 of 36 articles). Investigators need to become better acquainted with residual plots, which give insight into how well the fitted line models the data, and with confidence bounds for regression lines. Statistical computing packages enable investigators to use these techniques easily.

D ata analysis is largely a search for patterns — that is, for meaningful relations among the various items observed. Are older persons more susceptible to some disease than younger ones? Does the survival of cancer patients depend on the size of the cancer at the time of diagnosis? Is one metabolic variable related to another? Such patterns in the data can often be summarized by means of mathematical models.

One of the simplest patterns (or models) for data on a pair of variables is a straight line. The techniques for fitting lines to data and checking how well the line describes the data are called linear-regression methods. Using these methods, we can examine the relation between a change in the value of one variable and a change in the main variable of interest in the study. For example, Smith and Landaw,[1] in a study of patients with smokers' polycythemia, used a straight line to examine the relation between blood

---

*The editors have revised this chapter slightly from the original in the first edition.

levels of carboxyhemoglobin and P_{50}, the partial pressure of oxygen at which available hemoglobin is 50 percent saturated under normal conditions. The two variables have somewhat different roles in the interpretation of a linear regression. One of the variables, called the predictor variable or the independent variable (in this example, the blood level of carboxyhemoglobin), is regarded as a cause, predictor, or correlate of the other (here, P_{50}), the outcome variable, which is also known as the dependent or response variable.

SIMPLE LINEAR REGRESSION

In this chapter I shall consider only the simplest regression model, that involving only one predictor variable and one outcome variable. We usually refer to the predictor variable as x. For example, in a study of red-cell storage,[2] the authors examined the relation between length of storage and the level of extracellular potassium in samples of red-cell suspensions. The predictor variable was the length of storage (x), and the outcome variable, usually labeled y, was the extracellular potassium measurement.

In Volumes 298 through 301 of the *New England Journal of Medicine* (published in 1978 and 1979), 36 of the 332 Original Articles made use of simple linear-regression techniques (Table 1).[1-36] This chapter will not discuss the mechanics of calculating a regression. Many standard reference works, such as those by Armitage,[37] Bliss,[38] and Snedecor and Cochran,[39] provide formulas and examples. Instead, I examine how simple regression methods are used in the *Journal* and how they may be applied to the readers' own work.

The simple linear-regression model assumes that the relationship between x and y can be summarized as a straight-line graph. That is, for a unit change in x, y changes by a constant amount over the range of interest. The line itself can be represented mathematically by two numbers, the intercept (where the line crosses the y axis) and the slope. The actual values of the slope and intercept are rarely known, but they are easily estimated to fit a data set. A formula for simple linear regression is

$$\text{predicted outcome} = \text{intercept} + (\text{slope} \times \text{predictor value}).$$

The outcomes predicted by the equation are our estimates for the outcome variable based on specific values of the predictor variable. They will seldom exactly equal the values for the outcome variable found in the study, but we choose the slope and the intercept to make the two sets of values agree as well as they can in a sense described later. To distinguish between the values we observe for the variable y and those predicted by the equation, we use a different notation for the latter by placing a circumflex ($^\wedge$) over the y. The circumflex reminds us that these values are estimated from a line and not actually observed.

In the polycythemia study,[1] y is P_{50}, measured in millimeters of mercury, and x is the blood level of carboxyhemoglobin stated as a percentage. The fitted equation is

$$\hat{y} = 26.9 - 0.39x.$$

The intercept is 26.9, the predicted value of P_{50} for a person with no blood carboxyhemoglobin. The slope (-0.39) indicates that for every 1 percent drop in the concentration of carboxyhemoglobin, the P_{50} level increases on average by 0.39 mm Hg. This may or may not be a causal relationship.

Besides the slope and intercept, a third important number is associated with a regression equation: the variability of observed data points around the regression line. This number provides a measure of how closely the predicted values for the outcome variable correspond to the actual values of y. Like the slope and the intercept, this number can only be estimated from the data. It is often expressed as the estimated variance, also called the residual variance; another way to represent the variability is to give the square root of the variance, the standard deviation.

The difference between the observed value and the predicted value of the outcome variable ($y - \hat{y}$) is called the residual, and we usually label it e. This number is the vertical distance of the observed point from the fitted line. It will be positive if the point is above the line and negative if the point is below the line. A good regression line minimizes the residuals in some sense. In the most common method of fitting, called least-squares regression, we choose regression coefficients that minimize the sum of the squared residuals. As a byproduct of this process, the sum of the residuals is zero.

Table 1. Articles Using Simple Linear-Regression Techniques in Volumes 298 through 301 of the *New England Journal of Medicine*.*

Reference No.	Predictor Variable (x)	Outcome Variable (y)	Correlation?	Summary?	Regression Coefficients Intermediary?	Prediction?	Equation?	Figure?	Data Transformed?
1	Blood carboxyhemoglobin concentration (%)	Partial pressure (mm Hg) of oxygen at which available hemoglobin is 50 percent saturated under standard conditions (P_{50})	yes	yes	no	no	yes	yes	no
2	Number of days of blood storage	Extracellular potassium concentration in the supernatant of stored blood	yes	yes	no	no	yes	yes	no
3	Drop in plasma glucose induced by leg exercise	Disappearance of insulin from injection sites in the legs of patients with diabetes	yes	no	no	no	yes	yes	no
4	Percent inhibition of IgG complexes in patients with lupus nephritis and nephrotic syndrome	Percent inhibition of C1q binding complexes	yes	no	no	no	no	yes	no

5	Days from diagnosis of toxic effects of chlordecone	Log (blood concentration of chlordecone) (ng/ml)	yes	yes	yes	no	no	no	y
6	Alveolar oxygen tension (mm Hg)	Minute ventilation of oxygen (liters/min)	no	yes	yes	no	no	yes	x
7	Log (uninvolved renal-vein plasma renin activity)	Ratio of blood pressure after and before administration of nonapeptide converting-enzyme inhibitor in patients with hypertension	yes	no	no	no	yes	yes	x
8	Peroneal nerve-conduction velocity (mm/sec) in patients on dialysis	Serum parenchymal hormone level (pg/ml)	yes	no	no	no	yes	no	no
9	Hepatic clearance of indocyanine green (ml/min/kg)	Hepatic lidocaine clearance (ml/min/kg) in patients with congestive heart failure and controls	yes	yes	yes	yes	yes	yes	no
10	Number of cigarettes smoked per day by women under 50 with recent myocardial infarction and by controls	Log (relative risk of myocardial infarction)	yes	yes	no	no	yes	yes	y

(Table continues)

Table 1. (Continued)

Reference No.	Predictor Variable (x)	Outcome Variable (y)	Correlation?	Summary?	Regression Coefficients Intermediary?	Prediction?	Equation?	Figure?	Data Transformed?
11	Body fat as percentage of total fat in children with Turner's syndrome	24-Hour mean plasma concentration of luteinizing hormone	yes	no	no	no	yes	yes	no
12	Biochemical test and liver-biopsy features in patients with acute and chronic active hepatitis (piecemeal necrosis, serum bilirubin level, etc.)	Log (titers of antibody to liver-specific membrane lipoprotein)	yes	no	no	no	no	no	y
13	Mean arterial blood pressure in patients with renal parenchymal disease	Urinary kallikrein activity	yes	no	no	no	yes	yes	no
14	Leukocyte-inhibitor factor activity induced by type II collagens by cells from patients with rheumatoid arthritis	Leukocyte-inhibitor factor activity induced by type III collagens by cells from patients with rheumatoid arthritis	yes	no	no	no	no	yes	no

15	Blood mineral level (calcium, phosphate) in patients with hypocalcemia	Blood hormone level (parathyroid hormone, $1\alpha, 25(OH)_2D_3$, urinary cyclic AMP)	yes	yes	no	no	no	yes	no
16	Months from immunization with dodecavalent pneumococcal vaccine in patients with Hodgkin's disease	Log (mean pneumococcal-antibody concentration)	yes	no	no	yes†	yes	y
17	Mean carcinoembryonic-antigen level	Mean recurrence time in months of Dukes' B and Dukes' C cancer	yes	yes	no	no‡	yes	no
18	Ratio of bound to free platelet sites in binding of ^{3}H-dihydroergocryptine (DHE) to platelets in patients with essential thrombocythemia and controls	Bound platelet site rate (DHE sites/platelet)	yes	yes	no	no	yes	no
19	Log (minimum inhibitory concentration of penicillin G)	Drug-zone parameter obtained with penicillin and methicillin disks for 100 strains of pneumococci	no	no	yes	yes	yes	no
20	Serum cobalamin intrinsic factor in normal subjects	Serum cobalamin–binding protein	no	yes	no	yes	no	no

(Table continues)

Table 1. (Continued)

Reference No.	Predictor Variable (x)	Outcome Variable (y)	Correlation?	Summary?	Regression Coefficients Intermediary?	Prediction?	Equation?	Figure?	Data Transformed?
21	Renal clearance rate of creatinine in patients with burns	Renal clearance rate of insulin, iothalamate, tobramycin	yes	no	no	no	yes	yes	no
	Age in years	Renal clearance rate of creatinine and tobramycin	yes	no	no	no	yes	yes	no
22	Days of viral excretion in patients with initial episode of genital herpes simplex infection	Log (peak stimulation index with homologous herpes simplex virus antigen)	yes	no	no	no	yes	yes	y
23	Base-line plasma insulin level in patients with isolated growth hormone deficiency and controls	Specific binding of insulin to monocytes	yes	no	no	no	no	yes	no
24	Days to onset of graft-versus-host disease after bone marrow transplantation	Days to onset of pulmonary disease in patients with lymphocytic bronchitis	yes	no	no	no	yes	yes	no

#									
25	Urea production rate during treatment with Intralipid and sorbitol in patients with trauma	Change in urea production rate on transfer to treatment with glucose and insulin or glucose alone	yes	yes	no	no	yes	yes	no
	Urea production rate during treatment with glucose	Urea production rate during treatment with insulin alone	yes	yes	no	no	yes	yes	no
26	Age in days of neonates with respiratory syncytial virus infection	Log (quantity of syncytial virus shed in nasal washes)	yes	no	no	no	yes	yes	y
27	Oxygen saturation (ear) during sleep in normal subjects	Oxygen saturation (artery)	yes	no	no	no	yes	no	no
28	Age in years of Pima Indians	Percent bile cholesterol saturation	yes	no	no	no	no	yes	no
29	Frequency of documented coronary artery disease in patients	Likelihood of coronary artery disease after tests (electrocardiographic stress test, thallium scintigraphy)	yes	no	no	no	yes	yes	no
30	Days of hospitalization	Days of antimicrobial prophylaxis	yes	no	no	no	yes	yes	no

(Table continues)

Table 1. (Continued)

Reference No.	Predictor Variable (x)	Outcome Variable (y)	Correlation?	Regression Coefficients Summary?	Interme-diary?	Predic-tion?	Equa-tion?	Figure?	Data Trans-formed?
31	Left ventricular end-systolic volume	Left ventricular transmural pressure and systolic arterial pressure in patients with respiratory disease (asthma, etc.)	yes	yes	no	no	yes	yes	no
32	Minutes after injection of radioactively labeled iodine before and during treatment with L-thyroxine	Log (plasma-specific albumin activity in patients with hypothyroidism)	no	yes	no	no	no	yes	y
33	Years (date)	Log (pertussis incidence rate per 100,000 population in Massachusetts)	no	yes	no	no	yes	yes	no
34	Sine (time in minutes)	Plasma concentrations of insulin, glucose, and C-peptide	yes	no	no	no	yes	no	x

			correlation	summary	prediction	equation	figure	x	y
35	Percentage increase in systemic vascular resistance at peak rebound after nitroprusside withdrawal	Percentage change in heart rate during nitroprusside administration in patients with severe chronic heart failure	yes	no	no	no	yes	yes	no
36	Increase in maximal aerobic power in healthy adults	Increase in insulin-mediated glucose uptake after physical training	yes	no	no	no	no	yes	no

* The "correlation" column indicates whether the regression was used to calculate the correlation between x and y; the "summary" column, whether the regression was used to summarize the data; regression coefficients were considered intermediary if they were used to calculate final results in the analysis of the data; the "prediction" column indicates whether the regression equation was used to predict values not in the original data set; the "equation" column, whether a regression line was presented; the "figure" column, whether a figure showing data with the fitted line was presented. When data were transformed, x indicates that the predictor variable was transformed; y, that the outcome variable was transformed.

† The standard deviation about the regression line was reported.

‡ Complete data for the outcome and predictor variables in the regression were reported.

The study of residuals is an important part of a regression analysis. Residuals may point to possible outliers (unusual values) in the data or to problems with the regression model, and they can sometimes give us more important information about the data than does the regression equation itself. Residuals will be discussed later in this chapter.

A study by Slone et al.[10] makes clear the difference between observed values (y) and predicted values ($\hat{y}$) of the response variable. The authors measured cigarette consumption (x) and the logarithm of the relative risk of myocardial infarction (y) for a group of women; they then analyzed the relation between these variables by a simple linear regression of log(relative risk) of myocardial infarction on cigarette consumption. Figure 1, adapted from their original figure, shows both observed and predicted values of log(relative risk), along with the regression line itself.

The horizontal scale (cigarettes per day) shows categories of subjects according to cigarette use. The biologic rationale for the scale is not given; there is no clear reason to put exsmokers midway between nonsmokers and users of 1–14 cigarettes per day, or to make spacings of higher-use categories equal. However, this kind of arbitrary quantitation of categories that have a natural ordering often works well, as discussed by Moses et al. in Chapter 13 of this book.

Although all the predicted risk values (triangles) necessarily lie on the line, only one of the observed risk values (solid circles) does, and only by chance. The vertical deviations of the solid circles from the triangles on the dotted line are the residuals.

Simple linear regression can serve different purposes in analyzing data. I shall discuss three uses of the technique: prediction, correlation, and summarization. Table 1 summarizes the uses of regression made by the 36 individual papers.

PREDICTION

The most direct aim of regression is to use x to predict y. In the simple regression model, when the degree of correlation between y and x is high, so that the residual variance (which measures the error associated with the residuals) of the regression line is small, $\hat{y}$ will be a good predictor of y.

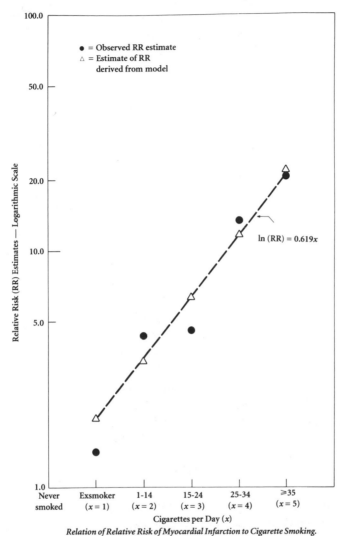

Figure 1.   Example of a Regression Line Fitted to Data.*
The solid circles mark the observed values. The dotted line is the predicted regression line for the data set, and the triangles mark the predicted values for the points in the data set. For this data set, the predictor variable (x) codes the number of cigarettes smoked per day; the outcome variable (y) is the relative risk of myocardial infarction (in the logarithmic scale).
*Data from Slone et al.[10]

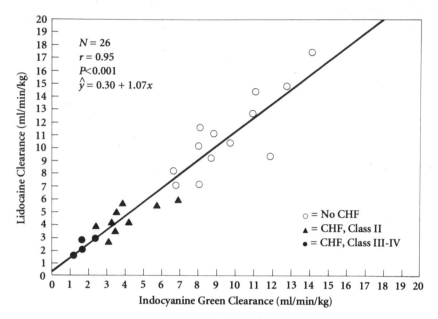

Individual Lidocaine and Indocyanine Clearances for Patients with and
without Congestive Heart Failure (CHF).

$$\text{Lidocaine clearance rate} = \frac{\text{lidocaine constant infusion rate } (\mu g/kg/min)}{\text{lidocaine level, steady state } (\mu g/ml)}$$

Figure 2.   The Regression Analysis of Zito and Reid.*
The figure gives the data, the fitted regression line, the equation of the fitted line,
and the correlation between the predictor and outcome variables.
*Data from Zito and Reid.[9]

For example, Zito and Reid[9] used measured hepatic clearance rates of
indocyanine green dye (x) and lidocaine (y) to fit a regression equation
where $\hat{y}$ was the predicted lidocaine-clearance rate. The object was to
see whether the indocyanine-clearance rate could be used to estimate
the lidocaine-clearance rate and thus to adjust lidocaine infusion so as
to avoid giving patients potentially toxic doses of lidocaine. The result was
$\hat{y} = 0.3 + 1.07x$, with x and y both measured in milliliters per minute per
kilogram of body weight. Figure 2 shows the results of their regression. The
authors reported that the intercept (0.3) was not significantly different

from zero and that the slope (1.07) was not significantly different from 1.00, as one would expect if the clearance rates for indocyanine green and lidocaine were identical. In this case, the authors used the regression to conclude that one rate might be used in place of the other. This finding has important clinical implications for the cost of treatment, the risks to patients, and the availability of information on clinical status. Figure 2 also suggests that points of high values are not in general as close to the fitted line as points of low values. This is common in biologic data, and is discussed below.

CORRELATION

The degree of correlation between the variables x and y is expressed by the correlation coefficient, which measures the degree of linear relation between the two. Its value does not depend on the units in which x and y are expressed. The slope of the fitted line does depend, however, on the units used to measure x and y, a point that should be kept in mind when one is judging the importance of the slope.

The slope and the correlation coefficient are not equivalent measures of association. There is a basic asymmetry in the regression equation, and y and x are not interchangeable. For this reason, the slope of the equation using x to predict y will not be the same as the slope of the equation using y to predict x, unless the standard deviations of the two variables x and y happen to be equal. The correlation coefficient can be calculated directly from the data without deriving the regression equation. Formulas are available in introductory statistical textbooks like Snedecor and Cochran's,[39] and statistical computing packages can also provide correlations.

In a study[4] of patients with lupus nephritis and steroid-responsive nephrotic syndrome, the authors used both regression analysis and the correlation coefficient (r) between C1q binding (y) and IgG-containing complexes (x) in their patients. Figure 3 shows their results. Although the regression line is indicated, the article provides no statistics other than the correlation coefficient itself.

A fitted regression is often used in this way in the *Journal*. More than half the papers (21 of 36) using simple regression techniques in Volumes 298 through 301 of the *Journal* presented correlations.

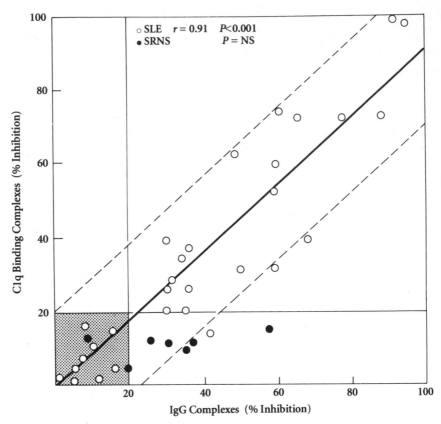

*Correlation between C1q Binding and IgG-Containing Complexes in Children
and Adults with Systemic Lupus Erythematosus (SLE) and Nephritis (○) and in
Children with Steroid-Responsive Nephrotic Syndrome (SRNS) (●).
The regression line and 95 percent confidence limits for the patients with
lupus are shown. The limit of normal for both systems is indicated.*

Figure 3.   An Example of the Use of Regression Analysis with the Correlation between the Outcome and Predictor Variables.*
The solid line is the fitted regression equation. The dotted lines are not the confidence interval curves described in the text.
*Data from Levinsky et al.[4]

SUMMARIZATION

Sometimes investigators want to know more than the degree of correlation between the x and y variables but do not intend to use the measure-

ments of x to predict (and perhaps to replace in the future) measurements of y. Instead, they want to know something about the relation between x and y, beyond the simple correlation. A regression can be used to summarize the relation between the two variables. For each value of x, there is an average value of y. When these averages lie on a line or close to it, the line summarizes all these averages and thus summarizes the relation between x and the average value of y at that value of x. A regression line is more specific than a correlation coefficient in that it can predict the change in y associated with a particular change in x. This prediction depends on the slope.

Smith and Landaw[1] use regression in this way in their study of patients with smokers' polycythemia. A regression equation summarizes the relation between the blood concentration of carboxyhemoglobin (x) and the partial pressure of oxygen at which available hemoglobin is 50 percent saturated under normal conditions (y). Figure 4 gives their results, along with the equation of the regression line. The fitted line summarizes the relation between the partial pressure and the blood concentration of carboxyhemoglobin, indicating that a 10 percent drop in the carboxyhemoglobin concentration produces an increase of about 4 mm Hg in the partial pressure needed to obtain 50 percent saturation.

Of the 36 articles using simple regression techniques in Volumes 298 through 301 of the *Journal*, 12 employed the method simply to summarize the data.

GUIDELINES FOR USING REGRESSION

LOOKING AT THE DATA

The first step in fitting a regression equation to the data is to examine the data. We should plot the data (y vs. x) for two reasons. First, the shape of the plot may suggest that a straight-line equation is not appropriate. Procedures for selecting and fitting curved lines are not discussed here, but they may be found in standard textbooks such as those by Snedecor and Cochran[39] and Bliss.[38] For example, in an investigation by Lang et al.[34] of the oscillations over time of plasma glucose and insulin concentrations, the periodic (wavelike) shape of the data suggested that the data be converted, or transformed, to the sine of time (x) before the investigators fitted a regression equation

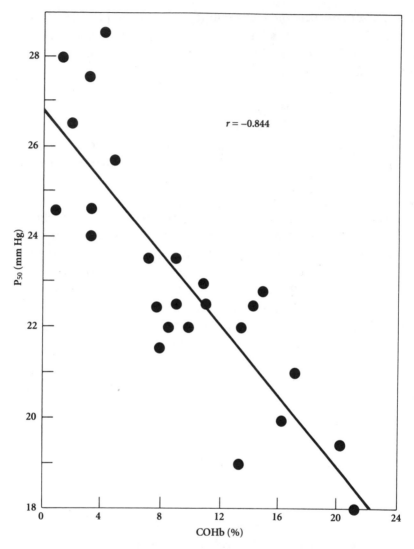

*Relation between Blood Carboxyhemoglobin (COHb) Concentration
and P_{50} in 20 of the 22 Smoking Polycythemic Subjects.
A number of subjects were studied at different COHb levels for
a total of 25 observations. The equation for the straight line is:
P_{50} (mm Hg) = −0.39 (percent COHb) + 26.9; and (r = −0.84, P<0.01).*

Figure 4.   The Regression Analysis Performed by Smith and Landaw.[*]
Unlike the lines in Figures 1 to 3, the line here has a negative slope.
[*]Data from Smith and Landaw.[1]

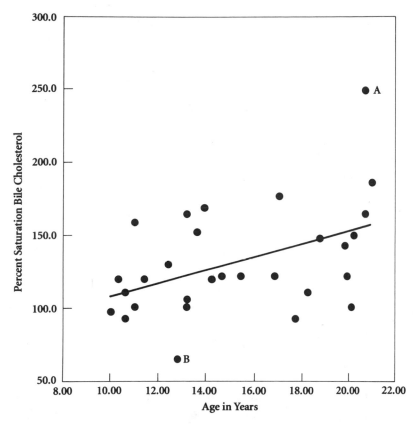

Figure 5.   Relation between Bile Cholesterol Saturation and Age in Young
Female Pima Indians.*
Note the two outliers in the data set, marked A and B. The regression line, giving
the saturation as a function of age, is $\hat{y} = 63 + 4.5x$. The correlation between x
and y is 0.47.
*Data from Bennion et al.[28]

Second, plots can indicate outlying data points. Figure 5, adapted from
a figure in a study by Bennion et al. on bile cholesterol levels in Pima
Indians,[28] shows bile cholesterol saturation (y) versus age (x) in females.
The plot shows a clear outlier (A) in the upper right-hand corner. This
value for bile cholesterol saturation was extremely high. Point B, in

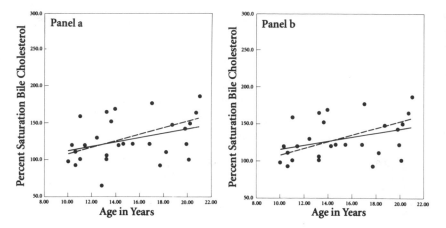

Figure 6.   Data from Figure 5 Reworked with Point *A* Removed (Panel a) and
Points *A* and *B* Removed (Panel b).
For each plot, the fitted regression line from Figure 5 is the broken line, and the
new regression line calculated after removal of the outlier or outliers is the solid
line. The regression equation for the plot in Panel a is $\hat{y} = 82 + 3.0x$, with a correlation of 0.38 between x and y. The regression equation for the plot in Panel b is
$\hat{y} = 89 + 2.7x$, with a correlation of 0.36 between x and y.

the lower part of the plot, may also be an outlier; it has a low value of y.
An outlier can be the result of an error in measurement, a true value for
an unusual member of the population, or an observation on a person (or
other subject of study) that does not really belong in the population under
study.

What can happen when the data are not examined before we begin the
regression? Even a single outlier can have a profound effect on the relationship derived from the regression line. If we remove the upper outlier (A)
from the data in Figure 5, the slope of the regression line will be reduced
from 4.5 to 3 (Fig. 6a). If we also remove the apparent outlier (B) in the
lower part of the plot, we flatten the line even more, reducing the slope to
2.7 (Fig. 6b). Although the sample includes 29 persons, points A and B account for nearly half the estimated slope of the relation between age and
percentage of saturation. We cannot tell from the paper whether those
points should be included or excluded in estimating the regression, but either way their influence should be recognized.

When should we remove an outlier from the data being analyzed? When an anomalous value appears and is determined not to be the result of an error in measurement or recording, the investigators have a decision to make. Should they discard the point as a fluke or consider it representative of the extreme range of the sampled population? In other words, do they want the regression to reflect the occasional outlier or to apply only to the majority of "usual" data points? Unfortunately, without outside information, remeasurement (if feasible), or a larger sample, it is often impossible to discover whether an outlier is more than a fluke. Thus, the investigator must rely on intuition and experience as well as on the data. An alternative way to handle outliers is to make use of a different method to fit a line to the data. Several methods, not described here, are relatively unaffected by outliers. Velleman and Hoaglin[40] discuss one such technique.

LOOKING AT RESIDUALS

The residuals contain the information in the data that is not explained by the fitted line. If the residuals have a discernible pattern, more is going on in the data than the model reveals, and we should consider changing to a model that incorporates more of the information. Thus, the residuals tell us not only how well a particular model fits the data but also whether a different model might fit the data better.

An analyst should always verify, at least roughly, the assumption that the regression equation accurately summarizes the data. After the regression line has been calculated, the investigator should examine the residuals $(y - \hat{y})$ by plotting them against the fitted values $\hat{y}$. As in the data plot, outliers may come to light. Ideally, the scatter plot of the regression residuals will be a patternless cloud of points with no perceptible slope or axis, centered about the horizontal axis. Figure 7 shows an artificial but typical pattern. Other patterns may suggest that the regression model being used does not provide the best fit to the data.

If the data do not seem to lie on a line, or if the residuals show a clear pattern, the linear model may not be consistent with the data. There are several ways to deal with this problem. If the data seem to follow a particular curve, such as a logarithmic, sinusoidal, or exponential curve, nonlinear regression, discussed by Bliss[38] and Snedecor and Cochran,[39] is an appropriate method. If part of the data seem to lie along one line and part along another line, the data can be stratified. Using this technique, we fit a different line to each segment of the data, as in Figure 8. In this case, also artificial, we assume that a linear relation holds within each stratum, although the exact relation may not be the same from stratum to stratum. Such a model is called a piecewise linear model.

A third solution to the problem of nonlinearity is transformation. By changing the measurements of the x or y variable to a different scale, such as the logarithmic, we may produce a better fit to the data; the result will be a formless residual plot when the regression model is refitted to the transformed data.

Figure 9, which plots hypothetical residuals, shows a wedgelike pattern, suggesting that the variance of the residuals (that is, the variance about the

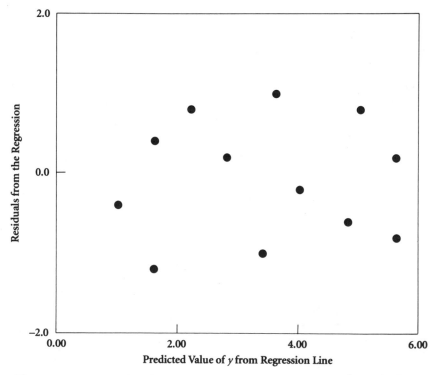

Figure 7. A Scatter Plot after Regression, Showing No Pattern in the Residuals. The residuals are plotted on the vertical axis. The fitted values of y are plotted on the horizontal axis.

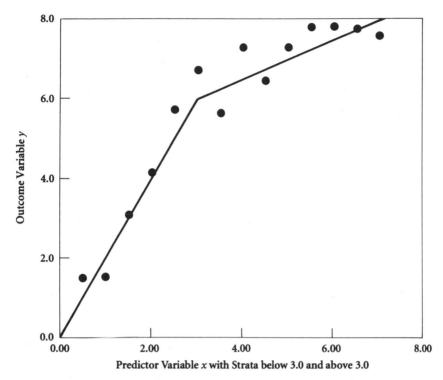

Figure 8.   Fitting a Piecewise Linear Model to Data.
The data are divided into two strata, below 3.0 and above 3.0, on the horizontal
axis. A separate regression line has been fitted for each stratum. Those regression
lines are the solid lines in the plot.

regression line) increases with $\hat{y}$. This pattern violates the usual assumption
that in a regression the residual variance will be the same for all values of $\hat{y}$.
The values in Figure 9 show a wedge flaring to the right. Taking square
roots or logarithms of y will reduce larger values more than smaller values
and thus can shrink the variance at the larger values. Similarly, a wedge
flaring to the left may suggest that squared values of y should be used.
These transformations sometimes, but not always, have an obvious meaning. If we take the reciprocal of the absorption time of a drug, for instance,
we have the rate of absorption.

A common purpose for transforming the data on a variable is to find the
"natural" units in which to measure the variable. For example, if we are

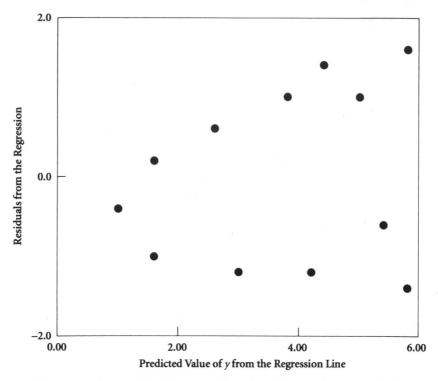

Figure 9. A Scatter Plot after Regression, Showing a Wedge Pattern in the Residuals.
The residuals are plotted on the vertical axis. The fitted values of *y* are plotted on the horizontal axis.

studying motor abilities by means of timed tests, we can record the time it takes a person to complete the test or the speed at which he or she performs the various tasks. The first measurement is in units of time; the second, in reciprocal units of time (1/time). If we want to use these measurements as the *y* variable in a regression against another variable, such as the age of the subjects, the plot of the speeds versus age may be linear, or the plot of time versus age may be linear. We usually choose the units for *y* that give a linear shape to the plot.

Histograms can also alert us to the possibility of using a transformation of *x* or *y*. If the histogram of one of the variables seems highly nonsymmetrical, we may want to transform that variable into another unit. For example, a histogram of a variable that is positively skewed (having a long tail in the direction of larger values) suggests a transformation like the logarithmic, to shrink the larger values of the variable.

In general, we transform the values for the y variable when the plot of the residuals indicates that their variance changes with $\hat{y}$, as in Figure 9. We transform the values for the x variable when their distribution appears to be nonsymmetrical. Of course, one can do both.

Transforming the variables, especially y, alters the standard regression model. For example, if taking logarithms of y gives a good linear fit, the residual $(\log(y) - \overline{\log(y)})$ will have no pattern or slope, but the fitted $\hat{y}$s, which we calculate as the antilogarithms of the fitted values in the regression of $\log(y)$, and the residuals $(y - \hat{y})$ calculated from this regression model, may look quite different. Another consideration when choosing a transformation is the ease with which the results can be interpreted. If the transformed measurement has no particular meaning, as is the case with the square root of age, the transformed model may be less useful even though the fit has been improved. Thus, it may be necessary to weigh a more accurate fit against a model that is more easily interpreted.

LOOKING AT THE FITTED LINE

An overall picture of the variability of y about the fitted line is provided by the confidence bounds for the fitted values. That is, for each value of x that we observe, we calculate the confidence interval for the mean value of y associated with that value of x. Then we draw a curve connecting all the lower confidence bounds and another connecting all the upper confidence bounds. To do this, of course, we need to calculate the variance of the fitted values of y. Assuming that the values of x and y are normally distributed, we can derive a formula to estimate this variance, using our regression results. The variance of the mean of $\hat{y}$ associated with a given observation x' of the predictor variable depends on x' and the variance of the x values. The variance of $\hat{y}$ is smallest when x' is the mean of the x values and increases as x' moves away from the mean of x in either direction. In such a case, the pair of curves drawn through the confidence bounds are hyperbolas that arch away from the regression line. Figure 10, from an article by Zerbe and Robertson,[41] shows such confidence bounds. When the variance of the fitted slope is small, the curves will be nearly straight (that is, almost

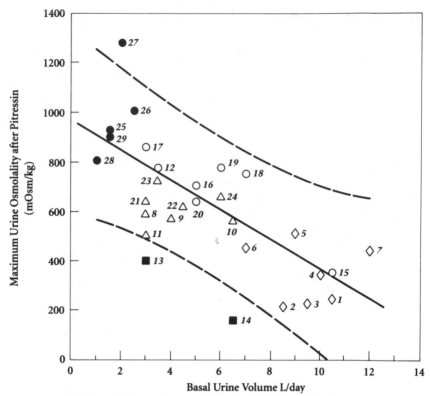

Relation of Urinary Concentrating Capacity to Basal Urine Output.

*The solid circles (●) represent healthy control patients. The other symbols represent patients
with various degrees of disease. The solid line depicts the regression function generated by the
least-squares method. The correlation was significant (r = −0.77; P<0.001). The broken lines
depict the 99 percent confidence limits for the regression of y on x.*

Figure 10.   An Example of a Fitted Regression Line and the Associated
Confidence-Interval Curves.*
*Data from Zerbe and Robertson.[41]

parallel to the fitted line). These curves give us a confidence interval for the
fitted line as a whole. The formula tells us that the uncertainty about the
entire fitted line is smallest near the center, where we often have the most
data values, and progressively greater as we move away from the center.
Also, the greater our uncertainty about the value of the slope, the greater
our uncertainty about the fitted values of *y*.

Figure 3 (page 216) shows confidence limits that are not adjusted for the value of x' and so are parallel to the fitted regression line. These are not true confidence bounds for $\hat{y}$, but if the variance of the fitted slope is small, they will be approximately correct, at least for values of x that are not higher or lower than those actually observed.

We have been calculating the variance of the means of fitted values of y for values of x. We can also calculate a confidence interval for the predicted value of y associated with a single value of x (for instance, a new value of x not observed in the original data). This variance is larger than that of the mean of the fitted y values. Again, the variance increases as we consider values of x further from the mean of x. Extrapolated data values — predictions for values of x outside the observed range of data — thus have greater uncertainty than those interpolated for values of x within this range. When we extrapolate, we also face uncertainty about how far the approximate linearity of the data extends; that is, it may not be clear whether the linear model holds for values of x outside the observed range.

REPORTING REGRESSION RESULTS

Readers of a journal article should usually be told enough about the authors' statistical procedures to be able to duplicate their results, given the data. Thus, the authors should include a brief but complete statement of their methods; one carefully worded sentence is often enough.

Results should also be reported in sufficient detail for the intended users. Readers may need the sample size, an equation of the fitted line, the fraction of the variation in y explained by the x variables (which can be measured as the square of the correlation between x and y), the variances of each of the fitted coefficients a and b (or their standard deviations), and the residual variance of y. With this information, readers can conduct their own significance tests of the fit of the regression. For example, in the lidocaine study,[9] although the correlation is high, the authors do not report the residual variance or the variances of the intercept and the slope. So even though the authors report that the intercept is not significantly different from zero, a reader cannot test whether it is significantly different from other values — for instance, 0.5. Other readers might want to calculate a confidence interval for the lidocaine-clearance rate of their own patients, given the patients' indocyanine-clearance rates. If they knew the variances of *a* and *b*, they could do this using formulas available in many introductory statistical textbooks.

Plots can give readers additional information about the fit of the regression. Figure 10 shows one way to do this. It includes two curves drawn through the upper and lower confidence limits for the fitted values of the outcome variable, based on the regression line. These are the confidence interval curves described earlier.

In general, articles in the *Journal* include a figure of the fitted regression line and the fitted equation and correlation. Only 1 article of 36, however, reported the standard deviation associated with the regression.[16]

STATISTICAL PACKAGES AND REGRESSION

Though software changes rapidly, I discuss methods of using regression in four popular computing packages: Minitab, SPSS, SAS, and BMDP.

Minitab was designed as an interactive program; the analyst submits commands one by one to the computer while sitting at a terminal, and these commands are executed at once. The other packages are intended for batch processing, with all commands submitted at once. Minitab is thus more flexible than the other packages, since the user can alter the direction of the analysis as it proceeds, instead of requesting all statistics in advance. Flexibility during processing comes at a price, however. The extensive program required for an interactive computer package makes it difficult to include large subprograms for specialized tasks. Thus, each of the other packages provides options that Minitab does not offer.

All four packages provide the regression coefficients and their standard errors, along with the standard error for the regression and the fraction of the variation in *y* explained by the regression. All four produce the *t* ratio for testing the significance of each regression coefficient, but Minitab does not calculate the *P* values for these *t* ratios.

Each of the programs will generate plots of the residuals, but the formats vary. Minitab and SPSS produce plots based on standardized residuals, where each residual has been divided by its standard deviation; the raw residuals, which we have been discussing here, may be calculated directly as the difference between the observed and the predicted values of *y*. The residual plots in SPSS also standardize the predicted values of *y*. SAS and BMDP use the raw residuals in producing plots. For further information on the options offered by the various packages, the reader should consult the appropriate reference manuals.[42-46]

Statistical packages provide plots and regression fits quickly but mechanically. A regression fit that maximizes the fraction of the variance that is explained may not be the best possible fit according to other criteria, such as ease of interpretation or cost of data collection. It is important to examine plots of the data for nonlinear shape or outliers before using regression routines. As in Figure 5 for the Pima data,[28] a single point can have a great deal of influence over the fit of the final regression. A plot will alert the investigator to possible problems.

Because of the way a computer plot is generated, it may mask some details of the data. If there is an extreme outlier (as in the Pima data), the scale of the plot will "shrink," since the axes must include the entire range of data, so that the main body of the data will be closely clustered. This condensation makes it more difficult to see the relation among those clustered points. In a situation like this, it may be profitable for investigators to obtain a second plot of the data excluding the outliers. Different computer packages may also use different algorithms to produce the axes of plots, so that the same plot may not be drawn to the same scale by two different packages.

CONCLUSIONS

Simple linear regression is a technique for studying the relation between a predictor variable (x) and a response variable (y). Regression analysis can summarize the relation between the response and predictor variables or produce a formula for predicting the response variable from the predictor variable. Authors often use simple linear regression to measure the correlation between y and x.

To make the best use of regression techniques, investigators should take three important steps. First, they should plot and examine the data beforehand and check for outliers or definite nonlinear patterns in the data. Second, they should plot residuals from the regression against the fitted values for $\hat{y}$. Once again, they should look for outliers and for patterns suggesting that the data should be transformed. Third, in reporting the study, they should give the reader enough information to follow and reproduce their results. Along with the regression equation itself, they should report the variances of the coefficients and the residual variance.

Regression is a powerful tool for providing a simple picture of data on two variables. If investigators take care to avoid possible pitfalls by examining the data and the residuals, simple linear regression can help to both summarize the data and communicate the results of this analysis clearly.

REFERENCES

1. Smith RJ, Landaw SA. Smokers' polycythemia. N Engl J Med 1978; 298:6-10.

2. Högman C, Hedlund K, Zetterström H. Clinical usefulness of red cells preserved in protein-poor mediums. N Engl J Med 1978; 299:1377-82.

3. Koivisto VA, Felig P. Effects of leg exercise on insulin absorption in diabetic patients. N Engl J Med 1978; 298:79-83.

4. Levinsky RJ, Malleson PN, Barratt TM, Soothill JF. Circulating immune complexes in steroid-responsive nephrotic syndrome. N Engl J Med 1978; 298:126-9.

5. Cohn WJ, Boylan JJ, Blanke RV, Fariss MW, Howell JR, Guzelian PS. Treatment of chlordecone (kepone) toxicity with cholestyramine: results of a controlled clinical trial. N Engl J Med 1978; 298:243-8.

6. Mountain R, Zwillich C, Weil J. Hypoventilation in obstructive lung disease: the role of familial factors. N Engl J Med 1978; 298:521-5.

7. Re R, Novelline R, Escourrou M-T, Athanasoulis C, Burton J, Haber E. Inhibition of angiotensin-converting enzyme for diagnosis of renal-artery stenosis. N Engl J Med 1978; 298:582-6.

8. Avram MM, Feinfeld DA, Huatuco AH. Search for the uremic toxin: decreased motor-nerve conduction velocity and elevated parathyroid hormone in uremia. N Engl J Med 1978; 298:1000-3.

9. Zito RA, Reid PR. Lidocaine kinetics predicted by indocyanine green clearance. N Engl J Med 1978; 298:1160-3.

10. Slone D, Shapiro S, Rosenberg L, et al. Relation of cigarette smoking to myocardial infarction in young women. N Engl J Med 1978; 298:1273-6.

11. Boyar RM, Ramsay J, Chipman J, Fevre M, Madden J, Marks J. Regulation of gonadotropin secretion in Turner's syndrome. N Engl J Med 1978; 298:1328-31.

12. Jensen DM, McFarlane IG, Portmann BS, Eddleston ALWF, Williams R. Detection of antibodies directed against a liver-specific membrane lipoprotein in patients with acute and chronic active hepatitis. N Engl J Med 1978; 299:1-7.

13. Mitas JA II, Levy SB, Holle R, Frigon RP, Stone RA. Urinary kallikrein activity in the hypertension of renal parenchymal disease. N Engl J Med 1978; 299:162-5.

14. Trentham DE, Dynesius RA, Rocklin RE, David JR. Cellular sensitivity to collagen in rheumatoid arthritis. N Engl J Med 1978; 299:327-32.

15. Bilezikian JP, Canfield RE, Jacobs TP, et al. Response of 1α,25-dihydroxyvitamin D₃ to hypocalcemia in human subjects. N Engl J Med 1978; 299:437-41.

16. Siber GR, Weitzman SA, Aisenberg AC, Weinstein HJ, Schiffman G. Impaired antibody response to pneumococcal vaccine after treatment for Hodgkin's disease. N Engl J Med 1978; 299:442-8.

17. Wanebo HJ, Rao B, Pinsky CM, et al. Preoperative carcinoembryonic antigen level as a prognostic indicator in colorectal cancer. N Engl J Med 1978; 299:448-51.

18. Kaywin P, McDonough M, Insel PA, Shattil SJ. Platelet function in essential thrombocythemia: decreased epinephrine responsiveness associated with a deficiency of platelet α-adrenergic receptors. N Engl J Med 1978; 299:505-9.

19. Jacobs MR, Koornhof HJ, Robins-Browne RM, et al. Emergence of multiply resistant pneumococci. N Engl J Med 1978; 299:735-40.

20. Kolhouse JF, Kondo H, Allen NC, Podell E, Allen RH. Cobalamin analogues are present in human plasma and can mask cobalamin deficiency because current radioisotope dilution assays are not specific for true cobalamin. N Engl J Med 1978; 299:785-92.

21. Loirat P, Rohan J, Baillet A, Beaufils F, David R, Chapman A. Increased glomerular filtration rate in patients with major burns and its effect on the pharmacokinetics of tobramycin. N Engl J Med 1978; 299:915-9.

22. Corey L, Reeves WC, Holmes KK. Cellular immune response in genital herpes simplex virus infection. N Engl J Med 1978; 299:986-91.

23. Soman V, Tamborlane W, DeFronzo R, Genel M, Felig P. Insulin binding and insulin sensitivity in isolated growth hormone deficiency. N Engl J Med 1978; 299:1025-30.

24. Beschorner WE, Saral R, Hutchins GM, Tutschka PJ, Santos GW. Lymphocytic bronchitis associated with graft-versus-host disease in recipients of bone-marrow transplants. N Engl J Med 1978; 299:1030-6.

25. Woolfson AMJ, Heatley RV, Allison SP. Insulin to inhibit protein catabolism after injury. N Engl J Med 1979; 300:14-7.

26. Hall CB, Kopelman AE, Doublas RG Jr, Geiman JM, Meagher MP. Neonatal respiratory syncytial virus infection. N Engl J Med 1979; 300:393-6.

27. Block AJ, Boysen PG, Wynne JW, Hunt LA. Sleep apnea, hypopnea and oxygen desaturation in normal subjects: a strong male predominance. N Engl J Med 1979; 300:513-7.

28. Bennion LJ, Knowler WC, Mott DM, Spagnola AM, Bennett PH. Development of lithogenic bile during puberty in Pima Indians. N Engl J Med 1979; 300:873-6.

29. Diamond GA, Forrester JS. Analysis of probability as an aid in the clinical diagnosis of coronary-artery disease. N Engl J Med 1979; 300:1350-8.

30. Shapiro M, Townsend TR, Rosner B, Kass EH. Use of antimicrobial drugs in general hospitals: patterns of prophylaxis. N Engl J Med 1979; 301:351-5.

31. Buda AJ, Pinsky MR, Ingels NB Jr, Daughters GT II, Stinson EB, Alderman EL. Effect of intrathoracic pressure on left ventricular performance. N Engl J Med 1979; 301:453-9.

32. Parving H-H, Hansen JM, Nielsen SL, Rossing N, Munck O, Lassen NA. Mechanisms of edema formation in myxedema — increased protein extravasation and relatively slow lymphatic drainage. N Engl J Med 1979; 301:460-5.

33. Koplan JP, Schoenbaum SC, Weinstein MC, Fraser DW. Pertussis vaccine — an analysis of benefits, risks and costs. N Engl J Med 1979; 301:906-11.

34. Lang DA, Matthews DR, Peto J, Turner RC. Cyclic oscillations of basal plasma glucose and insulin concentrations in human beings. N Engl J Med 1979; 301:1023-7.

35. Packer M, Meller J, Medina N, Gorlin R, Herman MV. Rebound hemodynamic events after the abrupt withdrawal of nitroprusside in patients with severe chronic heart failure. N Engl J Med 1979; 301:1193-7.

36. Soman VR, Koivisto VA, Deibert D, Felig P, DeFronzo RA. Increased insulin sensitivity and insulin binding to monocytes after physical training. N Engl J Med 1979; 301:1200-4.

37. Armitage P. Statistical methods in medical research. New York: John Wiley, 1971.

38. Bliss CI. Statistics in biology: statistical methods for research in the natural sciences. Vols. 1, 2. New York: McGraw-Hill, 1967, 1970.

39. Snedecor GW, Cochran WG. Statistical methods. 8th ed. Ames, Iowa: Iowa State University Press, 1989.

40. Velleman PF, Hoaglin DC. Applications, basics, and computing of exploratory data analysis. Boston: Duxbury Press, 1981.

41. Zerbe RL, Robertson GL. A comparison of plasma vasopressin measurements with a standard indirect test in the differential diagnosis of polyuria. N Engl J Med 1981; 305:1539-46.

42. Dixon WJ, Brown MB, Engelman L, et al. BMDP statistical software. Berkeley, Calif.: University of California Press, 1983.

43. SAS Institute Inc. SAS user's guide: statistics, version 5 edition. Cary, N.C.: SAS Institute, 1985.

44. SPSS Inc. SPSS X user's guide. 2nd ed. New York: McGraw-Hill, 1986.

45. Minitab reference manual Release 7. University Park, Pa.: Pennsylvania State University, 1989.

46. Ryan BF, Joiner BL, Ryan TA Jr. Minitab student handbook. 2nd ed. North Scituate, Mass.: Duxbury Press, 1985.

12

~

COMPARING THE MEANS OF SEVERAL GROUPS

Katherine Godfrey, Ph.D. *

ABSTRACT   To explain statistical methods for comparing the means of several groups, this chapter focuses on examples from 50 Original Articles published in the *New England Journal of Medicine* in 1978 and 1979 and includes a follow-up study of Volume 321 (July through December 1989). Although medical authors often present comparisons of the means of several groups, the most common method of analysis, multiple *t*-tests, is usually a poor choice because it does not take into account multiplicity. Which method of analysis is appropriate depends on what questions the investigators wish to ask. If the investigators want to identify groups under study that differ from the rest, they need a different method from the one required if they wish simply to decide whether the groups share a common mean. More complicated questions about the group means call for more sophisticated techniques. Of the 50 *Journal* articles examined, 27 (54 percent) could have used more appropriate statistical methods to analyze the differences between group means. Analysis of variance and multiple comparison methods including the Bonferroni, Duncan, Newman–Keuls, Scheffé, and Tukey methods offer useful methods for comparing means.

I n one simple form of experiment or observational study, the investigator compares sets of measurements taken from two groups to decide whether the group means differ. Emerson and Colditz[1] (Chapter 3 of this book) have reported that the *t*-test, the standard analysis for such an experiment, is the most commonly used statistical procedure in the *New England Journal of Medicine*.

When an experiment includes more than two groups, the choice of an appropriate statistical method for comparing group means depends on the

---

*The editors have revised this chapter slightly from the original in the first edition.

experimental design and on the questions asked. Although investigators have many options, few are appropriate, and the most frequently used method, multiple *t*-tests, is usually a poor choice. Moreover, even appropriate methods of analysis may not directly answer the investigators' questions. A poorly chosen analysis may generate misleading results by giving incorrect answers to the investigators' questions, by giving correct answers to the wrong questions, or by failing to use all the information available from the experiment.

Because computer programs offer ready numerical solutions to formerly obscure and difficult analytic problems, investigators can often match the type of data and the goal of the study to the method and obtain an adequate analysis. Using examples from the *Journal*, this chapter discusses some standard statistical methods for comparing group means and describes which questions about group means each method is designed to answer. Some standard methods of statistical analysis include multiple *t*-tests, analysis of variance, and multiple comparisons. More sophisticated techniques are available to answer complicated questions about group means.

In this chapter, I do not consider the mechanics of the calculations involved in the various techniques. Standard textbooks, such as those by Snedecor and Cochran,[2] Armitage,[3] Dixon and Massey,[4] Bliss,[5] Sokal and Rohlf,[6] Winer,[7] Kleinbaum and Kupper,[8] and Brownlee,[9] provide details of the calculations.

METHODS

The 332 Original Articles published in Volumes 298 to 301 (1978 and 1979) of the *Journal* provided the examples used in this chapter. In these four volumes, 50 Original Articles[10-59] included a comparison of several group means. This list of articles was checked against a similar list prepared for Emerson and Colditz.[1] I have not included in this total articles comparing observed proportions and have considered only articles dealing with means of measurements. Table 1 provides information on the analyses reported in each of these articles. These articles deal almost exclusively with testing.

WHY NOT t-TESTS?

Investigators comparing three or more group means frequently examine each possible pair of groups separately, using the t-test to examine each pair. For example, Toft et al.[10] examined estimated thyroid-remnant weight in patients in four postoperative states: euthyroidism with normal serum thyrotropin levels; euthyroidism with raised serum thyrotropin levels; temporary hypothyroidism; and permanent hypothyroidism. Examining each pair of groups would involve six tests: Groups 1 versus 2, 1 versus 3, 1 versus 4, 2 versus 3, 2 versus 4, and 3 versus 4. In general, any experiment with k groups has $k(k-1)/2$ different pairs available for testing.

Suppose that all the groups share the same underlying population mean. We pick two groups — for instance, Groups 1 and 2 — and test whether their group means are equal. In this kind of problem, when we are testing the equality of group means, the standard assumption that the means are equal is called the null hypothesis. If the outcome of the test falls beyond a certain value, called the critical value, the test result is considered statistically significant, and we reject the null hypothesis that the two means are equal. We choose the critical value so that when the null hypothesis is true (that is, when the group means are equal), there is only a small probability (such as 5 percent) that the test result will be as large as the critical value. This is the probability that we will reject the null hypothesis when it is true and reach the false conclusion that the means are unequal, when they are in fact equal. This probability is called the α-level of the test. By changing the critical value, we change this probability; conversely, a test using a different α-level will have a different critical value. The value 5 percent is a common choice for the α-level, but 10 percent, 1 percent, or other values could be selected. All the examples in this chapter will use the critical value 5 percent.

In testing by pairs, problems arise because of the number of tests made and because of lack of independence. We first discuss the situation when we do have independence from test to test. We can calculate the probability of obtaining statistically significant test results for two independent tests simultaneously by multiplying the probabilities that each test individually will have a significant result. If each test result has the probability 0.05 of being significant under the null hypothesis, then the probability that both test results will be significant is $0.05 \times 0.05 = 0.0025$. The probability that

Table 1. Articles Comparing Group Means in Volumes 298 through 301 of the *New England Journal of Medicine.*

Reference No.	Variable Studied	Basis for Grouping	No. of Groups in Study	Multiple-Comparison Analysis?	Analysis of Variance?	Figure?	Comments
10	Thyroid-remnant weight in thyrotoxic patients	Postoperative status after subtotal thyroidectomy	4	no	no	no	1
11	Sodium excretion in urine specimens	Diet and drug treatments for hypertension	3	no	yes	no	2
12	Platelet monoamine oxidase activity	Chronic or paranoid schizophrenia; controls	5	no	no	no	1
13	Plasma creatinine levels	Normal or phosphate-restricted diet schemes	4	yes	no	no	3, 4
14	Completeness of history taking	Type of hospital staff (nurse, pediatrician, house officer)	3	no	yes	no	
15	Chlordecone concentration in body tissues	Cholestyramine or control	2	no	yes	no	5
16	Circulating immune-complex levels	Cellular and fibrotic interstitial lung disease; controls	4	no	no	yes	1
17	Age at diagnosis of pernicious anemia	Ethnic group; sex	6	no	no	yes	1

18	Lymphocyte blast transformation to antigens	Kidney disease (glomerulonephritis, nonglomerular renal disease); controls	3	no	no	yes	6
19	Serum levels of total protein and complement components	Shigellosis type (hemolytic, hemolytic uremic, leukemoid); uncomplicated dysentery; controls	5	no	no	yes	6
20	Motor-nerve conduction velocity	Uremia type (diabetic, nondiabetic) by serum parathyroid hormone levels	4	no	no	yes	
21	Lung-function test measurements	Total pathology score for lung disease	4	yes	no	yes	7
22	Antibody titers against liver-specific membrane lipoprotein	Liver disease (chronic and acute hepatitis, hemochromatosis, hepatic necrosis)	6	no	no	yes	6
23	Heart function (heart rate, blood pressure)	Passive smoking exposure; controls	3	no	no	no	8
24	Months between ischemic event and entry into trial	Stroke therapy (aspirin, sulfinpyrazone, both, neither)	4	?	?	no	9
25	Breast-cancer tumor diameter	Frequency of breast self-examination	4	yes	no	yes	10, 11

(Table continues)

Table 1. (Continued)

Reference No.	Variable Studied	Basis for Grouping	No. of Groups in Study	Multiple-Comparison Analysis?	Analysis of Variance?	Figure?	Comments
26	Response of leukocyte-inhibitory factor to collagens	Arthritis type (rheumatoid, osteo-, inflammatory); controls	4	no	no	yes	1
27	Duration of survival of second kidney graft	Duration of survival of first kidney graft	4	no	no	no	1
28	Plasma glucose levels	Insulin-infusion protocol	3	no	no	no	1
29	Serum levels of calcium, phosphate, etc.	Time from infusion of mithramycin	6	no	no	yes	12
30	Antibody reactions to pneumococcal polysaccharides	Treatment for Hodgkin's disease (chemotherapy, radiotherapy, both); controls	6	yes	no	no	3, 4
31	Plasma and red-cell electrolyte levels	Congestive heart failure; controls	3	no	no	no	1
32	Binding ratios (triiodothyronine, thyroxine, thyrotropin)	Liver disease (cirrhosis, hepatitis)	4	no	no	no	13
33	Total circulating lymphocytes	Amyotrophic lateral sclerosis; parkinsonism-dementia; controls	4	no	no	no	
34	Kidney-graft survival rate	Number of blood transfusions	6	no	no	yes	1, 11

35	Plasma growth hormone levels	Treatment and disease status (controls, isolated growth hormone deficiency before and after treatment)	3	no	no	no	14
36	Red-cell uroporphy-rinogen activity	Porphyria type (cutanea tarda, variegate, others)	6	no	no	yes	1, 6
37	Type of heterosexual or autosexual activity	Contraceptive method and segment of menstrual cycle	3	no	yes	table instead	15, 16
38	Serum lipoprotein levels	Controls; severity of hypoproteinemia	5	no	no	no	1
39	Serum cortisol concentrations	Asthma treatment groups (prednisone, beclomethasone, others); controls	5	no	no	yes	1
40	Adult height	Age at treatment for congenital virilizing adrenal hyperplasia	4	no	no	yes	1
41	Apolipoprotein levels	Kin phenotype of relatives of patient with apolipoprotein C-II deficiency	4	no	no	no	1, 11

(Table continues)

Table 1. (Continued)

Reference No.	Variable Studied	Basis for Grouping	No. of Groups in Study	Multiple-Comparison Analysis?	Analysis of Variance?	Figure?	Comments
42	F-reticulocyte levels	Repeated sample of F-reticulocyte levels in patients with sickle-cell anemia	15	no	yes	no	17
43	Glycoprotein and albumin levels	Disease groups (cirrhosis, arthritis, Crohn's disease, etc.); controls	6	no	no	no	18
44	Segment length; myocardial thickness	Repeated echocardiographic images	2	no	yes	no	17
45	Plasma glucose levels	Treatment stage for juvenile diabetes	3	no	no	no	11
46	Phenylephrine concentration	Asthma; rhinitis; controls	3	no	no	no	11
47	Systolic blood pressure and heart rate	Beta-blocking drugs (propranolol, etc.); controls	6	no	yes	yes	13
48	Visual threshold of dark-adapted eyes	Pancreatitis with or without steatorrhea; controls	3	?	?	yes	9

49	Age of high-cost patients	Hospital type (community, county, etc.)	4	no	yes	no	19
50	Age/length of stay in hospital	Type of bleeding lesion (with or without visible vessel, etc.)	3	no	yes	no	
51	Blood pressure; serum cholesterol levels; etc.	Hypertension therapy (ticrynafen, hydro-chlorothiazide, etc.)	4	yes	yes	no	3, 4
52	Age of adult lymphoma patients	Histoimmunologic classes (B, T, null)	4	no	yes	no	4
	Age of adult lymphoma patients	Immunologic classes (B, T, null)	3	no	yes	no	
	Age of adult lymphoma patients	Histologic classes (size of lymphoma, etc.)	6	no	yes	no	
53	Blood pressure	Extent of involvement of arteries in coronary artery disease	4	yes	yes	no	11, 20
54	Protease activity	Fibrosis, sarcoidosis; controls	3	no	no	yes	1, 6
55	Age; blood chemistry	Anticoagulant therapy (aspirin and dipyrid-amole, warfarin); controls	3	?	?	no	9
56	Plasma insulin levels	Repeated measures of insulin levels in normal subjects	10	no	yes	no	17

(Table continues)

Table 1. (Continued)

Reference No.	Variable Studied	Basis for Grouping	No. of Groups in Study	Multiple-Comparison Analysis?	Analysis of Variance?	Figure?	Comments
57	Change in ejection fraction	Number of transfusions in anemic patients; controls	3	no	no	yes	1, 11
58	Heart rate; blood pressure	Time after administration of nitroprusside	4	yes	yes	yes	3, 11
59	Psychological-test scores of drug addicts (IQ, personality tests)	Type of drug abused (opiate, psycho-stimulant, psycho-depressant)	3	no	yes	no	15, 17

1. Multiple t-tests were the only method of analysis performed to compare groups.
2. Student's t-tests were performed as a supplementary analysis.
3. Multiple comparisons were made by Duncan's method.
4. Two-way analysis of variance could have been performed.
5. Analysis of variance with two groups is equivalent to a t-test.
6. Results of the analysis were obvious in the accompanying figure.
7. Multiple comparisons were made by Dunnett's method.
8. Crossover design was used.
9. It was unclear what method was used to compare groups.
10. Multiple comparisons were made by Scheffé's method.

11. There was an apparent trend in the grouping variable.
12. Time analysis would be appropriate for this study.
13. Analysis of covariance could be performed for treated groups.
14. There was correlation between the disease and treatment groups.
15. Two-way analysis of variance was performed.
16. Three-way analysis of variance could be performed.
17. Repeated-measures analysis of variance was performed.
18. Nonparametric analysis of variance would be useful.
19. A specific contrast was examined separately from the analysis of variance.
20. Multiple comparisons were made by the Newman–Keuls method.

neither test result will be significant is $0.95 \times 0.95 = 0.9025$. The probability that at least one of the two test results will be significant is $1 - 0.9025$, or 0.0975. Thus, the probability of incorrectly deciding that the members of either one or both pairs of means are unequal using two tests is nearly twice the probability of making the same error for a single test (0.0975 vs. 0.05). If we add a third group, the probability that none of the three tests will be significant is $0.95 \times 0.95 \times 0.95 = 0.8574$, so the probability that at least one test will be significant is about 14 percent or nearly three times 5 percent. In general, if we make n tests, the probability of finding at least one spuriously significant independent result can be calculated as follows:

$$\text{Probability of at least one spurious test result} = 1 - (1 - \alpha)^n.$$

As n increases, this probability becomes much larger than 0.05, our original α-level for a single test; indeed for n, it is approximately $0.05n$. As Table 2 shows, the probability of at least one false-positive result grows quickly with the number of independent tests.

The formula above assumes that the tests are statistically independent. If this were the case, there would be no relation between any of the differences between pairs of group means, but obviously the difference between the means of Groups 1 and 2 is related to the difference between the means of Groups 2 and 3, and to the difference between the means of Groups 1 and 4, and so on. Because the formula makes the false assumption of independence, it is a conservative one, and so it may overestimate the probability of a false-positive result. If we assume that the group means are normally distributed, we can calculate the probability that the difference between the largest and smallest of the group means in an experiment of k groups would be large enough to be judged statistically significant if those two groups were the only ones in the study, without making the assumption of independence. This probability is the same as the probability of finding at least one significant result. These probabilities are smaller than those for the independent case, but they still increase rapidly with the number of groups. The corresponding probabilities for both the independent and the normal, nonindependent situations are shown in Table 2.

Of the 50 articles comparing group means, more than half (27) used only multiple t-tests to make comparisons among groups. As we have just seen, however, multiple t-tests introduce a large and misleading increase in the probability of finding at least one significant test result when no real population differences exist. In comparing five groups, if one of the 10 t-tests turns out to be significant, we wonder whether it means that the null hypothesis is not true after all or whether this is a spurious positive test result, as will occur in about 30 percent of such experiments (Table 2). When we have 10 groups, even when the population means of all groups are equal, we expect the 45 comparisons to generate on the average $45 \times 0.05 = 2.25$ results that are significant at the 5 percent level. Thus, when we find two or three significant results, we are not impressed. Although the technique of using many t-tests appears simple, the results are hard to interpret.

Table 2. Probability of Finding at Least One Comparison Significant at the 0.05 Level for Various Numbers of Independent Tests and for Testing All Possible Pairs of Group Means for 2 to 7 Groups.

No. of Tests	Tests Independent	Tests Nonindependent (Normal Distribution)
1	0.05	0.05 (2 groups)
3	0.14	0.11 (3 groups)
6	0.26	0.21 (4 groups)
10	0.40	0.30 (5 groups)
15	0.54	0.39 (6 groups)
21	0.64	0.47 (7 groups)

One method for decreasing the risk of an incorrect test result adjusts the testing so that the simultaneous risk of finding one or more spurious significant results in all the t-tests combined is the chosen α-level — for instance, 0.05. That is, we alter the nominal α-level for individual comparisons so that the probability of finding at least one significant test result for the entire experiment (looking at all the t-tests simultaneously) is 0.05 when all the underlying group means are equal. This approach controls the chance of error per experiment. A common way to approximate this method is to divide the desired simultaneous α-level for the experiment as a whole by the number of tests being made; the result is a new α-level to be used for each of the individual t-tests. This is called the Bonferroni method. For an experiment with five groups, there are 10 tests of pairs, so with a simultaneous α-level of 0.05, the Bonferroni method leads us to conduct each of the 10 individual tests with an α-level of $0.05/10 = 0.005$. In a follow-up study of usage, versions of the Bonferroni method appeared twice in Volume 321 of the *New England Journal of Medicine* (July through December, 1989).

The Bonferroni method gives an approximation when applied to either independent tests or to simultaneous comparisons of many pairs. We might prefer to arrange the calculations to get a pre-set α-level for the whole investigation. For independent trials the approximation is rather good. If we want $\alpha = 0.05$ overall for 10 independent trials, then the α-level to use for each independent trial is 0.00512 instead of the approximate 0.005 (which leads to an overall 0.0489 instead of 0.05). We discuss above and show results in Table 2 for trials that are not independent, and we discuss below other approaches when trials are not independent.

ANALYSIS OF VARIANCE

An analysis of variance indicates whether there are differences among the population means of the groups being compared, but it does not pinpoint which groups, if any, differ from the others. For example, Perrin and Goodman[14] compared the way in which three groups of medical personnel (house officers, practicing pediatricians, and nurse practitioners) handled telephone calls from the parents of acutely ill children. Each worker received a score for completeness of telephone interviewing based on how much information he or she collected over the phone. The authors tested for differences among the three groups by using an analysis of variance, which indicated that differences in interviewing ability did exist.

Analysis of variance generalizes the t-test from two groups to three or more groups. It replaces multiple t-tests with a single F test of the assumption that the underlying group population means are all equal. Because only one test is performed, the multiple-comparison problem does not arise, as it does with several t-tests. The logic of the F test is as follows: if all the groups have a common population mean m, then the observed group means should all lie near m, and we are not likely to find a significant difference. If they are sufficiently dispersed, then the F statistic is likely significant, and we conclude that at least one of the population means for the groups differs from the others. By itself, however, analysis of variance will not tell us which groups differ from which others. That question can be treated with a multiple-comparison analysis.

In some cases, analysis of variance is a very useful technique. When comparing several groups to evaluate treatment effects, for example, investigators want to make sure that the groups are similar in "covariates," characteristics other than treatment that may affect the outcome. For instance, age and sex often affect the way patients react to treatment, and investigators should assure themselves that the groups being compared are similar in age and of the same sex when these factors may influence the outcome. Analysis of variance is well suited for testing the usual assumption that a number of treatment groups have comparable population means for variables other than the main variable of interest. For instance, in the study of medical personnel described above,[14] we may wish to see whether the three groups had comparable mean years of experience. Analysis of variance provides a simple way to test for differences of this sort.

Analysis of variance offers a standard method for comparing various groups when there is no presumption beforehand that they differ. We can use this method to compare drug therapies or disease groups that we consider to be comparable or "about the same." For example, Griffiths et al.[50] performed an analysis of variance to test whether the mean ages of patients in three endoscopic categories were similar. When this test proved nonsignificant, they went on to make another analysis of variance to look for differences among the three groups in transfusion requirements, without adjusting for age.

The original investigators used analysis of variance to compare several group means in about 30 percent (14 of 50) of the *Journal* articles. In the 1989 follow-up study, at least 11 articles in Volume 321 of the *New England Journal of Medicine* used one- or two-way analysis of variance.

MULTIPLE COMPARISONS

Before the study begins, an investigator may well have an idea what the results will be or which questions are most important. For example, when comparing the efficacy of several treatments with that of a placebo, he or she may expect that the active treatments will have similar effects but that each will be more effective than the placebo. An example is an experiment by Thadani et al.,[47] who compared the performance of five beta-adrenergic–blocking drugs in the treatment of angina with that of a placebo. The authors found that the responses of the patients given the placebo seemed quite different from those of each of the other treatment groups, but that the drug-treated groups were all similar.

What is an appropriate analysis for such an experiment, with one placebo group and several treated groups? In the context of analysis of variance, we ask a single question of the data: Do the groups have equal means? An ordinary analysis of variance simultaneously compares all six groups and, in this instance, confirms differences among the groups, but it does not tell us which groups differ from which others. If we want more information than analysis of variance can provide, we need to ask more questions. Multiple-comparison methods allow us to do this and still control the overall α-level. We can find all the pairwise differences among the group means and test them individually for significance, or we can separate the group means into clusters of like means, so that all the means in one cluster differ significantly from all the means in any other cluster. Multiple-comparison methods can also test certain weighted sums of the group means; using this technique, we multiply each group mean by its own constant (a weight) and then add these products together. A difference between two group means is only one example of such a weighted sum; the weights for such a difference are 1 and -1 for the two group means in the difference and 0 for each of the other means. More complicated examples are described below.

MULTIPLE-COMPARISON TECHNIQUES

The Bonferroni method, described above, adjusts t-tests to make each test more stringent, but it is conservative and may have low statistical power. Methods that are less conservative have been developed for making multiple comparisons among pairs of group means. These methods attempt to keep the overall α-level of the experiment at the intended level (for instance, 0.05) while making it less likely that a true difference between two groups will be missed. Several of these methods are discussed briefly below, in order of decreasing conservatism: those of Scheffé, Tukey, Newman–Keuls, and Duncan. Both the SAS[60] and SPSS[61] computing packages provide calculations for each of these methods. Miller[62] discusses methods for making multiple comparisons in detail. Each method assumes that the data are normally distributed and that the true (but unknown) variance within each group is the same. Games et al.[63] survey developments in multiple-comparison techniques. Of the 50 *Journal* articles cited in Table 1, 7 used a multiple-comparison technique. In the 1989 follow-up study of Volume 321 carried out by the editors, 4 articles used multiple comparison techniques in addition to 2 using versions of Bonferroni. The Scheffé, Newman–Keuls (twice), and Duncan methods were used.

Scheffé's test allows the investigator to examine the data and then to choose one or more weighted combinations of means to test, without bias to the results. To do this, the test must allow for many different weighted combinations, because investigators may be imaginative in choosing a weighted sum. The weights must add up to zero, so that the value of the sum will be zero when all the group means are equal. Such sums are called contrasts. For example, in a study of motor ability in children in grades one, two, and three, we might ask whether the second-graders' performance differed from the average performance of the children in the first and third grades. In the usual notation, we are asking about

$$\bar{x}_2 - \frac{1}{2}(\bar{x}_1 + \bar{x}_3).$$

The weights for this contrast are $-\frac{1}{2}$ for $\bar{x}_1$, 1 for $\bar{x}_2$, and $-\frac{1}{2}$ for $\bar{x}_3$. To be able to look at any contrast, the method must allow for the variability of all possible contrasts, even one, such as $0.2\bar{x}_1 + 0.7\bar{x}_2 - 0.9\bar{x}_3$, that few investigators are likely to consider. One of the 50 *Journal* articles in my survey used Scheffé's method for multiple comparisons.[25]

Tukey's method exemplifies those that test for differences among group means by using the difference between the largest and smallest means, often called the range, as a measure of their dispersion. This difference replaces the usual variance calculated from the sum of squares. Instead of the t distribution, which is based on the sum of squares, these methods employ the q distribution, or Studentized range, which is based on the range. Tables of the critical points for the Studentized range are available in many textbooks, including that by Snedecor and Cochran.[2] Like the F distribution, the q distribution depends on the number of groups.

Tukey's method is the most conservative of those based on the q distribution. It uses for a multiplier of the standard deviation the q statistic based on the α-level and the total number of groups. Like Scheffé's method, Tukey's method allows the investigator to look at all possible contrasts; however, it requires that all the groups be the same size. The confidence intervals for Tukey's method are narrower than those for Scheffé's method for simple contrasts such as differences between two group means; thus, Tukey's method is less conservative in comparing group means. For more complicated contrasts involving several groups, Tukey's method may give wider confidence intervals than Scheffé's. Tukey's method is useful for testing all possible pairs of group means when the investigator does not wish to examine any contrasts other than differences in group means. None of the articles I examined used Tukey's method.

The Newman–Keuls and Duncan multiple-range tests create clusters of like group means, which may overlap. The analyst begins by comparing the largest and the smallest group means. If this test of the interval spanning all k groups is not significant, the analysis is complete and it may be concluded that there are no significant differences among any of the group means. Otherwise, the analyst goes on to examine the two intervals spanning $k - 1$ group means, then the intervals spanning $k - 2$ group means, and so forth. The process continues until all pairs of means have been tested. In an experiment with five groups, where we number the means 1 through 5 in order of increasing size, the first test compares mean 1 and mean 5. If this test is significant, means 2 and 5 and means 1 and 4 are compared. If any two means are declared similar, all the means that lie between them are also similar. Thus, if means 2 and 5 are similar, we need only compare mean 1 with the others to complete our multiple-range test.

The Newman–Keuls and Duncan methods differ in the way in which the appropriate q statistic is chosen for each test; this difference makes Duncan's test much less conservative than the Newman–Keuls test. One of the 50 articles cited in Table 1 used the Newman–Keuls method.[53] Duncan's test, the most popular of the multiple-comparison methods, was included in 4 of the 50 articles (8 percent).[13,30,51,58]

Although multiple-comparison methods usually consider all possible pairs of differences among the groups, they are well suited to examining a few selected differences. In the experiment reported by Thadani et al.,[47] we may be interested only in the 5 differences between each drug and the placebo and not in all 15 possible differences among the six study groups. We can use multiple-comparison techniques to test these preselected differences without calculating the others. For example, we can use Bonferroni-adjusted *t*-tests, taking as our divisor for the α-level the number of tests we make — five. When we test only a few differences, the Bonferroni method works very well. Note that we are testing for differences chosen for study before the experiment begins, not those suggested by the collected data. The latter case is discussed below.

QUESTIONS SUGGESTED BY THE DATA

Ideally, investigators should decide which statistical tests they will perform, including which groups to compare, before they examine the data in even a cursory fashion. In practice, however, the data in hand may suggest comparisons that were not originally planned. If groups that the investigators expected to be similar have quite different means, for instance, there may be reasons to test this apparent difference. Investigators who are careful in choosing multiple-comparison methods can make this new comparison without changing the overall α-level for the experiment.

If we choose to test only the significance of the larger observed mean differences in an experiment, the probability of finding apparently significant differences will be greater than if we test pairs of means chosen at random before the experiment began. In such a case, the overall α-level for the experiment will actually be higher than we determined beforehand. The more conservative multiple-comparison methods allow investigators to select which group means to test after the experiment begins, by allowing for all possible comparisons. If group means are chosen for comparison testing on the basis of their apparent differences after the data have been collected, it is important to use these conservative methods. Scheffé's method, in particular, was designed to permit this sort of data dredging.

In summary, a more conservative method gives broader confidence intervals, is less likely to report a difference between means when none exists, and is more likely to miss a real difference. Thus, it has lower statistical power than a less conservative method. A less conservative method, in turn, has a better chance of detecting a small difference but also has a greater chance of reporting a difference that is not real. Choosing a multiple-comparison method requires the analyst to balance these two kinds of errors.

TWO-WAY ANALYSIS OF VARIANCE

A one-way analysis of variance assumes that the groups are at the same level in some hierarchy. In other words, the groups should be distinct, comparable, and of equal stature. This is not always true for the groups we wish to compare, however. There may be several different levels within a group. For instance, three different drug treatments, A, B, and C, administered at dosages A1, B1, and C1, respectively, may be taken as being at the same level, but an experiment that includes three additional treatments — drug A at dosage A2, drug B at dosage B2, and drug C at dosage C2 — is more complicated in that it includes two different dosages of each of the drugs in the design. A one-way analysis of variance will not allow us to test the effect of different dosages of the same drug; it allows us only to compare all six groups at once. A multiple-comparison method may help answer questions about how the effect of a single drug varies with dosage, but the comparability of these six treatment groups remains unclear. A better method is an analysis of variance that permits the dose level to be included as a variable in the analysis. This method allows for one F test of the effect of differing dosage in all the groups simultaneously, as well as a separate F test comparing the three drugs. An analysis of variance that compares group means in this way, across two separate categorizing variables, is called a two-way analysis of variance.

In a study by the Veterans Administration Cooperative Study Group on Antihypertensive Agents,[51] each of two antihypertensive drugs (ticrynafen and hydrochlorothiazide) was administered at two different dosages. The authors conducted a one-way analysis of variance comparing all four treatment groups. A two-way analysis of variance would have allowed the authors to look at the effects of dosage as well as of drug.

Carmel and Johnson[17] considered three groups of patients with anemia (European, black, and Latin American) and also examined the effect of sex. The authors used more than a dozen t-tests to try to get at the effects of ethnic group and sex. A two-way analysis of variance would have allowed them to estimate the effects of sex and ethnic group separately, looking at all the data at once, as well as to measure the joint effect of sex and ethnic group over and above their separate effects.

Two of the 50 articles in my survey used a two-way analysis of variance. McLellan et al.[59] analyzed patients in three drug-abuse groups by means of psychological testing and retesting. The two-way analysis permitted the investigators to use scores on the test and the retest simultaneously in comparing the three groups. It also compared the test and retest scores themselves. Adams et al.[37] studied levels of female sexual activity according to contraceptive method and segment of the menstrual cycle (first or second half). This analysis could have been extended to a three-way analysis of variance if the authors had included another variable analyzed elsewhere in the article, type of sexual activity. This approach would have allowed them to examine the effects of all three variables simultaneously.

OTHER METHODS

Depending on what questions researchers wish to answer, they can design their experiments in different ways. In the thyroid experiment described earlier,[10] the groups might have fallen into two, three, or four separate categories, depending on the way the researchers viewed the data. Each group could have been considered separately, giving four categories. Another possible design would be two categories — euthyroidism and hypothyroidism — each with two subcategories. There could also be three categories — euthyroidism with two subcategories, temporary hypothyroidism, and permanent hypothyroidism. Each different design calls for its own analysis. We have discussed only three types of analysis-of-variance designs: one-way, two-way, and three-way. Brownlee[9] and Winer[7] discuss other analysis-of-variance designs and give examples of the calculations involved.

POSSIBLE DIFFICULTIES IN COMPARING GROUP MEANS

Both the multiple-comparison methods and analysis of variance assume equality of variances in the different groups. This assumption allows for a pooled estimate of the common group variance, based on the variance of each group, that is more precise than any of the individual group variances. Although the required F test is not greatly affected by small differences in group variances, the data should be checked for large differences in group variances before the means are compared. Several tests have been devised for this purpose, including Hartley's F-max, a test based on the ratio of the largest group variance to the smallest group variance. Details are available in textbooks such as Snedecor and Cochran's.[2] The BMDP computing package[64] provides another such method, Levene's test, also described by Snedecor and Cochran.

The variances of groups compared in the *Journal* articles I surveyed often differed substantially, raising questions about the accuracy of the analysis. If one group variance is much larger than the others, it will increase the estimate of the pooled variance, making it more difficult to detect differences among the groups with small variances. Sometimes converting, or transforming, the measurements to another scale can make the variances more nearly equal. For example, if the variance of the group increases as the mean of the group increases, taking logarithms or square roots of the original data may make the variances more nearly equal in the transformed scale. The analysis then proceeds in the new scale. Adams et al.[37] used the *F-max* test to compare the variances of their study groups. The disparity in variances led them to take square roots of the data before performing an analysis; this technique corrected the disparity.

One possible drawback to the use of such transformations is that all the results of the analysis refer to the transformed data, not the original data. Sometimes the units of the transformed data cannot be interpreted clearly. For example, taking reciprocals of data on the time required to complete some action gives the speed of completion, but the square roots of such data have a less obvious meaning.

When we have evidence that the assumptions of normality and equal variance of analysis-of-variance or multiple-comparison methods do not hold, we may want to use the nonparametric techniques available to test for differences among means. Nonparametric methods do not require the usual assumption that the data are normally distributed and so are appropriate when it is likely that the data do not meet that assumption. One such method is the Kruskal–Wallis test, available on SAS, SPSS, and BMDP. Sokal and Rohlf[6] give an example.

No matter how the data are analyzed, it is almost always useful to provide a plot of the results. Often, a plot speaks so clearly that the message is obvious regardless of the method of analysis. Differences among groups are instantly apparent, and outlying values are easily recognized.

REFERENCES

1. Emerson JD, Colditz GA. Use of statistical analysis in the *New England Journal of Medicine*. N Engl J Med 1983; 309:709-13. [Chapter 3 of this book.]

2. Snedecor GW, Cochran WG. Statistical methods. 8th ed. Ames, Iowa: Iowa State University Press, 1989.

3. Armitage P. Statistical methods in medical research. New York: John Wiley, 1971.

4. Dixon WJ, Massey FJ. Introduction to statistical analysis. 4th ed. New York: McGraw-Hill, 1983.

5. Bliss CI. Statistics in biology: statistical methods for research in the natural sciences. Vol. 1. New York: McGraw-Hill, 1967.

6. Sokal RR, Rohlf FJ. Biometry: the principles and practice of statistics in biological research. 2nd ed. San Francisco: WH Freeman, 1981.

7. Winer BJ. Statistical principles in experimental design. 2nd ed. New York: McGraw-Hill, 1971.

8. Kleinbaum DG, Kupper LL, Muller KE. Applied regression analysis and other multivariable methods. 2nd ed. Boston: PWS-Kent, 1988.

9. Brownlee KA. Statistical theory and methodology in science and engineering. 2nd ed. Melbourne, Fla.: Krieger, 1984.

10. Toft AD, Irvine WJ, Sinclair I, McIntosh D, Seth J, Cameron EHD. Thyroid function after surgical treatment of thyrotoxicosis: a report of 100 cases treated with propranolol before operation. N Engl J Med 1978; 298:643-7.

11. Reisin E, Abel R, Modan M, Silverberg DS, Haskel HE, Modan B. Effect of weight loss without salt restriction on the reduction of blood pressure in overweight hypertensive patients. N Engl J Med 1978; 298:1-6.

12. Potkin SG, Cannon HE, Murphy DL, Wyatt RJ. Are paranoid schizophrenics biologically different from other schizophrenics? N Engl J Med 1978; 298:61-6.

13. Ibel LS, Alfrey AC, Haut L, Huffer WE. Preservation of function in experimental renal disease by dietary restriction of phosphate. N Engl J Med 1978; 298:122-6.

14. Perrin EC, Goodman HC. Telephone management of acute pediatric illnesses. N Engl J Med 1978; 298:130-5.

15. Cohn WJ, Boylan JJ, Blanke RV, Fariss MW, Howell JR, Guzelian PS. Treatment of chlordecone (kepone) toxicity with cholestyramine: results of a controlled clinical trial. N Engl J Med 1978; 298:243-8.

16. Dreisin RB, Schwartz MI, Theofilopoulos AN, Stanford RE. Circulating immune complexes in the idiopathic interstitial pneumonias. N Engl J Med 1978; 298:353-7.

17. Carmel R, Johnson CS. Racial patterns in pernicious anemia: early age at onset and increased frequency of intrinsic-factor antibody in black women. N Engl J Med 1978; 298:647-50.

18. Fillit HM, Read SE, Sherman RL, Zabriskie JB, van de Rijn I. Cellular reactivity to altered glomerular basement membrane in glomerulonephritis. N Engl J Med 1978; 298:861-8.

19. Koster F, Levin J, Walker L, et al. Hemolytic-uremic syndrome after shigellosis: relation to endotoxemia and circulating immune complexes. N Engl J Med 1978; 298:927-33.

20. Avram MM, Feinfeld DA, Huatuco AH. Search for the uremic toxin: decreased motor-nerve conduction velocity and elevated parathyroid hormone in uremia. N Engl J Med 1978; 298:1000-3.

21. Cosio M, Ghezzo H, Hogg JC, et al. The relations between structural changes in small airways and pulmonary-function tests. N Engl J Med 1978; 298:1277-81.

22. Jensen DM, McFarlane IG, Portmann BS, Eddleston ALWF, Williams R. Detection of antibodies directed against a liver-specific membrane lipoprotein in patients with acute and chronic active hepatitis. N Engl J Med 1978; 299:1-7.

23. Aronow WS. Effect of passive smoking on angina pectoris. N Engl J Med 1978; 299:21-4.

24. Canadian Cooperative Study Group. A randomized trial of aspirin and sulfinpyrazone in threatened stroke. N Engl J Med 1978; 299:53-9.

25. Foster RS Jr, Lang SP, Costanza MC, Worden JK, Haines CR, Yates JW. Breast self-examination practices and breast-cancer stage. N Engl J Med 1978; 299:265-70.

26. Trentham DE, Dynesius RA, Rocklin RE, David JR. Cellular sensitivity to collagen in rheumatoid arthritis. N Engl J Med 1978; 299:327-32.

27. Opelz G, Terasaki PI. Absence of immunization effect in human-kidney retransplantation. N Engl J Med 1978; 299:369-74.

28. Raskin P, Unger RH. Hyperglucagonemia and its suppression: importance in the metabolic control of diabetes. N Engl J Med 1978; 299:433-6.

29. Bilezikian JP, Canfield RE, Jacobs TP, et al. Response of $1\alpha,25$-dihydroxyvitamin D_3 to hypocalcemia in human subjects. N Engl J Med 1978; 299:437-41.

30. Siber GR, Weitzman SA, Aisenberg AC, Weinstein HJ, Schiffman G. Impaired antibody response to pneumococcal vaccine after treatment for Hodgkin's disease. N Engl J Med 1978; 299:442-8.

31. Loes MW, Singh S, Locke JE, Mirkin BL. Relation between plasma and red-cell electrolyte concentrations and digoxin levels in children. N Engl J Med 1978; 299:501-4.

32. Schussler GC, Schaffner F, Korn F. Increased serum thyroid hormone binding and decreased free hormone in chronic active liver disease. N Engl J Med 1978; 299:510-5.

33. Hoffman PM, Robbins DS, Nolte MT, Gibbs CJ Jr, Gajdusek DC. Cellular immunity in Guamanians with amyotrophic lateral sclerosis and parkinsonism-dementia. N Engl J Med 1978; 299:680-5.

34. Opelz G, Terasaki PI. Improvement of kidney-graft survival with increased numbers of blood transfusions. N Engl J Med 1978; 299:799-803.

35. Soman V, Tamborlane W, DeFronzo R, Genel M, Felig P. Insulin binding and insulin sensitivity in isolated growth hormone deficiency. N Engl J Med 1978; 299:1025-30.

36. Felsher BF, Norris ME, Shih JC. Red-cell uroporphyrinogen decarboxylase activity in porphyria cutanea tarda and other forms of porphyria. N Engl J Med 1978; 299:1095-8.

37. Adams DB, Gold AR, Burt AD. Rise in female-initiated sexual activity at ovulation and its suppression by oral contraceptives. N Engl J Med 1978; 299:1145-50.

38. Rapoport J, Aviram M, Chaimovitz C, Brooks JG. Defective high-density lipoprotein composition in patients on chronic hemodialysis: a possible mechanism for accelerated atherosclerosis. N Engl J Med 1978; 299:1326-9.

39. Wyatt R, Waschek J, Weinberger M, Sherman B. Effects of inhaled beclomethasone dipropionate and alternate-day prednisone on pituitary-adrenal function in children with chronic asthma. N Engl J Med 1978; 299:1387-92.

40. Urban MD, Lee PA, Migeon CJ. Adult height and fertility in men with congenital virilizing adrenal hyperplasia. N Engl J Med 1978; 299:1392-6.

41. Cox DW, Breckenridge WC, Little JA. Inheritance of apolipoprotein C-II deficiency with hypertriglyceridemia and pancreatitis. N Engl J Med 1978; 299:1421-4.

42. Dover GJ, Boyer SH, Charache S, Heintzelman K. Individual variation in the production and survival of F cells in sickle-cell disease. N Engl J Med 1978; 299:1428-35.

43. Piafsky KM, Borgå O, Odar-Cederlöf I, Johansson C, Sjöqvist R. Increased plasma protein binding of propranolol and chlorpromazine mediated by disease-induced elevations of plasma α_1 acid glycoprotein. N Engl J Med 1978; 299:1435-9.

44. Eaton LW, Weiss JL, Bulkley BH, Garrison JB, Weisfeldt ML. Regional cardiac dilatation after acute myocardial infarction: recognition by two-dimensional echocardiography. N Engl J Med 1979; 300:57-62.

45. Tamborlane WV, Sherwin RS, Genel M, Felig P. Reduction to normal of plasma glucose in juvenile diabetes by subcutaneous administration of insulin with a portable infusion pump. N Engl J Med 1979; 300:573-8.

46. Henderson WR, Shelhamer JH, Reingold DB, Smith LJ, Evans R III, Kaliner M. Alpha-adrenergic hyper-responsiveness in asthma: analysis of vascular and pupillary response. N Engl J Med 1979; 300:642-7.

47. Thadani U, Davidson C, Singleton W, Taylor SH. Comparison of the immediate effects of five β-adrenoreceptor-blocking drugs with different ancillary properties in angina pectoris. N Engl J Med 1979; 300:750-5.

48. Toskes PP, Dawson W, Curington C, Levy NS, Fitzgerald C. Non-diabetic retinal abnormalities in chronic pancreatitis. N Engl J Med 1979; 300:942-6.

49. Schroeder SA, Showstack JA, Roberts HE. Frequency and clinical description of high-cost patients in 17 acute-care hospitals. N Engl J Med 1979; 300:1306-9.

50. Griffiths WJ, Neumann DA, Welsh JD. The visible vessel as an indicator of uncontrolled or recurrent gastrointestinal hemorrhage. N Engl J Med 1979; 300:1411-3.

51. Veterans Administration Cooperative Study Group on Antihypertensive Agents. Comparative effects of ticrynafen and hydrochlorothiazide in the treatment of hypertension. N Engl J Med 1979; 301:293-7.

52. Bloomfield CD, Gajl-Peczalska KJ, Frizzera G, Kersey JH, Goldman AI. Clinical utility of lymphocyte surface markers combined with the Lukes–Collins histologic classification in adult lymphoma. N Engl J Med 1979; 301:512-8.

53. Okada RD, Pohost GM, Kirshenbaum HD, et al. Radionuclide-determined change in pulmonary blood volume with exercise: improved sensitivity of multigated blood-pool scanning in detecting coronary-artery disease. N Engl J Med 1979; 301:569-76.

54. Gadek JE, Kelman JA, Fells G, et al. Collagenase in the lower respiratory tract of patients with idiopathic pulmonary fibrosis. N Engl J Med 1979; 301:737-42.

55. Pantely GA, Goodnight SH Jr, Rahimtoola SH, et al. Failure of antiplatelet and anticoagulant therapy to improve patency of grafts after coronary-artery bypass: a controlled, randomized study. N Engl J Med 1979; 301:962-6.

56. Lang DA, Matthews DR, Peto J, Turner RC. Cyclic oscillations of basal plasma glucose and insulin concentrations in human beings. N Engl J Med 1979; 301:1023-7.

57. Leon MB, Borer JS, Bacharach SL, et al. Detection of early cardiac dysfunction in patients with severe beta-thalassemia and chronic iron overload. N Engl J Med 1979; 301:1143-8.

58. Packer M, Meller J, Medina N, Gorlin R, Herman MV. Rebound hemodynamic events after the abrupt withdrawal of nitroprusside in patients with severe chronic heart failure. N Engl J Med 1979; 301:1193-7.

59. McLellan AT, Woody GE, O'Brien CP. Development of psychiatric illness in drug abusers: possible role of drug preference. N Engl J Med 1979; 301:1310-4.

60. SAS Institute Inc., SAS user's guide: statistics — Version 5 edition. Cary, N.C.: SAS Institute, 1985.

61. SPSS Inc., SPSS user's guide, 2nd ed. Chicago, Ill.: SPSS, 1986.

62. Miller R Jr. Simultaneous statistical inference. New York: Springer-Verlag, 1981.

63. Games PA, Keselman HJ, Rogan JC. A review of simultaneous pairwise multiple comparisons. Stat Neerland 1983; 37:53-8.

64. Dixon WJ, Brown MB, Engelman L, et al. BMDP statistical software 1983. Berkeley, Calif.: University of California Press, 1983.

13

~

ANALYZING DATA FROM ORDERED CATEGORIES

Lincoln E. Moses, Ph.D., John D. Emerson, Ph.D.,
and Hossein Hosseini, Ph.D.

ABSTRACT   Clinical investigations often involve data in the form of ordered categories — e.g., "worse," "unchanged," "improved," "much improved." Comparison of two groups when the data are of this kind should not be done by the chi-square test, which wastes information and is insensitive in this context. The Wilcoxon–Mann–Whitney test provides a proper analysis. Alternatively, scores may be assigned to the categories in order, and the *t*-test applied. We demonstrate both approaches here.

Sometimes data in ordered categories are reduced to a 2 × 2 table by collapsing the high categories into one category and the low categories into another. This practice is inefficient; moreover, it entails avoidable subjectivity in the choice of the cutting point that defines the two super-categories. The Wilcoxon–Mann–Whitney procedure (or the *t*-test with use of ordered scores) is preferable.

A survey of research articles in Volume 306 (1982) of the *New England Journal of Medicine* shows many instances of ordered-category data and no instance of analysis by the preferred methods presented here. Seven years later, Volume 321 (1989) shows several instances of analysis by the preferred methods that use the ordering.

The clinical investigator must at times rely on quantitative information that is intrinsically imprecise. The clinical response (worse, unchanged, improved) is such a variable; the response can be qualitatively ordered, but it often cannot be expressed on a precise numerical scale. Input variables of the same sort arise, too; stage of disease (I, II, III, IV) is an example. Effective ways to analyze such information are gradually gaining in use. Sometimes inefficient methods of analysis are applied; this is equivalent to ignoring part of the data. This chapter points out ways to go wrong, but its primary emphasis is on methods that use such information efficiently, particularly for comparing two groups of observations expressed in an ordered classification. A randomized controlled clinical trial comparing two combination drug treatments for advanced Hodgkin's disease offers an illustration. If the response is three levels of tumor remission (complete, partial, none), then the methods described here are well suited to the analysis of the 2×3 table of counts.

An initial survey of Volume 301 of the *New England Journal of Medicine* identified 13 (of 71) Original Articles containing 31 instances of ordered categorical data. A second, more detailed survey of Volume 306 of the *Journal* considered all 168 articles and revealed that 32 articles contained 47 instances of ordered categorical data. A survey of Volume 321 found that 45 of 163 articles contained 89 ordered categorical variables. In some articles, the authors used established classification systems, such as the New York Heart Association scale for patients with congestive heart failure. In others, the ordered classes arose from the lumping of measurements into ranges of measured values. Some authors devised original systems of ordered categories as vehicles for reporting their observations.

The central feature of a set of ordered categories is that they express in increasing or decreasing order the extent or the degree of intensity or complexity of some observable phenomenon. Each class used has a definite place in the order of the set of classes; if numbers are used, as on a four-point scale, they exhibit the order, but it may not be obvious just how they are to be interpreted numerically. Pain scored at 4+ is more severe than that scored at 2+, but is it twice as severe?

Inappropriate analysis of such data (for example, reducing them to a two-point scale indicating presence or absence) can sacrifice much of the information in the data — information that has been obtained with considerable effort, expense, and perhaps patient cooperation. Often, ordered categories are the best available way to capture important information, such as the stage of disease or the severity of symptoms. Efficient statistical methods for analyzing such data have considerable value.

Some medically important categorical variables are not ordered; ABO blood types are an example. Chapter 15 of this book[1] considers the analysis of unordered categorical variables. This chapter does not deal with the analysis of such data but with problems in which an approximate quantification is attributable to the categories. We assume also that over the range of data under consideration, "more" is either always better than "less" or always worse. This condition generally applies to variables such as degree of recovery, remission, or level of physical functioning. But there are some variables, such as levels of arousal (described by such terms as torpid, normal, and hyperactive) or trust in strangers (from too trusting to overly suspicious), that are optimal at intermediate values and for which "more" is not always better (or always worse) than "less." The methods described in this chapter may not be suitable for analyzing such data. This chapter also does not address the analysis of tables whose entries are measurements.

A SUITABLE METHOD OF ANALYSIS

The Wilcoxon or Mann–Whitney test is appropriate for comparing two sets of results that are combined and then scored in terms of an ordered classification.[2] In 1945 Wilcoxon[3] reported a method for comparing two samples, taking account of only the relative order of the observations. He arranged the two samples in a single rank order from smallest to largest and devised a test that compared the average ranks of the two samples in that ordering. He made use of the fact that large observations produce larger ranks, so that if one of the treatments tends to produce larger observations than the other, its average rank will be large. Thus, comparison of the average ranks in the two samples could replace comparison of the average values, as in the ordinary t-test. In addition to its uses described in this book, the Wilcoxon test is well suited to comparing two sets of precisely ordered measurements; it gives a nonparametric analysis that parallels the two-sample t-test.

Soon after Wilcoxon's method appeared, Mann and Whitney[4] devised a different version of Wilcoxon's test, a second form of it. The two forms are exactly equivalent; sometimes one is more convenient to use, and sometimes the other. The computations for the two forms of the test are different, and lead to consulting different tables, although a knowledgeable person can easily deduce one table from the other. The Mann–Whitney form of the test compares each individual in the first group with each individual in the second group, recording how many times this comparison favors the individual from the second group. This chapter uses the Mann–Whitney form of the calculations because of their greater convenience for the $2 \times k$ table.

Table 1. Scoring System for Outcomes of Treatment for Chronic Oral
Candidiasis.*

Clinical Findings	Laboratory Findings	Score
Absent	Negative	1
Improved	Negative	2
Improved	Positive	3
Unimproved	Positive	4

*From Kirkpatrick and Alling.[5]

In an article published in the *Journal* in 1978, Kirkpatrick and Alling[5]
applied the Mann–Whitney test in a clever way to assess the results of
a randomized clinical trial dealing with the treatment of chronic oral candidiasis. Every subject was scored two to seven days after treatment,
as shown in Table 1. This scoring scheme captures two kinds of outcomes
and combines them in such a way that the larger of any two scores
connotes the poorer outcome; these scores define an ordered classification. (If in any of the subjects in the study there had been no improvement but negative laboratory tests, or an absence of clinical findings but
positive laboratory tests, it would not have been obvious how to put their
results into an ordered classification; but since only the four outcomes
shown did occur, the investigators were able to use this ordered classification to analyze their data.) The authors presented the data reproduced in Table 2.

Visual inspection of the tables suggests that the treatment led to pre-
dominantly smaller scores — i.e., to more favorable outcomes. The Mann–
Whitney test offers strong statistical support for this observation.

In the example, each of the 10 patients receiving treatment is compared
with each of the 10 control patients, for a total of 100 comparisons. (If the
sample sizes had been 15 and 20, there would have been $15 \times 20 = 300$
comparisons.) If the two treatments were equivalent in their effects, about
half of such pairwise comparisons should favor the control group and half
the treatment group. In the instance before us, the outcome is far from
being half and half, as we shall see by doing the counting. Some pairs favor
the placebo, some the treatment, and in some the two are tied; the frequen-

Table 2. Outcomes after Two to Seven Days of Treatment in 20 Patients with Chronic Oral Candidiasis.*

Treatment	Outcome Category				Total
	1	2	3	4	
Clotrimazole	6	3	1	0	10
Placebo	1	0	0	9	10

*Data from Kirkpatrick and Alling.[5]

cies of these three situations must add up to 100, the total number of pairs. The data provide six pairs in which a subject from the treatment group is tied with a subject from the placebo group; these six pairs are the ones comprising the placebo-group subject who was scored 1 and each of the six treatment-group subjects who also were scored 1 (if two placebo-group subjects had been scored 1, we would have had $2 \times 6 = 12$ ties from the pairings of the 1s). The data in the example provide four pairs in which the subject from the placebo group has a better (lower) score than the subject from the treatment group; these are the pairs involving the placebo-group subject who was scored 1, the three treatment-group subjects who were scored 2, and the one treatment-group subject who was scored 3. We have found six tied pairs and four pairs favoring the placebo; the remaining 90 ($= 100 - 6 - 4$) pairs must favor the treatment. This is readily confirmed: All 10 subjects who received treatment have scores of less than 4, and when each of those 10 is paired with each of the 9 placebo-group subjects who were scored 4, we have 90 pairs in which the treatment-group subject has the better score.

To summarize the results of this counting, in 90 pairs the treatment was better, and in 6 it was tied with the placebo; we credit half the ties to the treatment, as if it had been superior, and the other half to the placebo. Our summary statistic is then $90 + 3 = 93$. This statistic was named U by Mann and Whitney.[4] In general, U is calculated as we have done, by counting the pairs favoring one of the groups and adding half the tied pairs.

The calculated value of U allows us to do two useful things. First, we can construct an informative descriptive statistic, as follows: divide U by

the total number of pairs — 100 in our example. This ratio is the proportion of pairs favoring the treatment, and for the data in Table 2 we get
93/100 = 0.93. The ratio estimates the probability that a subject chosen at random (from a group like those used in this randomized trial) and
given the treatment will have a better outcome than will a subject similarly chosen and given the placebo. This index offers a numerical estimate
in answer to the question "For this population, what is the probability
of a better response to treatment than to placebo?" (In this question, we
must understand that the "better" of two tied observations is to be determined by the flip of a coin.) The ratio cannot address the question "How
much better?" This question may not even make sense when the scale
is ordinal.

In a second use of U, we can assess statistical significance. In these data a
value of about 50 would be expected for U if the placebo and treatment
were equivalent in effect. Is the observed value of 93 convincingly different
from 50? Computer programs[6,7] allow us to answer the question. The exact
two-sided significance level is 0.00041, and we conclude with a high degree
of confidence that the treatment, clotrimazole, is superior to the placebo in
terms of producing more favorable outcomes.

If the two sample sizes, m and n, are both large, the value of U is computed as we have done. The descriptive interpretation of U/mn continues
to hold, but significance can be assessed without recourse to special computer programs, by the use of a normal approximation. With equivalent
treatments, U is nearly normally distributed, with a mean of $mn/2$ (50 in
this example) and with a standard deviation that is calculated approximately by

$$\sigma = \sqrt{mn(m + n + 1)/12}$$

(13.23 for this example). When some of the ranks are tied, as in this example, the approximation can be improved by multiplying σ by a *correction
factor for ties* (0.931 for the example); we refer to Armitage[2] for the algebraic details of the correction. We then express U as a departure from its
mean ($mn/2$), measured in units of corrected standard deviation (σ_U), obtaining

$$z = \frac{U - mn/2}{\sigma_U} = \frac{93 - 50}{(13.23)(0.931)} = 3.49.$$

From tables of the normal distribution we find that 3.49 corresponds to a two-sided significance level of 0.0005. Again, we conclude with a high degree of confidence that clotrimazole is superior to the placebo.

Considerable numerical investigation of the normal approximation shows that a clear indication of significance or nonsignificance can be relied on provided that each sample comprises at least 10 observations and that no one category contains more than half the combined set of observations. If the normal approximation gives a result near the threshold of significance, it is wise to compute the exact P value with an appropriate computer program. Emerson and Moses[8] give further recommendations on the use of exact methods for carrying out calculations for Mann–Whitney.

APPROXIMATE METHODS OF ANALYSIS

An alternative way to analyze data comparing two samples scored in ordered categories is to assign ordered numerical values to those categories and then compute the two sample averages, comparing them by means of the two-sample t-test. This procedure has advantages and drawbacks. The main advantage is the familiarity of the t-test, although computing it from data presented in the format of Table 2 may seem strange at first. A clear example is given in Snedecor and Cochran.[9] The primary drawback is that the choice of numerical values for the ordered categories has an essential arbitrariness, so that different analyses of a set of ordered data may generate different P values and thus lead to different conclusions.

This drawback is mitigated by two considerations, however. First, in any particular data set, the Mann–Whitney test can be shown necessarily to yield essentially the same result as that obtained from the t-test when some set of ordered scores (chosen for that data set) is used. Second, t-tests using different sets of ordered scores for a given data set ordinarily produce similar results.[9] From these two principles it follows that the t-test will serve as an approximation of the Mann–Whitney test.

To illustrate how the choice of scores may have very little effect (so long as the order is not disturbed), let us again consider the data presented in Table 2. If the results are analyzed by computing the t-statistic with numerical scores (1, 2, 3, and 4 or 1, 2, 8, and 9) for the four outcome categories, then the results are as follows: $t = 5.88$ with a P value of 0.000014, and $t = 5.89$ with a P value of 0.000014, respectively. The agreement is close.

Both these values of t (with 18 degrees of freedom) give two-sided P values that are considerably smaller than the exact value, 0.00041. Two points are illustrated rather sharply: the change in choice of scores had little effect, and the evaluation of significance is qualitatively in accord with that of the Mann–Whitney result (clotrimazole being judged definitely better than the placebo), but the numerical agreement is not exact.

A second and inferior approximate method of treating such ordered data is to collapse the set of categories into only two. The imposition of such a coarse dichotomy on data that are intrinsically more fine-grained gives an appearance of simplicity but nearly always at a cost in information; the cost can be great, though it is not always so.

With the data in Table 2 we could reduce the ordered categories to a dichotomy without violating the order, by dividing them into two groups at the boundary between 1 and 2, between 2 and 3, or between 3 and 4. These three approaches yield the reduced data shown in Table 3.

Not only does collapsing the table waste data; it also forces us to be arbitrary in choosing where to divide the ordered categories. We can, as before, compute U/mn and estimate the probability that treatment will be better than placebo. Its values for A, B, and C in Table 3 are 0.75, 0.90, and 0.95, respectively. This kind of variation suggests a danger of imposing a dichotomy — that the choice of where to divide the categories may reflect the investigator's wish to maximize (or minimize) the appearance of difference between the two samples. To some extent, this same danger may be feared in the assignment of scores to the ordered categories, but the danger is greater in the case of dichotomization because that is an extreme instance of assigning scores. Imposing a dichotomy is equivalent to defining all the scores for categories on either side of the cutting point as being equal to each other. The Mann–Whitney approach has the

Table 3. The Three Possible Dichotomized Versions of the Data in Table 2.

Treatment	A		B		C	
	1	2–4	1, 2	3, 4	1–3	4
Clotrimazole	6	4	9	1	10	0
Placebo	1	9	1	9	1	9

property of using exactly the information about order — and nothing else — in the analysis.

In summary, we have discussed two kinds of approximate analysis. One, the use of the t-statistic, is usually not objectionable; the other, collapsing the categories into a dichotomy, has important drawbacks.

At this point it is worth reviewing the advantages of the U statistic. First, it is applicable to very small samples (e.g., $m = 3$, $n = 5$), to which we would apply the t-test at our peril. Second, it yields a descriptive statistic that is much more valuable than is widely known; that statistic, U/mn, does not depend on the units of measurement. (Indeed, we have applied it where there are no units of measurement, only an underlying order.) The value of U/mn is interpretable directly in terms of the probability that one treatment will be more effective than the other in individual application to subjects of the kind used in the study furnishing the data. Third, the computation of U/mn is simple, once the data are arranged in order. Many standard computer packages include the Mann–Whitney test and report a P value, suitable when m and n are not small, that is based on the normal approximation. However, advances in computing and in algorithms for implementing exact statistical methods now make the exact Wilcoxon–Mann–Whitney procedure available. Several microcomputer statistics packages incorporate exact methods; we have used StatXact.[10]

ORDERED INPUT VARIABLES

In the preceding two sections we have shown how to compare the responses of two groups, expressed in an ordered classification. But ordered classifications arise in medicine not only as outcomes but also as input variables. Among the examples found in our survey of Volumes 301, 306, and 321 of the *Journal*, many involved ordered input variables; these included the extent of disease in entering patients, their socioeconomic status, the elapsed interval before the initiation of treatment, and the patients' nutritional status. Table 4 from Baker et al.[11] provides an example of the last sort, showing the nutritional status of 48 hospitalized patients and the occurrence of infections among these patients. These data have columns bearing a natural and meaningful order, but they describe a condition, not a response; the outcomes correspond to "infection" or "no infection" — that is, to the row variable. Here our attention naturally fastens on the

Table 4. **Occurrence of Infection in 48 Hospitalized Patients Classified According to Their Nutritional Status.***

| | | Nutritional Status | |
| | | | |
Outcome	Normal	Mild Malnutrition	Severe Malnutrition
Infection	4	3	11
No infection	21	4	5

*Data from Baker et al.[11]

percentage with infection in each of the three columns. Before, when columns represented outcomes, we compared the rows. (In using the *t*-test we looked at two means, one for each row.)

So this problem, with ordered columns corresponding to levels of an input variable, has a different structure. We naturally attend to the trend we see, in which the percentage with infection increases as the nutritional status worsens. Is this trend statistically significant?

Answering this question is made easy by a pleasant, surprising mathematical fact: We can assess the significance of such a trend in a table with two rows by exactly the same methods presented earlier. This statement is formally justified by Armitage,[12] who shows that Kendall's rank-correlation coefficient (a measure of correlation that is based only on ranks and is suitable for testing trend) reduces to U when there are two rows. Furthermore, although it is far from obvious, it can be proved that the ordinary correlation coefficient, in which numerical scores are used for the columns and there are only two rows, yields as a test statistic the very *t*-statistic that we have already described. Though we do not pursue these issues here, they justify our using the same statistical methods as those already described.

We have, then, two ways to test the trend of the percentage having infection (for example) with regard to an ordered characteristic of treatment (intensity, duration) or of subject material (nutritional status, stage of disease). The first method is to apply the U test. This will do for assessing significance, but the descriptive interpretation of U is less direct than before. U now estimates the probability that if one infected patient and one uninfected patient are drawn randomly from the population that this sam-

ple represents, then the infected patient will be the one with the worse nutritional status. Alternatively, one can, as before, assign numbers to the columns and apply the *t*-test to the two rows. This will also do for assessing
approximate significance; with the score for each column used to define the
nutritional status in the column, the difference between the row averages
serves as an estimate of the average difference in nutritional status between
infected and uninfected patients.

We now apply these two methods, using U and t, to the data in Table 4.
The value of U is 413.5. If there were no association between infection and
nutritional status, U would have an expected value of

$$\frac{18 \times 30}{2} = 270.$$

The value of σ, corrected for ties, is 42.5; so the significance is assessed in
terms of

$$z = \frac{413.5 - 270}{42.5} = 3.38,$$

which, according to a table of the normal distribution, corresponds to a
two-sided P value of 0.00072. (The exact P value is also 0.00072.) The value
of U found from Table 4 is large enough that we can conclude with near
certainty that the trend of increasing infection with worsening nutritional
status is not an artifact of chance.

Alternatively, we can apply the two-sided t-test to these data. The first
task is to assign scores to the columns. The value of t, with the scores 1,
2, and 3, is 3.76, which has a two-sided P value of 0.00048, according to the
t distribution with 46 degrees of freedom. We conclude, as before, that the
trend is too marked to be regarded as a chance phenomenon.

Before turning from this example, we pause to show the weakness of
analyzing data with ordered categories using the usual chi-square test for
the $2 \times k$ table. The chi-square statistic for the infection–nutrition data
is significant at 0.0029 with two degrees of freedom. Notice first that this
P value is considerably less critical than the two-sided P value of 0.0007
that was attained by using U; the difference is not damaging here, but the
lesser sensitivity of chi-square for identifying trends is evident, and it can
be decisive in some data sets. Less obvious is the underlying principle: The
chi-square statistic takes no account of the order.

Panel A of Figure 1 plots the data from Table 4 as the percentage of patients with infection at each of the three nutritional levels. There is a clear trend, and chi-square is 11.68 ($P = 0.0029$). Now, if Table 4 had contained frequencies of infection and noninfection that were the same but were differently associated with nutritional status, then other dispositions of the three percentages 0.16, 0.43, and 0.69 would have resulted. The figure shows all six possibilities. They have very different trends and very different values for z based on the Mann–Whitney test, but chi-square is the same for all — 11.68. When trend is the issue, then order matters, and the chi-square test is inappropriate. Using a version of Fisher's exact test for a $2 \times k$ table is not a remedy; it too fails to take ordering into account and so has the same conceptual drawbacks as the chi-square test.

AN ASSESSMENT OF PRACTICE

To assess the occurrence of ordered categorical variables in the research published in the *Journal* and to learn about the ways in which these variables are used in statistical analyses, we surveyed 168 articles in Volume 306 (January through June, 1982). We included all the articles belonging to the following categories: Original Articles, Special Articles, Medical Progress, and Medical Intelligence.

Using the Rand Corporation's 1955 table of random digits,[13] we randomly ordered all 25 issues of Volume 306. These issues were then assigned in random sequence for review to members of the Study Group for Statistical Methods in the Biomedical Sciences; all the members, who are listed in the acknowledgment at the end of this chapter, were faculty members, postdoctoral fellows, or graduate students trained in the quantitative sciences; each had considerable acquaintance with statistical applications to medical research. One of these persons reviewed each issue, without blinding, and the numbers of issues reviewed varied considerably among them. They searched the assigned articles for occurrences of ordered categorical variables, and for each variable identified they completed a 12-question checklist. We then briefly reviewed the articles a second time to try to identify additional variables that might have been overlooked by the assigned readers. This additional review also enabled us to improve the accuracy and consistency of the responses to the questionnaires.

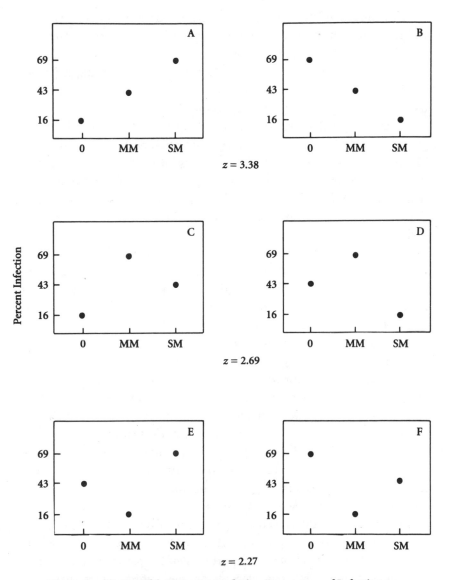

Figure 1.   Six Possible Data Sets Relating Percentage of Infection to
Nutritional Status: Normal (0), Mild Malnutrition (MM), and Severe
Malnutrition (SM).
Chi-square is 11.68 for all panels. Panel A shows the data observed.

We included in our study only the ordered variables, with three or more categories, for which frequencies were reported or clearly could have been reported. We included some ordered variables that were used by authors in analysis of variance or in logistic regressions and not in contingency tables, but in these situations we thought that the variables could also have been used for grouping the frequencies in a contingency table.

Of the 168 articles reviewed, 32 contained 47 instances of ordered categorical variables. Table 5 summarizes information derived from the checklists for the 47 variables. The table provides information of three general types: descriptive information about the ordered variable itself, information about whether and how the variable was incorporated into statistical analyses, and information about the clarity of reporting on the uses of the variable.

Twenty-three of the ordered categorical variables we identified were derived from measured variables (such as age) whose values were partitioned into consecutive ranges. Twenty-eight of the 47 ordered categorical variables described patient characteristics such as age or weight; these are among the variables we have referred to as input variables.

Although 22 of the ordered categorical variables were used in at least one formal statistical analysis, there were eight situations in which we could not be reasonably certain whether the variable was used in this way. Authors did not indicate, for example, whether a formal test for trend was made or whether they were simply reporting an empirical discovery made by inspection. Of course, many ordered categorical variables are used for descriptive purposes and not in statistical computation.

More frequently, we were unable to determine from the article which statistical testing procedure (e.g., the chi-square test, Fisher's exact test, or the Mann–Whitney test) was used to obtain a reported P value or a finding of nonsignificance. Our difficulty often arose from the tendency of authors to list together, in a Methods section, all the statistical techniques that were used for their study. When the results are reported later, the particular statistical test that was used cannot always be inferred from the context. In these articles, the extent of nonreporting or inadequate reporting of the results of statistical analyses and of the techniques used is somewhat greater than that determined by DerSimonian and her colleagues[14]; among reports of clinical trials published in the *Journal*, they found that 92 percent included results of statistical analyses and the same percentage specified techniques.

Table 5. Data for 47 Ordered Categorical Variables in 32 Articles Published in Volume 306 of the *New England Journal of Medicine*.

Item	Response Categories	No. of Variables
Type of classification	Established system	9
	Range of measured values	23
	Apparently original	15
Role of variable	Treatment variable	9
	Response variable	10
	Other (e.g., descriptive)	28
Used in statistical analysis?	Yes	22
	No	17
	Unclear	8
Statistical technique identified?	Clearly identified	9
	Can be inferred	8
	Unclear	13
	None used	17
P value or confidence interval reported?	Yes	17
	No	30
Did authors collapse categories for a contingency-table analysis?	Yes	9
	No, they analyzed full table	8
	Unclear	10
	No contingency-table analysis	20

P values (or, in one article, confidence intervals) were reported for analyses that made use of 17 of the 47 ordered variables we identified. When no statistical analysis was done, then there was, of course, no P value to report.

The authors of the 32 articles apparently did not carry out any statistical analysis of a contingency table for 20 of the 47 ordered variables we found. It appears that the authors analyzed 27 tables using some statistical method; they collapsed categories in 9 instances, and they analyzed the full table using methods that ignore order in 8 instances. It also appears that the authors may have collapsed categories in 10 other tables in which a statistical analysis was made. We found no indication that any of the authors of these articles in Volume 306 used an analysis, such as the Mann–Whitney test, that takes the ordering into account; however, in Volume 299, the article by Kirkpatrick and Alling[5] analyzed the data on oral candidiasis in Table 2 correctly, using a Mann–Whitney test.

UPDATING THE ASSESSMENT

To assess whether changes in statistical practice have occurred since 1982, we reviewed 163 articles in Volume 321 (July through December, 1989) belonging to the same four *Journal* sections (Original Articles, Special Articles, Medical Progress, and Medical Intelligence). We found an increase in the use of ordered categorical variables, with 45 articles containing 89 instances. Twenty articles carried out at least one statistical analysis of a contingency table that incorporated an ordered variable, and each of these reported a *P* value or reported that the test was not significant.

Eight of the twenty articles used analyses that took the ordering into account; the remaining twelve articles used either chi-square or Fisher's exact test, often after collapsing categories. Although these findings leave room for further improvement, we find encouragement in the increased use of statistical analyses that properly accommodate the ordering of a categorical variable. We are optimistic that, as the availability of computer software with the appropriate methods continues to increase, improvements in statistical use also may continue.

REANALYSIS OF ORDERED-CONTINGENCY TABLES

A Mann–Whitney analysis can give results that differ substantially from a chi-square analysis. The authors who reported on the details of their statistical analyses for ordered-contingency tables most often used a chi-square test. To compare the results obtained by these authors with those that would be obtained from the use of a Mann–Whitney test to analyze the tables, we narrowed our attention to 11 of the 47 ordered variables we studied from Volume 306. The authors of the articles that used these variables were more complete in their statistical reporting than most of the others; thus, we were able to compare alternative methods of statistical analysis.

We identified all ordered-contingency tables from Volume 306 of the *Journal* that met the following criteria: (1) frequencies for the table were given or could be inferred from the article; (2) the format of the table was $2 \times k$ or $k \times 2$ (where k was greater than two), and the variable with more than two categories was ordered; (3) the authors used a method of statistical analysis that was clearly identified or could be inferred from the article; and (4) the authors reported a P value or a significance level with a boundary (e.g., $P<0.02$), or they reported both nonsignificance and the level of the test used. For a table to be included in this part of our study, we did not require that a P value be identified as either one-sided or two-sided (all but one were apparently two-sided). Even so, we were dismayed that incomplete reporting prevented many interesting tables from being included in this phase of our study.

We discovered only six contingency tables, involving five ordered categorical variables, that met all the criteria for inclusion in this part of our investigation; these tables appeared in five different articles. (In one article, two tables made use of the same ordered categorical input variable, but the response variables differed.) To these we added five tables whose reanalysis we thought might be instructive. Table 6 summarizes the analyses and reanalyses for the 11 tables. For each table the authors apparently used either chi-square analyses or Fisher's exact test for 2×2 tables; both of these methods fail to take full account of the order implicit in $k \times 2$ or $2 \times k$ tables.

For all 11 tables selected, we first attempted an exact replication of the analysis done by the authors. We also reanalyzed the tables using the Mann–Whitney test; for this part of the study we used both an exact analysis and a normal approximation corrected for ties. For most of the tables, we performed a variety of additional analyses on both uncollapsed and collapsed versions of the data. For these statistical analyses, we used three widely available statistical software packages: SPSS,[15] BMDP,[16] and Minitab.[17]

In four of the first six tables, we were able to replicate the results reported by the authors, using the methods of analysis that they reported using. In two instances our results seemed not to confirm those reported by the authors; in one case the authors apparently reported a one-tailed P value, and in another they probably reported the results for an uncorrected rather than a corrected chi-square analysis. In the remaining five tables, the authors did not report fully on their statistical analyses. Still, we were able to support their general conclusions in three of the five tables, using the chi-square analyses that we suspect they adopted. Lack of complete information about what analyses the authors performed

Table 6. Reanalysis of 11 Ordered-Contingency Tables from Articles Published in Volume 306 of the *New England Journal or Medicine*.*

Table No.	In Authors' Report	Two-Sided *P* Values		
		In Reanalysis Using Authors' Method	Exact	Mann–Whitney Normal Approximation
1	<0.0005	0.0002	0.0010	0.0021
2	0.094	0.188	0.106	0.100
3	0.02, 0.005	0.04, 0.005	0.010	0.012
4	<0.005	0.0017	0.0007	0.0008
5	<0.005	0.0029	0.0007	0.0007
6	0.04	0.044	0.044	0.044
7	Not significant	0.267	0.378	0.357
8	<0.01	—	0.00003	0.0001
9	Nonsignificance implied	—	0.119	0.113
10	0.009, 0.043	—	0.010	0.0077
11	Not significant, but differences cited	—	0.949	0.948

*When two *P* values are reported, they correspond to two different collapsed versions of a table. The first six tables were selected because the authors were relatively complete in their statistical reporting; the remaining five tables were selected for their statistical interest and instructive value.

may have prevented us from confirming their findings for the two other tables.

Table 6 compares our results with those reported by the authors. The normal approximation to the Mann–Whitney test, corrected for ties, is almost always in very good agreement with the exact two-sided *P* value. But the results shown in this table, and those from many other analyses, indicate that a Mann–Whitney analysis can generate a *P* value that is substantially different from the value that results from a chi-square analysis. In 6 of these 11 tables, the authors collapsed the data to a 2 × 2 table, and in 3 of these cases (data sets 3, 10, and 11 in Table 6) they analyzed two different collapsed versions of the original table. In several instances, the authors may have believed that low cell frequencies in the original table required that the data be collapsed before a chi-square analysis could be performed. This issue is, of course, circumvented when a Mann–Whitney test is used; the method even applies to contingency tables without ties, in which each cell has an entry of either 0 or 1.

Because few of the P values produced by an exact Mann–Whitney analysis contradict any of the authors' general conclusions about significance or nonsignificance (only two change at the 1 percent level, none at the 5 percent level, and one at the 10 percent level), we may ask just how important the choice of an appropriate statistical method really is. This choice can, in fact, make a practical difference in one's conclusion. For example, in two papers that analyzed two different collapsed versions of an original table, the authors reached conclusions that differed at common levels of significance, leaving the meaning of the data necessarily ambiguous; the Mann–Whitney test resolves the ambiguity once and for all. The difficulties can be more dramatic than the data in Table 6 suggest, especially when the authors collapse a table and perform a chi-square analysis on a 2×2 table. For example, the inappropriate use of a chi-square test to analyze one collapsed version of the 11th data set in Table 6 reduced a P value from 0.95 to 0.06. These illustrations emphasize the importance of using a method of statistical analysis that is suitable in the context of the particular data being considered.

SUMMARY AND RECOMMENDATIONS

Ordered classifications are often used in medicine to indicate individual responses to treatment. Both intended effects (e.g., improvement, lack of change, or deterioration) and side effects (e.g., whether they are absent, slight, moderate, or severe) are often scored in ordered classifications. Ordered variables appear in about 20 percent of the research articles recently published in the *Journal*, and they often define the categories for contingency tables that undergo statistical analysis.

Comparing two groups in which the data come in this form is naturally and readily done by the Mann–Whitney U statistic. This statistic is easy to compute, and it provides a direct estimate of the probability that a patient will do better if treated with method A rather than with method B. The P values are accurate even in very small samples, and a normal approximation for Mann–Whitney is often suitable.

A satisfactory alternative method when neither sample is very small is to apply numerical values to the categories and compare the two groups by means of a t-test. It is generally not satisfactory to collapse the ordered categories into only two so as to apply methods for comparing two proportions. Data are wasted by doing this.

Either U or t also applies when the rows give two possible outcomes (live or die, conceive or do not conceive) and the columns correspond to some ordered input variable, such as the stage of disease or the intensity of treatment. Now the issue is to assess any trend in the percentage of patients responding favorably, across the ordered input variable. It is a happy mathematical accident that this kind of problem can be treated by the methods natural to the first problem.

In contexts in which reasonable interpretation of the data should take the order of categories into account, it is a mistake to use the chi-square statistic, which has often been adopted by *Journal* authors for tables with two rows and more than two columns, as the tool of analysis. That statistic completely ignores the order of the columns. As a consequence, the chi-square test is not as responsive to trends that are truly present; the Mann–Whitney test or the t-test is more likely to find that a trend is statistically significant. For contingency tables with ordered variables on two margins and for tables with three or more rows and three or more columns in which at least one variable is ordered, special techniques are available[18] and investigators should probably seek the assistance of a professional statistician.

With current reporting practices, it is often difficult or impossible to determine precisely what statistical technique was used in a particular analysis. For an analysis of a two-way contingency table, adequate information should be reported in order to allow independent verification of the analysis. The authors should clearly identify the table of frequencies analyzed; whether the data were collapsed and, if so, in what way; what test statistic was employed; the P value of the test; and whether the P value is one-sided or two-sided.

The adoption of suitable statistical methods, together with adequate reporting of the methods used and the results obtained, can enhance the credibility and value of the published report of a scientific investigation. The high costs of medical investigation, whether measured in time, effort, or money, mandate our careful attention to these issues.

We are indebted to John Bailar, Graham Colditz, Katherine Godfrey, Katherine Taylor Halvorsen, Robert Lew, Thomas Louis, Frederick Mosteller, and John Williamson, all members of the Study Group for Statistical Methods in the Biomedical Sciences, who assisted us in reviewing articles published in Volume 306 of the *Journal;* to Cyrus Mehta and David Tritchler; and to Cleo Youtz, who provided useful suggestions on an earlier version of the manuscript.

REFERENCES

1. Zelterman D, Louis TA. Contingency tables in medical studies. [Chapter 15 of this book.]

2. Armitage P. Statistical methods in medical research. New York: John Wiley, 1971:section 3.3.

3. Wilcoxon F. Individual comparisons by ranking methods. Biometrics 1945; 1:80-3.

4. Mann HB, Whitney DR. On a test of whether one of two random variables is stochastically larger than the other. Ann Math Stat 1947; 18:50-60.

5. Kirkpatrick CH, Alling DW. Treatment of chronic oral candidiasis with clotrimazole troches: a controlled clinical trial. N Engl J Med 1978; 299:1201-3.

6. Klotz J, Teng J. One-way layout for counts and the exact enumeration of the Kruskal–Wallis H distribution with ties. J Am Stat Assoc 1977; 72:165-9.

7. Mehta CR, Patel NR, Tsiatis AA. Exact significance testing to establish treatment equivalence with ordered categorical data. Biometrics 1984; 40:819-25.

8. Emerson JD, Moses LE. A note on the Wilcoxon–Mann–Whitney test for $2 \times k$ ordered tables. Biometrics 1985; 41:303-9.

9. Snedecor GW, Cochran WG. Statistical methods. 8th ed. Ames, Iowa: Iowa State University Press, 1989.

10. Cytel Software Corporation. StatXact. Cambridge, Mass., 1989.

11. Baker JP, Detsky AS, Wesson DE, et al. Nutritional assessment: a comparison of clinical judgment and objective measurements. N Engl J Med 1982; 306:969-72.

12. Armitage P. Tests for linear trends on proportions and frequencies. Biometrics 1955; 11:375-86.

13. Rand Corporation. A million random digits with 100,000 normal deviates. New York: The Free Press, 1955.

14. DerSimonian R, Charette LJ, McPeek B, Mosteller F. Reporting on methods in clinical trials. N Engl J Med 1982; 306:1332-7. [Chapter 17 of this book.]

15. SPSS Inc. SPSS user's guide, 2nd ed. Chicago, Ill.: SPSS, 1986.

16. Dixon WJ, Brown MB, Engelman L, et al. BMDP statistical software 1983. Berkeley, Calif.: University of California Press, 1983.

17. Minitab reference manual Release 7. University Park, Pa.: Pennsylvania State University, 1989.

18. Agresti A. Analysis of ordinal categorical data. New York: John Wiley, 1984.

14

~

STATISTICAL ANALYSIS OF SURVIVAL DATA

Stephen W. Lagakos, Ph.D.

ABSTRACT   A primary end point in many medical studies is the time until the occurrence of an event such as disease progression, death, or discharge from a hospital. The statistical analysis of such data, commonly referred to as failure-time data or survival data, requires the use of special methods because the event may not yet have occurred in all subjects by the time the data are analyzed. This paper reviews standard statistical methods for analyzing survival data. We describe the Kaplan–Meier method of estimating a survival distribution and the log-rank test for assessing the equality of two or more survival distributions. We also review Cox's proportional-hazards regression model, which is a popular methodology for assessing the simultaneous association between multiple baseline factors and survival. These techniques can be very helpful in identifying factors that influence survival time and in describing the survival experience of a population.

In clinical trials and other medical studies, interest often centers on an assessment of the times until the participants experience some specific event. This may be a clinical outcome such as disease progression, disappearance of a tumor, or death; or it may be an event related to the patient's clinical course, such as discharge from the hospital, discontinuation of study medication, or development of an adverse reaction to treatment.

Data that arise in this way are usually referred to as *survival data* or *failure-time data*. Implicit in the definition of survival time is that survival is measured from some well-defined starting point. For example, in a study of survival in patients with AIDS, the starting point might be the time of diagnosis of AIDS. Similarly, in a clinical trial evaluating the use of chemotherapy to prolong remission in women who have received a mastectomy, the starting point might be the time of surgery or the time of randomization into the trial.

A common aspect of nearly all such studies is that some of the subjects will not experience the event defining the end of survival by the time the data are analyzed. For example, if three years have elapsed since an individual was diagnosed with AIDS, his survival time is known only to be something in excess of three years. This is commonly denoted as a survival of 3+ years and referred to as a *censored survival time.*

The use of survival methods in the *New England Journal of Medicine* has increased dramatically in recent years.[1] The purpose of this chapter is to provide an introduction to the most commonly used statistical methods for analyzing survival data. These are the Kaplan–Meier estimator of the survival distribution, the log-rank test for comparing two or more survival distributions, and Cox's proportional-hazards regression model for assessing the association between multiple base-line explanatory variables and survival.

SURVIVAL AND HAZARD FUNCTIONS

The most common way to describe the survival characteristics of the population is to report the survival function, denoted $S(t)$, which is defined as the proportion of individuals who survive beyond time t. For example, if time is measured in years, then $S(5) = 0.35$ means that 35 percent of the population survive beyond 5 years. By specifying the value of $S(t)$ for each t, we can fully describe the survival characteristics of the population. This is often done graphically and is referred to as a "life table" or "survival plot." To illustrate, Figure 1 gives a hypothetical survival plot describing survival after the diagnosis of AIDS. Like any other life table, the curve takes the value 1 at time 0 and then decreases, indicating that the proportion of survivors decreases with time. The median survival is the time at which 50 percent of the subjects have failed. Thus, in Figure 1, the median survival for this population is 1.7 years.

Suppose that someone with AIDS has already survived for one year since being diagnosed. What is this person's prognosis? An alternative to the survival function that provides a response to this question is the hazard function, which we will denote $h(t)$. Essentially, the hazard function at time t is the proportion of persons who fail at time t from among those who have not failed previously. Formally, this function is related to the survival function by the expression below.

$$h(t) = -\frac{d}{dt}[\log S(t)]$$

That is, the hazard function is just the negative of the slope of the survival curve when the latter is plotted on a logarithmic scale. In epidemiologic studies, $h(t)$ is closely related to the concept of age-specific mortality. The main difference is that in epidemiologic studies "t" usually refers to the age of a subject, while in clinical trials or other medical investigations it usually refers to time under study.

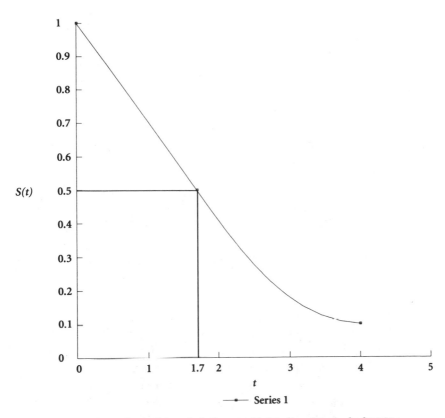

Figure 1.   Hypothetical Survival Curve with Median Survival of 1.7 Years.

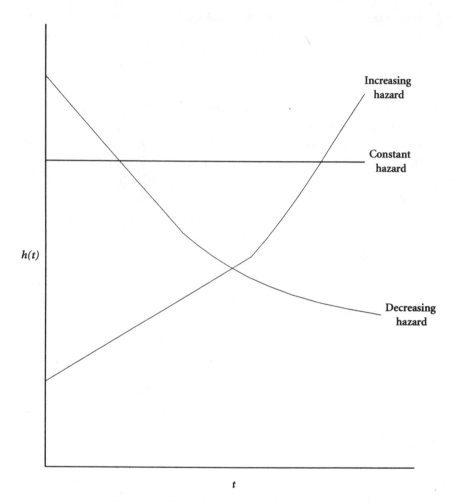

Figure 2.   Hypothetical Hazard Functions.

To illustrate how hazard functions can be useful in describing the survival characteristics of a population, Figure 2 gives three hypothetical hazard functions. The increasing hazard function reflects a population in
which the proportion of persons who fail in a particular year (from among
those who have not failed previously) increases with time. This type of distribution appears to describe the time between infection with HIV and the
development of AIDS.[2] As an infected person survives another year with-

out developing AIDS, his chances of AIDS developing in the subsequent year are greater than in the previous year. In contrast to this, the decreasing hazard function in Figure 2 reflects a population in which the conditional probability of failing becomes less each year; that is, the longer an individual survives, the less his or her chances of failing in the subsequent year. Finally, the constant hazard in Figure 2 describes a population in which the proportion of subjects who are expected to fail in a given year, given that they haven't failed previously, is the same as in any other year. For example, survival from the time of diagnosis of inoperable lung cancer has been shown by several studies to be well approximated by a constant hazard function.[3]

By displaying survival curves on a logarithmic vertical axis, one can also get a qualitative feel for the shape of the hazard function. For example, note that the survival plots in Figure 3 are approximately linear. Because $S(t)$ is plotted on a logarithmic axis, this means that the hazard functions for these populations are approximately constant.

CENSORING

A unique feature of survival data is the presence of censored observations. Because of limited periods of follow-up in observational studies, only a portion of the survival time is observed for some participants. Thus, the data available for analysis will typically consist of some uncensored observations as well as some that are censored, usually at different time points. To illustrate, Table 1 gives the durations of remission among 42 leukemia patients who were either untreated or treated with the drug 6-mercapto-purine (6-MP).[4] Note that the censored durations of remission, which are denoted with a "+," are smaller than some of the uncensored times and larger than others. Because patients are usually followed for different times, this is typical of censored survival data. Note also that the censoring does not occur equally in the two treatment groups; in fact, in Table 1 there are no censored data in the control group.

All the methods described in this chapter, and almost all in regular use, assume that censoring is noninformative.[5] Loosely speaking, this means that the mechanism causing censoring does not act selectively on subjects who are more (or less) likely to fail than those who are not censored. Censoring that arises simply because some subjects have not yet failed by the

Table 1.   Durations of Remission (in Weeks) of 42 Leukemia Patients.[4]

6-MP Group ($n = 21$):	6, 6, 6, 6+, 7, 9+, 10, 10+, 11+, 13, 16, 17+, 19+, 20+, 22, 23, 25+, 32+, 32+, 34+, 35+
Control Group ($n = 21$):	1, 1, 2, 2, 3, 4, 4, 5, 5, 8, 8, 8, 8, 11, 11, 12, 12, 15, 17, 22, 23

time the data are analyzed is noninformative because it implies nothing about the subjects' future status. In other situations, however, it is less obvious that censoring is noninformative. For example, suppose censoring occurs because individuals are lost to follow-up. The noninformative assumption means that subjects who are lost are representative, with respect to ultimate survival, of those who remain under observation. Yet if individuals become lost to follow-up because their disease has progressed, their survival prognosis would be poorer than that of persons not lost to follow-up. Such censoring would be informative and would render standard statistical methods invalid, though perhaps useful for approximation.

ESTIMATING $S(t)$: THE KAPLAN–MEIER ESTIMATOR

Many methods[6] have been proposed for estimating the survival function $S(t)$. One approach, called the Kaplan–Meier or Product Limit estimator, enjoys widespread acceptance and use in medical applications.[6]

   Calculation of the Kaplan–Meier estimator of $S(t)$ is relatively simple and now is included as an option in many statistical software packages. At the heart of the method is the fact that the survival function $S(t)$ can be expressed as a product of conditional probabilities, and that each observation, whether it is censored or uncensored, can help to estimate some of these conditional probabilities. To illustrate, suppose that time is measured in years and consider $S(10)$, the probability of surviving at least 10 years. This can be expressed as:

$$S(10) = S(1) \times S(2|1) \times S(3|2) \times \ldots \times S(10|9),$$

where $S(t|t-1)$ denotes the conditional probability of surviving t years, given that the individual has survived $t - 1$ years. For example, the first two terms on the right-hand side are the probability of surviving 1 year and the conditional probability of surviving 2 years, given survival to 1 year. When multiplied together, they equal the probability of surviving 2 years. Similarly, the product of the first three terms is the probability of surviving 3 years. The Kaplan–Meier estimator of $S(t)$ is constructed by estimating the individual terms, say $S(t|t-1)$, by the ratio n_t/N_t, where N_t denotes the number of subjects in the sample who are still being followed at time t, and n_t is the number of these who survive beyond time t. That is, the conditional probability of surviving t years, given that the individual has survived at least $t-1$ years, is estimated by the proportion of subjects at risk at t years who in fact survive beyond t years.

An important feature of the Kaplan–Meier estimator is that it makes no assumptions about the shape of $S(t)$ or corresponding hazard function $h(t)$. The method assumes only that censoring occurs noninformatively.

Figure 3 gives the Kaplan–Meier estimators of $S(t)$ for the two samples described in Table 1. Note the similarities in the shapes of the two Kaplan–Meier estimators. Each consists of a sequence of vertical and horizontal lines — called a step function — that begin at the value $S(t) = 1$ at $t = 0$ and decrease. It is the nature of Kaplan–Meier estimators that the jumps occur at observed times of failure. The greater frequency of jumps at earlier times is typical of survival data. Similarly, the larger jumps at later times are also typical. Because these are estimates, the curves are subject to statistical error. The variability in the Kaplan–Meier estimates tends to be greatest at the largest survival times since there are usually only a few subjects still under observation (and few failures) at these times.

Another feature of Kaplan–Meier survival curves is that they do not always drop completely to zero. In fact, a curve will reach zero only if the largest observation is uncensored. This phenomenon reflects the fact that there is insufficient follow-up to describe the complete survival experiences of the population. Thus, for the 6-MP group in Figure 3, the data provide no information about remission times longer than 35 weeks, and hence the Kaplan–Meier curves do not extend beyond this point.

COMPARISON OF TWO GROUPS: THE LOG-RANK TEST

Suppose that we wish to compare the survival distributions of two groups. For example, we may wish to compare the age-specific mortality rates

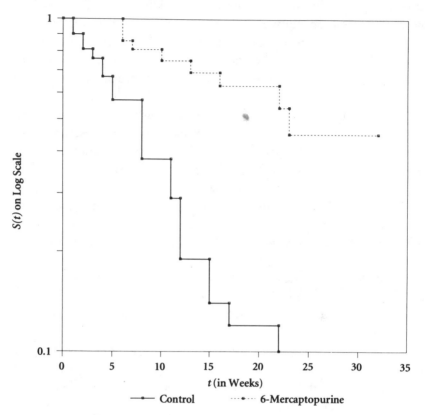

Figure 3.   Kaplan–Meier Plots for Data in Table 1.

of males and females, the age-specific rates of stroke for black and white males, or two experimental treatments being evaluated in a clinical trial. An informal way to carry this out is by plotting the Kaplan–Meier estimators for the groups to be compared. For example, the curves in Figure 3 suggest that 6-MP may prolong the time in remission. To more formally compare these groups, a statistical test is often employed. The most commonly used test for this purpose is the log-rank (or Mantel–Haenszel) test.[6]

The log-rank test is constructed by comparing estimates of the hazard functions of the two groups at each observed time of failure. Specifically, analogous to N_t and n_t, let M_j and N_j denote the numbers of subjects at risk in the 6-MP and control groups, respectively. And of these, let m_j and n_j denote the number of failures that occur at time t_j. For example, consider the data in Table 1 and suppose we let $t_j = 22$. It can be verified that $M_j = 7$, $N_j = 2$, $m_j = 1$, and $n_j = 1$. That is, 1 of the 7 subjects at risk in the 6-MP group failed at week 22, compared to 1 of the 2 in the control group. If we knew that a total of two failures occurred, and if the two groups had equal hazard functions at this time point, we would expect $E_j = 2(2/9) = 0.44$ of the failures to be in the control group because 2/9 of the persons at risk at this time point are in the control group. In fact, we observed $O_j = 1$ failure in the control group at this time point. The log-rank test is constructed by computing the observed (O_j) and expected (E_j) number of failures in one of the groups at each time and then summing these to obtain an overall observed and expected number of failures. The totals O and E are given as

$$O = \sum_j O_j \text{ and } E = \sum_j E_j.$$

In a similar way, a variance, say V, is computed by summing variance contributions for each time point. The resulting statistic, say

$$Z = (O - E)/\sqrt{V},$$

should behave approximately like a standard normal deviate when the two groups have equal survival distributions, and will tend to be large (either positive or negative) when they differ. For example, the data in Table 1 result in $Z = 4.1$ ($P<0.001$), suggesting that the control group has a higher failure rate (that is, shorter durations of remission) than the 6-MP group.

In constructing the log-rank test, O_j and E_j provide a comparison of the hazard functions of the two groups at time t_j. By summing these quantities, the test gives an overall test of the equality of the groups that gives equal emphasis to possible differences at each time point. Alternatively, if we wish to place greater emphasis on differences at earlier time points, a variant of the log-rank test, often called the generalized Wilcoxon test, is used.[6] This is constructed similarly to the log-rank test, but gives greater weight to the observed (O_j) and expected (E_j) number of failures at earlier time points when summing.

ASSESSING MULTIPLE EXPLANATORY VARIABLES: COX'S MODEL

In addition to estimating the survival distribution of a population and comparing two groups, it is often of interest to assess the simultaneous effect of several independent variables on survival time. For example, we may be interested in the combined effects of age and gender on survival. Or we might want to compare two treatment groups while simultaneously controlling for other factors that could influence survival. The most common approach to this type of analysis is Cox's proportional-hazards regression model.[6] In ordinary regression models, the dependent variable is usually expressed as a linear function of several independent variables, and the coefficients of these independent variables are estimated to assess the relation between the independent and dependent variables. Such methods do not easily adapt to censored data. Cox's model overcomes these technical problems by assuming that the independent variables are related to survival time by a multiplicative effect on the hazard function. For example, suppose we wanted to analyze the data in Table 1 while controlling for gender. If we let X_1 denote treatment group ($X_1 = 0$ for 6-MP and $X_1 = 1$ for control) and X_2 denote gender ($X_2 = 0$ for females and $X_2 = 1$ for males), then Cox's model would assume that the hazard function for a subject is of the form

$$h_0(t) \cdot e^{\beta_1 X_1 + \beta_2 X_2}.$$

Here $h_0(t)$ is some underlying hazard and β_1 and β_2 are unknown regression coefficients that can be estimated from the data. Expressed another way, the hazard functions for the four types of subjects are assumed to be

Female, 6-MP:	$h_0(t)$
Female, Control:	$h_0(t) \cdot e^{\beta_2}$
Male, 6-MP:	$h_0(t) \cdot e^{\beta_1}$
Male, Control:	$h_0(t) \cdot e^{\beta_1 + \beta_2}.$

Note that these hazard functions are proportional to one another, and differ only because of the multiplicative exponential term. Thus it is not necessary to know the underlying hazard $h_0(t)$ in order to compare the four groups using ratios. The hypothesis that treatment group is not associated with survival is given by $\beta_1 = 0$ while the hypothesis that gender is not associated with survival corresponds to $\beta_2 = 0$. An attractive feature of Cox's model is that it allows the coefficients β_1 and β_2 to be estimated from the data without any assumptions concerning $h_0(t)$.

The analysis of survival data using Cox's model consists of first estimating the parameters β_1 and β_2 and their standard errors. This allows the estimation of relative risks, the construction of confidence intervals for the parameters or relative risks, and tests of hypotheses about the βs. The computation of the estimators is somewhat complicated and must be done via computer programs, but these are available in many of the standard statistical packages. The interpretation of the results of such an analysis has many similarities to the interpretation of ordinary regression analyses.

There are several generalizations of Cox's model that are useful in the analysis of survival data involving multiple explanatory variables. These include the use of stratification by a categorical variable, the inclusion of explanatory variables that can vary with time, such as the value of a laboratory marker, and interaction terms. A thorough discussion of these extensions is given in Cox and Oakes.[6]

This work was supported by Grant AI24643 from the National Institute of Allergy and Infectious Diseases.

REFERENCES

1. Emerson JD, Colditz GA. Use of statistical analysis in the *New England Journal of Medicine*. N Engl J Med 1983; 309:709-13. [Chapter 3 of this book.]

2. De Gruttola V, Lagakos S. The value of AIDS incidence data in assessing the spread of HIV infection. Statistics in Medicine 1989; 8:35-43.

3. Stanley K. Prognostic factors for survival in patients with inoperable lung cancer. J Natl Cancer Inst 1980; 65:25-32.

4. Freireich EJ, Gehan E, Frei E, et al. The effect of 6-mercaptopurine on the duration of steroid-induced remissions in acute leukemia: a model for evaluation of other potentially useful therapy. Blood 1963; 21:699-716.

5. Lagakos SW. General right censoring and its impact on the analysis of survival data. Biometrics 1979; 35:139-56.

6. Cox DR, Oakes D. Analysis of survival data. Chapman and Hall, 1984.

15

~

Contingency Tables in Medical Studies

Daniel Zelterman, Ph.D., and Thomas A. Louis, Ph.D.

ABSTRACT   Categorical data take the form of counts of individuals falling into two or more discrete classifications. Data of this type are common in medical studies and are presented in contingency tables that show the number of persons belonging to each classification. Seventy percent of the articles in our survey of the *New England Journal of Medicine* used contingency tables as a means of describing characteristics of the patients under study. Therefore, consumers and producers of medical research reports should understand the concepts and issues that arise in interpreting contingency tables. We present basic concepts and analyses and describe the scope of scientific questions addressable by contingency analysis. Small 2 × 2 tables of counts are easily analyzed using either Fisher's exact test or Pearson's chi-square. Tables with three or more dimensions are often in original data, but do not appear often in the *Journal*. Injudiciously collapsing tables can create or eliminate interaction between factors. Simpson's paradox may occur if proper randomization is not performed in the allocation of subjects to all possible treatments. Logistic regression (including multiple logistic regression) is a powerful tool in analysis and has become common in the *Journal*.

C ontingency tables show the numbers of persons or other units that belong simultaneously to two or more distinct categories — for example, surgical patients cross-classified according to both gender and age group (two axes of classification) or bacterial cultures classified according to sensitivity to three different antibiotics (three axes). Hsia et al.[1] studied the accuracy of diagnosis related groups (DRG) coding for Medicare patients in hospitals. Table 1 presents data on the coding errors detected and illustrates the general form of a contingency table. It cross-classifies errors requiring changes in the coding by size of hospital and the reason for reclassification. (We will discuss Table 1 at length throughout this chapter.) Such tables provide useful descriptive information about dis-

Table 1.   Reasons for DRG Changes, According to Hospital Size.*

Reason	Number of Hospital Beds			Total
	<100	100–299	≥300	
Changes by Physician	268	298†	195	661
Changes by Hospital				
Miscoding	56	63	45	164
Resequencing	167	110	96	373
Other	47	70	59	176
Total	538	441	395	1374

    * Data from Hsia et al.[1]

    † This count is incorrect. See the section entitled "Collapsing Tables" for an explanation.

crete variables and can be used to make inferences about the relations among these variables. In this chapter we give examples of contingency tables, explain the types of statistical questions they can help answer, and identify some of the potential pitfalls in their construction and analysis. We use examples from Volume 318 (January–June, 1988) of the *New England Journal of Medicine*.

Contingency tables always deal with counts, but sometimes they can be confused with cross-classification of other kinds of data. In a study of endometriosis, Henzl et al.[2] described mean base-line laparoscopic scores by disease stage and daily treatment regime. Though the general form of Table 2 is similar to that of Table 1 in that row and column categories are cross-classified, Table 2 is not a contingency table because it presents summaries

Table 2.   Mean Base-Line Laparoscopic Scores for Stage III and IV Endometriosis Patients (Includes Mean S.E.).*

Treatment	Stage III	Stage IV
Nafarelin (800 mg)	25.6 (1.9)	73.8 (2.8)
Nafarelin (400 mg)	26.9 (1.8)	59.0 (2.8)
Danazol (800 mg)	24.6 (1.9)	55.1 (2.9)

    * Data from Henzl et al.[2]

of continuous measurements. This difference is important because one must generally use different statistical methods for counts and for continuous variables.

Two features of contingency tables define much of the analysis: the number of axes of classification and the number of classes in each axis. Table 1 demonstrates a variable with four categories: four reasons for changes in DRG. Notice the lack of any necessary ordering of these four categories. They could have been presented in any order, and the ordering in Table 1 is arbitrary. Contrast the structure of this attribute with that for the size of the hospital. The size of the hospital is a discrete measured variable, reporting the number of beds. There is an obvious ordering to these categories. But how should they be defined? Here the choice is arbitrary, but there are a few rules of procedure. Investigators commonly pick both the number of categories and their end points to balance several competing objectives; for example, simplicity, or avoiding categories that have low frequency. Table 1 presents hospital size in three categories, each with approximately the same number of DRG changes, but the data could have been presented as four ordered categories (<100 beds, 100–199, 200–299, ≥300 beds, for example), or as only two. Chapter 13 of this book describes the issues of ordered categories.

In the *Journal*, categorical data commonly describe characteristics of the patients under study. These tables sometimes present an important outcome but more often are used to determine whether the subjects represent the overall population. Table 3, extracted from Veronesi et al.,[3] presents characteristics of 612 patients who were randomized to one of two surgical procedures for malignant melanoma of the skin. The table gives three distinct contingency tables: extent of surgery cross-classified once with sex, once with site, and once with age. These give a good summary of the patient population. This informative type of display is common. Our survey of Volume 318 of the *Journal* revealed that 70 percent of all original articles included tables of this form.

Table 3 shows that the patients randomized to either narrow or wide excision are comparable. Overall, for example, there were more than twice as many women as men in the survey and 80 percent of all melanomas occurred on the trunk or lower limbs. The table indicates that randomization resulted in a well-balanced distribution of sites and sexes across treatments. We might wish to know additional details such as the relation between the

Table 3. Characteristics of 612 Patients with Primary Cutaneous Malignant Melanoma.*

Characteristic	Narrow Excision	Wide Excision
Sex		
Male	93	96
Female	212	211
Site		
Trunk	121	119
Upper limbs	60	61
Lower limbs	124	127
Age (yr)		
0–20	6	0
21–40	101	116
41–50	84	75
51–65	114	116

* Data from Veronesi et al.[3]

site and the sex of the patients. This information would be useful but cannot be gathered from the present table. To take another example from Table 3, notice that all of the youngest age group were randomized to the same treatment (narrow excision). Are the individuals in this youngest group otherwise different from all other subjects? Additional data would be needed to answer this question.

Questions such as those raised by Table 3 are commonly asked when one analyzes categorical data. The very nature of such data requires that one think simultaneously in terms of several axes of classification, with the goal of describing the relations between two, three, four, or more factors. The form of collapsed table given in Table 3 will be discussed later in this chapter.

BASIC ANALYSES

Many aspects of contingency tables are illustrated in the analysis of the most basic form of categorical data reporting: two classifications, each with just two classes. This basic 2×2 table classifies each subject into one of four unique categories. Consider the 2×2 table shown in Table 4. A letter

Table 4. A 2 × 2 Table from a Letter to the *New England Journal of Medicine*.*

Treatment	Response		Totals
	Early Death	Other	
Placebo	8	12	20
Propranolol	2	28	30
Totals	10	40	50

* Data from Smythe.[4]

to the *Journal*[4] described a group of subjects with esophageal varices, cross-classified by treatment (propranolol treatment or placebo control) and response (early death by hemorrhage before admission to a hospital intensive care unit, or other). Table 4 presents the counts and shows that 40 percent (20/50) of the patients were on placebo and 60 percent (30/50) on propranolol. Similarly, 20 percent (10/50) of all patients were classified as "early death." These descriptions of individual variables are available from the row and column totals (margins) of the table.

An obvious clinical question asks whether the death rates differ by treatment. The data permit calculation of these rates. For the rows, the probability of early death in the control group is estimated by: 0.40 (= 8/20), and the probability of early death in the propranolol group is estimated by 0.07 (= 2/30). These numbers are estimating the true probabilities P_c and P_t, respectively.

First we compare these response rates and ask whether the difference can be readily accounted for by random fluctuation. Contingency table methods can be used to produce a confidence interval for the difference between these rates or to test their equality.[5] The test can be performed in many ways, but the most direct method compares two estimated death rates. This comparison can be cast as investigating the degree of dependence between treatment group and death, with statistical independence equivalent to equality of the probabilities. Therefore, it is instructive to consider alternative approaches to testing equality based on the cross-classification. The row totals in Table 4 are fixed by the design of the experiment, whereas the column totals result from the occurrence of early death in the two treatment groups.

A direct comparison of the estimated death rates for binomial proportions depends on the unknown values of the P_c and P_t. To understand the approach and establish notation, consider the schematic table of counts:

Row
totals

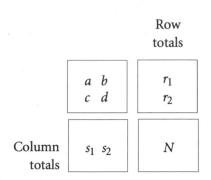

Our goal is to compare the row-specific probabilities of falling in the first column: a/r_1 and c/r_2. Equivalently, we can compare the row-specific odds for row 1 to the row-specific odds for row 2, where the odds are simply, in row 1, the ratio of the count for column 1 to the count for column 2, and similarly for row 2. We obtain a/b for row 1 and c/d for row 2. Algebraic work will show that the row probabilities will be equal if the odds are equal; and further, the odds are equal if the *odds ratio*, $(a/b)/(c/d)$, or (ad/bc) equals 1. Statistical tests investigate the hypothesis that the true, rather than the observed, odds ratio is 1.

An odds ratio of 1 is equivalent to independence of row and column variables (treatment group and death), and odds ratios larger or smaller than 1 can be interpreted as positive and negative dependence. The odds ratio provides one measure of the degree of dependence between categorical variables. In our example in Table 4, the estimated odds ratio of $ad/bc = (8 \times 28)/(12 \times 2) = 9.3$ indicates a high degree of dependence — here, that propranolol treatment greatly reduces the chances of early death. For the hypothesis test we need to compute the probability of observing an odds ratio equal to or more extreme than 9.3 under the null hypothesis that the true odds ratio is 1.

Fisher's exact test considers the 2×2 table with both the row and column totals fixed and describes the conditional probability of observing the four interior counts (a,b,c,d), assuming $P_c = P_t$ (that the odds ratio is 1). For Table 4 the test produces an extremely small P value ($P<0.006$). This conditional probability means that if $P_c = P_t$ and the tables are constructed at random conditional on matching the observed totals for both rows and columns, the probability of a result at least this extreme and in this direction is about 0.006. This result is strong evidence that P_c differs from P_t and that the true odds ratio differs from 1.

When the row and column totals are fixed, any one count inside the table will determine the other three. The remaining three cells are not free because of the fixed row and column sums. That only one cell determines the table gives rise to the expression "one degree of freedom." For tables with large counts and many categories or dimensions, the computation necessary for the exact test can be excessively lengthy, although advances[6] have extended its use. Alternative analyses, based on approximate methods such as Pearson's goodness-of-fit chi-square, can often be used to get very nearly the same results with much less computation.

The concept of statistical independence is important in the analysis of this table because $P_c = P_t$ is equivalent to statistical independence of treatment status and response category. Treatment status is independent of response when the treatment has no effect. Independence implies that *expected counts* for the four entries in Table 4 are given by products of row and column proportions times the table's total sample size. For example, the probability of classifying a patient into row 1, column 1 should be the product of the respective proportions, that is $20\!/\!50 \times 10\!/\!50 = 0.08$ if the treatment and response are independent. This is the proportion of the sampled individuals we expect to fall into this cell, so the expected count requires that we multiply this proportion by the total sample size. Table 5

Table 5. Expected Counts of Table 4 under the Hypothesis of Independence of Treatment and Response.

Treatment	Response		Totals
	Early Death	Other	
Placebo	$\dfrac{20}{50} \times \dfrac{10}{50} \times 50 = 4$	$\dfrac{20}{50} \times \dfrac{40}{50} \times 50 = 16$	20
Propranolol	$\dfrac{30}{50} \times \dfrac{10}{50} \times 50 = 6$	$\dfrac{30}{50} \times \dfrac{40}{50} \times 50 = 24$	30
Totals	10	40	50

presents the expected counts from Table 4 under the hypothesis of independence. While Table 4 is limited to integers, expected counts (as in Table 5) can be non-integers. However, the row and column sums of the expected counts (Table 5) are the same whole numbers as for the original observations (Table 4).

To avoid some of the computational difficulties encountered with Fisher's exact test with large samples, the Pearson chi-square is often used. When all of the row and column totals are large, these two tests yield approximately the same P values. There often is some confusion here because the Pearson statistic and its approximate distribution are both called chi-square. There is an approximation because the approximate distribution is continuous but the Pearson statistic takes on discrete values. This approximation to the distribution of the Pearson chi-square clarifies the use of the name "exact" when talking about Fisher's test.

The Pearson statistic summarizes the discrepancies between the observed and expected counts and is compared to a table of chi-square values. For each of the four cells we square the difference (observed minus expected) and divide by the expected number, then add the four results. In mathematical notation,

$$\chi^2 = \sum_{\text{all cells}} \frac{(\text{observed count} - \text{expected count})^2}{\text{expected count}}.$$

For Tables 4 and 5 this equals

$$\frac{(8-4)^2}{4} + \frac{(12-16)^2}{16} + \frac{(2-6)^2}{6} + \frac{(28-24)^2}{24} = 8.33.$$

When $P_c = P_t$ (that is, when the hypothesis of independence is true) and the row and column totals are large, then this statistic behaves approximately as the chi-square distribution with 1 degree of freedom. This chi-square test also concludes that P_c is very different from P_t in this example. The P value corresponding to $\chi^2 = 8.33$ is 0.004, found from a table of the chi-square distribution. This is in close agreement with the P value (0.006) obtained using Fisher's exact test.

The ideas of this section are easily generalized to larger tables. For example, a two-way table with R rows and C columns produces a test of independence (or equality of all population row- or column-specific percentages). There are $(R-1)(C-1)$ degrees of freedom, because if all row and column totals are fixed, then the last row and column of the table are determined by the other cells in the table. Additionally, association in the R-by-C table can be measured by $(R - 1)(C - 1)$ odds ratios, and the test for independence is equivalent to testing that they are all equal to 1. Tables with three or more axes of classification can be represented in a similar manner, and we give examples below.

SURVEY RESULTS AND OTHER ANALYSES

A survey of Volume 318 (January–June, 1988) of the *Journal* indicates that an extremely high proportion (70 percent, i.e., 78/109) of the original research articles, and many of the letters as well, contain some reference to tables of counts and/or the above mentioned techniques of analyzing them. Therefore, readers need at least a basic understanding of the analysis of categorical data if they are to understand a large proportion of papers in the *Journal*.[7] Chapter 3 of this book will be useful in this regard. A common use of a table of categorical data in the *Journal* articles surveyed was to describe demographic or other characteristics of a patient population under study, such as the age and sex distributions in Table 3. Substantially fewer contingency tables included the primary response to therapy (e.g., lived/died, rejected/didn't reject transplanted organ, etc.), perhaps because most outcomes are measured (e.g., change in blood pressure) rather than counted (e.g., deaths from hypertension). For tables where the response variable is categorical, a technique called logistic regression has been developed, which will be discussed at the end of this chapter.

Volume 318 of the *Journal* contained a total of 109 Original Articles. Among these we found 46 (42 percent) that made some reference to either Fisher's exact test or the Pearson goodness-of-fit chi-square test, described in the previous section. The use of Fisher's exact test was always limited to 2×2 tables but the chi-square test was also used to analyze them. We found only seven original articles with contingency tables in three variables, and none in more than three. Another group of 49 (45 percent) articles presented one- and two-dimensional tables (as our Table 3) that resulted from collapsing higher-dimensional tables. As we mentioned earlier in this chapter, such tables can be informative but are typically not analyzed. Our survey also revealed 27 articles (25 percent) that contained virtually no statistical data, but rather presented a list of subjects and their individual characteristics. These lists were typically rather short (rarely exceeding 20 patients).

The presentation of tables of categorical data in the *Journal* is usually limited to two or at most three factors at a time, and analysis tends to be even more limited. Notice that the two axes in Table 1 could have been extended by further cross-classification to show age group of patient (over or under 40 years), initial primary diagnosis (traumatic injury or other), size of hospital (small, medium, large), and reason for change in a four-way table of counts. Such a table would provide much greater information, but such high-dimensional tables are rarely seen in print because few people are able to visualize the complex interactions of more than two factors at one time. Too much information means more confusion for the reader. Models relating odds ratios in complex tables (commonly referred to as log-linear models[8,9]) were developed as a means of finding and understanding the complex relationships that are present in high-order dimensional tables of cross-classified data, but these models have met greater acceptance in the social sciences and epidemiology than in the clinical medical literature. Logistic regression, with categorical regressors (discussed at the end of this chapter), is an example of a log-linear model.

COLLAPSING TABLES

Another, perhaps more important, reason for the dearth of high-dimensional tables in medical journals may be a concern for succinctness and the desire to present a single point. Succinctness serves to draw the reader's attention to the point being made and the individual point of view may discourage the investigator from presenting alternative analyses at length. Obviously, a decision on how much to present must be made — it is impossible and undesirable to publish all the information available. Regardless of the author's motivation, tables that have been reduced (or *collapsed*) to two dimensions occur frequently in the *Journal*. These collapsed tables retain much of the original information, yet they can mislead the reader.

For an example, we will look more closely at Table 1 which cross-classifies DRG coding changes with hospital size and reason for change, but first we need to fix an apparent problem. Each marginal sum (total for a

Table 6.   Who Changes the DRG, with Corrected Counts.

Who Changes the DRG?	Hospital Size			Totals
	<100 beds	100–299 beds	≥300 beds	
Physicians	268	198	195	661
Hospitals	270	243	200	713
Totals	538	441	395	1374

row or column) should equal the sum of the counts in the corresponding column or row of the table. While this is generally true, we notice that the first row sums to 761 rather than 661 as given in Table 1. Similarly, notice that the second column, corresponding to the middle sized hospitals, sums to 541 rather than 441 as indicated. One way to repair this inconsistency is to change the count of the first row, second column to 198, from the 298 as reported in the original article. There is a good reason to suppose one error (likely typographical) occurred rather than two or more. It is much more likely that one error was made by prudent authors of a carefully prepared manuscript than two simultaneous errors of the same magnitude and direction — in this instance, in the two marginal totals. (Personal communication with the authors indicated that there was in fact one error rather than two.) The important lesson to be learned here is that tables of counted data usually have an internal consistency check. Few other methods of displaying the data would allow us to detect this kind of error.

Let us use this corrected table and return to the issue of collapsing large tables into smaller ones. Consider the issue of who changes the DRG and look at the reduced table given by Table 6. If we check to see if a hospital's size is related to who did the change, we obtain a Pearson's $\chi^2 = 2.70$ on 2 degrees of freedom, with a significance level of about 0.3. It seems reasonable to conclude, subject to considerations of statistical power, that the source of the change is the same for all sizes of hospitals.

Suppose now that we want to simplify the table by having two columns rather than three. Collapsing the second and third columns produces Table 7. After combining the counts in each row to get the entries in Table 7, we compute the proportion in each column and multiply by 100 to get the percentages. For example, the percentage of changes made by

Table 7. One Method of Collapsing Table 6 (Column Proportion × 100 ± S.E.).

Who Changes the DRG?	Number of Hospital Beds	
	<100	≥100
Physicians	268 (49.81 ± 2.16)	393 (47.01 ± 1.73)
Hospitals	270 (50.19 ± 2.16)	443 (52.99 ± 1.73)

physicians is $[268/(268 + 270)] \times 100 = 49.81$, with estimated standard error $100 \sqrt{(0.4981 \times 0.5019)/538} = 2.1$.

The numbers in parentheses represent column percentages plus or minus their estimated standard error. It appears that the proportion of errors made by physicians is smaller in the larger hospitals than in the smaller hospitals, although the difference of almost two points is not statistically significant. Conversely, by pooling results for hospitals with less than 300 beds, we can collapse Table 6 to obtain Table 8, again with column percentages plus or minus their standard errors in parentheses. As in Table 7, the differences are not statistically significant, but the suggestion is just the opposite: the proportion of errors made by physicians is greater in the larger hospitals than in the smaller hospitals. Clearly there is a lot of room for error and misunderstanding. Care is needed in forming and interpreting categories.

We have just shown how collapsing tables can produce associations. The opposite is also true: collapsing can make significant associations disappear, increase in magnitude, or even reverse their direction. Consider a fictitious setting in which a patient is given a certain therapy and we

Table 8. Another Method of Collapsing Table 6. (Column Proportion × 100 ± S.E.)

Who Changes the DRG?	Number of Hospital Beds	
	<300	≥300
Physicians	466 (47.60 ± 1.60)	195 (49.37 ± 2.52)
Hospitals	513 (52.40 ± 1.60)	200 (50.63 ± 2.52)

Table 9.   An Example of Simpson's Paradox.

	Women		Men	
	Treated (%)	Control (%)	Treated (%)	Control (%)
Success	8,000 (62)	4,000 (57)	12,000 (44)	2,000 (40)
Failure	5,000 (38)	3,000 (43)	15,000 (56)	3,000 (60)

judge its success or failure for that patient. The data might appear as in Table 9. The treatment is beneficial to women (62 percent treated successes versus 57 percent control successes) and to men (44 percent treated successes versus 40 percent control successes). These treatment effects are statistically significant for both men and women. The odds ratio for women ($[8,000 \times 3,000]/[5,000 \times 4,000] = 1.2$) is the same for men ($[12,000 \times 3,000]/[15,000 \times 2,000] = 1.2$). By the odds ratio measure of association, treatment effects are the same for men and women. We conclude then, that the therapy is beneficial and has the same statistically significant effect on women as on men.

The real shock comes when the data on separate sexes are combined as displayed in Table 10: the treatment effect is lost. That is, the odds ratio for the combined data is unity. The treatment works for women and it works for men, but it seems not to work for people! To expose this apparent paradox notice that, overall, men do not do as well as women (whether one considers treated or untreated patients) and the treated group is heavily weighted with men. Thus treated patients in Table 10 have a poor success rate, not because treatment is ineffective, but because most of them are male. This example shows the importance of proper stratification and of

Table 10.   Simpson's Paradox Realized.

	Both Sexes Combined	
	Treated	Control
Success	20,000	6,000
Failure	20,000	6,000

Table 11. Simpson's Paradox Explained.

	Women	Men
Treated	13,000	27,000
Control	7,000	5,000

combining data in a way that maintains differences. We could also have constructed examples in which the individual odds ratios were averaged, or greatly increased in the resulting 2 × 2 table, when two tables are combined.

Table 9 is an example of what is known as Simpson's paradox. The problem is not uncommon and can be traced to the fact that there is a strong relation between choice of therapy and gender, as demonstrated in Table 11.

In this study, men were much more likely to receive the treatment (as opposed to control) than the women subjects. If proper randomization to treatment or control had been part of the protocol, this degree of imbalance would have been very unlikely. In this study the relation between gender and choice of therapy was much stronger than the relation between therapy and outcome. If we collapse the gender variable, much information is lost and the therapy is incorrectly seen as useless, or worse.

Simpson's paradox is a well documented phenomenon, but the user of statistics should not live in constant fear of its appearance. Randomization reduces the chances of major problems, and we can usually collapse variables that are independent of the treatment and outcome, though the stratified analysis may provide important information. The problem with this example stems from the close relation between gender and the choice of treatment.

LOGISTIC REGRESSION

Greenberg et al.[10] retrospectively examined the records of lung cancer patients. Their article considers how a decision to perform surgery was related to the patient's marital status, medical insurance, or proximity to a major cancer-treatment center. For these types of data, the outcome categories (the decision to operate or not) are viewed as a response to covariates that may be continuous (proximity to a cancer center) or discrete (marital status). Binary-valued covariates are usually given some arbitrary numerical coding such as zero for male and one for female. Such data can be analyzed by logistic or probit regression, depending on the mathematical form of the response function that is considered.

Logistic regression is used to model the probability of occurrence of a binary (yes/no) outcome. Ordinarily we use linear regression to predict the value of a dependent variable (in this case, the decision to operate or not) from values of the independent covariates (e.g., marital status, proximity to the center, and enrollment in an insurance plan). Chapter 11 of this book discusses linear regression at greater length. We cannot use linear regression when the outcome is binary valued because the probability being modeled must lie between zero and one. Ordinary linear regression does not guarantee these limits. Specifically, if we tried to model P, the probability of surgery, as a function of covariates $(x_1, x_2, ..., x_k)$ in the form

$$P(x_1, x_2, ..., x_k) = \alpha + \beta_1 x_1 + \beta_2 x_2 + ... + \beta_k x_k,$$

then a patient who lives a very large distance from the medical center might have an estimated probability greater than one or less than zero depending on the sign of the β coefficient in front of this covariate. One solution to this problem might be to say that all negative estimates are replaced by zero, and those over one are to be replaced by one. This is not entirely satisfactory because we usually would like a relation something like Figure 1. For extreme values of the covariate x the probability $P(x)$ should be near zero or one, but not exactly, and should keep getting closer for even more extreme values of x. A better solution to this problem with linear regression lies in a non-linear response function. There are many such response functions in use but the one in most common use is called the logistic function, or logit.

The logit model is in terms of the odds $\frac{P}{1-P}$. The odds must be positive, so linear regression on the odds does not completely work either. The solution is to model the logarithm of the odds or *logit* as a linear model:

$$\log_e \left\{ \frac{P(x_1, x_2, ..., x_k)}{1 - P(x_1, x_2, ..., x_k)} \right\} = \alpha + \beta_1 x_1 + \beta_2 x_2 + ... + \beta_k x_k$$

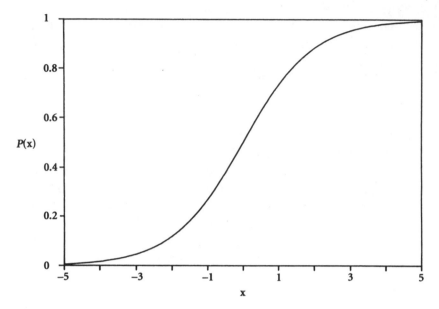

Figure 1. The Logistic Response Function $P(x)$.

for parameters $\alpha, \beta_1, \beta_2, \ldots, \beta_k$ that must be estimated from the data. This model is referred to as *logistic regression* and always predicts a value of P between zero and one. If there is only one covariate, x, and we set $\alpha = 0$ and $\beta_1 = 1$, then $P(x)$ is the non-linear function of x plotted in Figure 1. For further discussion of this model, standard references are available.[8] A typical use of logistic regression would be to assess the risk, P, of stroke as a function of diastolic blood pressure, x.

When the predictor variables as well as the dependent variables are binary valued, logistic regression can be performed using standard 2×2 contingency analysis. If x takes on only the values 0 and 1 corresponding to treatment and control, then the difference of the logits at these two different values is the slope β. The difference of two logits is the log of the odds ratio described in the Basic Analyses section of this chapter. For example, in the artificial example relating treatment effectiveness for men and women given in Table 9, we have a constant odds ratio of 1.2 for each gender. The logarithm of this value is the slope β on the treatment variable *x*. Similarly, the 2 × 2 table relating outcome to gender for the two treatment groups (Table 11) produces an odds ratio that determines the slope on gender in a logistic regression where *x* is 0 or 1 depending on the patient's gender. The logistic regression approach affords the flexibility of being able to model the effects of both discrete and continuous predictors with a binary or yes/no response.

SUMMARY

Categorical data are used to report counts of subjects rather than to summarize continuous measurements, but both types of tables may have a similar appearance. The choice and number of categories is arbitrary but usually dictated by a desire to focus the reader's attention. A large number of the Original Articles in our survey of the *Journal* used contingency tables but never with more than three variables at a time. The basic analysis of a table is to compare rates via Fisher's test or Pearson's chi-square. Collapsed tables can create or remove interactions due to Simpson's paradox, but proper randomization reduces the chances of this occurring. Contingency tables often have an internal error check when they are printed with their row and column totals. Logistic regression is useful for estimating the probability of a yes or no response from other covariates.

This research was supported by NIH grants P50-NS14538, P01-AGO8761, and NSF grant GMS 8702402. We thank Deb Sampson for preparing the manuscript.

REFERENCES

1. Hsia DC, Krushat WM, Fagan AB, et al. Accuracy of diagnostic coding for Medicare patients under the prospective-payment system. N Engl J Med 1988; 318:352-5.

2. Henzl MR, Corson SL, Moghissi K, et al. Administration of nasal Nafarelin as compared with oral Danazol for endometriosis: a multicenter double-blind comparative clinical trial. N Engl J Med 1988; 318:485-9.

3. Veronesi U, Cascinelli N, Adamus J, et al. Thin stage I primary cutaneous malignant melanoma. N Engl J Mcd 1988; 318:1159-62.

4. Smythe WR. Propranolol and hemorrhage from esophageal varices. N Engl J Med 1988; 318:994.

5. Kuzma JW. Basic statistics for the health sciences. Mountain View, Calif.: Mayfield Publishing Co., 1984.

6. Mehta CR, Patel NR. A network algorithm for performing Fisher's exact test in R × C contingency tables. J Am Stat Assoc 1983; 78:427-34.

7. Moses LE. Statistical concepts fundamental to investigations. N Engl J Med 1985; 312:890-7. [Chapter 1 of this book.]

8. Fienberg SE. The analysis of cross-classified categorical data. 2nd ed. Cambridge, Mass.: MIT Press, 1980.

9. Bishop YMM, Fienberg SE, Holland PW. Discrete multivariate analysis. Cambridge, Mass.: MIT Press, 1975.

10. Greenberg ER, Chute CG, Stukel T, et al. Social and economic factors in the choice of lung cancer treatment: a population-based study in two rural states. N Engl J Med 1988; 318:612-7.

SECTION IV

Communicating
Results

16

~

GUIDELINES FOR STATISTICAL REPORTING IN ARTICLES FOR MEDICAL JOURNALS*

AMPLIFICATIONS AND EXPLANATIONS

John C. Bailar, III, M.D., Ph.D., and Frederick Mosteller, Ph.D.

ABSTRACT   The 1991 edition of the *Uniform Requirements for Manuscipts Submitted to Biomedical Journals* includes guidelines for presenting statistical aspects of scientific research. The guidelines are intended to aid authors in reporting the statistical aspects of their work in ways that are clear and helpful to readers. We examine these guidelines for statistics using 15 numbered statements. Although the information presented relates to manuscript preparation, it will also help investigators in earlier stages make critical decisions about research approaches and protocols.

I n 1979, the group now known as the International Committee of Medical Journal Editors first published a set of uniform requirements for preparing manuscripts to be submitted to their own journals. These uniform requirements have been revised several times,[1] and have been widely adopted by other biomedical journals. In the 1988 revision,[2] the Committee added guidelines for presenting and writing about statistical aspects of research. The purpose of these guidelines is to assist authors in reporting statistical aspects of their research in ways that will be responsive to the queries of editors and reviewers and helpful to readers. The statistical guidelines remain unchanged in the 1991 revision.[1]

We present the statistical guidelines as a sequence of 15 numbered statements, and amplify and explain some of the reasoning behind the guide-

---

*Modified and reprinted from the *Annals of Internal Medicine,* 1988; 108:266-73, with the permission of the publisher.

lines. Our text focuses on manuscript preparation, but it should also be helpful at earlier stages when critical decisions about research approaches and protocols are made. We emphasize a few important aspects of reporting statistical work in publications, and we provide references to general statistical texts. The International Committee is not responsible for these amplifications; we have tried, however, to present the spirit of the Committee's discussions as well as our own views.

The International Committee's statistical guidelines are as follows:

Describe statistical methods with enough detail to enable a knowledgeable reader with access to the original data to verify the reported results. When possible, quantify findings and present them with appropriate indicators of measurement error or uncertainty (such as confidence intervals). Avoid sole reliance on statistical hypothesis testing, such as use of P values, which fails to convey important quantitative information. Discuss eligibility of experimental subjects. Give details about randomization. Describe the methods for, and success of, any blinding of observations. Report treatment complications. Give numbers of observations. Report losses to observation (such as dropouts from a clinical trial). References for study design and statistical methods should be to standard works (with pages stated) when possible, rather than to papers where designs or methods were originally reported. Specify any general-use computer programs used.

Put general descriptions of methods in the Methods section. When data are summarized in the Results section specify the statistical methods used to analyze them. Restrict tables and figures to those needed to explain the argument of the paper and to assess its support. Use graphs as an alternative to tables with many entries; do not duplicate data in graphs and tables. Avoid nontechnical uses of technical terms in statistics, such as "random" (which implies a randomizing device), "normal," "significant," "correlation," and "sample." Define statistical terms, abbreviations, and most symbols.

Our general approach is that scientific and technical writing should be comprehensible at the first reading for the average reader who is knowledgeable about the general area but is not a subspecialist in the specific topic of investigation.

1. *Describe statistical methods with enough detail to enable a knowledgeable reader with access to the original data to verify the reported results.*

Authors should report which statistical methods they used, and why. In many instances they should also report why other methods were not used, although this is rarely done.

Readers must be told about weaknesses in study design and about study strengths in enough detail to form a clear and accurate impression of the reliability of the data, as well as any threats to the validity of findings and interpretations. Such details are often omitted, although investigators probably would know them.[3,4]

The researcher must decide which statistical measures and methods are appropriate, given that a statistical goal has been defined. Investigators often have a choice: Mean or median? Nonparametric test or normal approximation? Adjustment, matching, or stratification to deal with confounding factors? Choosing statistical methods generally requires an appreciation of both the problem and the data, and an experienced biostatistician, statistician, or epidemiologist can often provide substantial help. This help should begin well before the study is started, because the foundation for reporting one's findings is laid before the study even begins.

Trying several reasonable statistical methods with collected data is often appropriate, but this strategy must be disclosed so that readers can make their own adjustments for the authors' industriousness or skill in fishing through the data for a favorable result. Whatever statistical task is defined, it is inappropriate, and indeed unethical, to try several methods and report only those results that suit the investigator. Results of overlapping methods need not be presented separately when they largely agree, but authors should state what additional approaches were tried, and that they did agree. Of course, results that do not agree also should be given, and investigators may sometimes find that such disagreements arise from important and unexpected aspects of the data.

Units should always be specified in text, tables, and figures, although not necessarily every time a number appears if the unit is clear to the reader. Often, careful choice of units of measurement can help clarify and unify the study question, biological hypotheses, and statistical analysis. Careful reporting of units can also help to avoid serious misunderstanding. Are quantities in milligrams or millimoles? Are rates per 10,000 or per 100,000? Does a figure show the number of different patients, the number of myocardial infarcts among those patients (including second infarcts), or the number of admissions to a given hospital (including readmissions)? Research investigators often use abbreviated language that is clear to their colleagues, but they may have to make a special effort to assure that such usage will not confuse nonspecialists, or even other experts.

2. When possible, quantify findings and present them with appropriate indicators of measurement error or uncertainty (such as confidence intervals).

Investigators have to choose a way to report their findings. The most useful ways give information about the actual outcomes, such as means and standard deviations as well as confidence intervals. The tendency to report a test of significance alone — rather than with this additional information — should be resisted, although a significance test in the context of other information may be helpful.

Readers have many reasons for studying a research report. One is to find out how a particular treatment does in its own right, not just in comparison with another treatment. At a minimum, readers should be offered the mean and the standard deviation for every appropriate outcome variable. Significance levels (P values), such as $P = 0.03$, are often reported to show that the difference seen or some other departure from a standard (a null hypothesis) had little probability of occurring if chance alone was the cause. Merely reporting a P value from a significance test of differences omits the information about both the average level of performance and the variability of individual outcomes for the separate treatments.

Exact P values rather than statements like "$P<0.05$" or "P not significant" should be reported where possible so that readers can compare the calculated value of P with their own choice of critical values. In addition, other investigators may need exact values of P if they are to combine results of several separate studies.

In independent samples, information about means, standard deviations, and sample sizes can often be readily converted to a significance test and thus into a P value. When only the P value is reported,[5,6] none of these quantities can be reconstructed, and important information is lost.

Make clear whether a reported standard deviation is for the distribution of single observations, for the distribution of means (standard errors), or for the distribution of some other statistic such as the difference between two means. If the standard deviation for single observations is given, together with sample sizes, then in independent samples the reader can compute the other standard deviations.

Each statistical test of data implies both a specific null hypothesis about those data (such as "The 60-day survival rate in Group A equals that in Group B," so that the difference is zero) and a specific set of alternative hypotheses (such as "The survival rate is different in Group B," which allows for a range of values for the difference). It is critical that both the null hypothesis and the alternatives be clearly stated, although many authors fail to do so. Clear reporting will not only help readers, it is also likely to reduce the frequency of abuse of P values.

It is critical also that authors specify how and when they developed each null hypothesis in relation to their consideration of the data. Statistical theory requires that null hypotheses be fully developed before the data are examined — indeed, before even the briefest view of preliminary results. Otherwise, P values cannot be interpreted as meaningful probabilities.

Authors should always specify whether they are using two-tail or one-tail tests.

3. Avoid sole reliance on statistical hypothesis testing, such as the use of P values, which fails to convey important quantitative information.

Confidence intervals offer a more informative way to deal with the significance test than does a simple P value. Confidence intervals for a single mean or a proportion provide information about both magnitude and its variability. Confidence intervals on a difference of means or proportions provide information about the size of an effect and its uncertainty, but not about component means, so these should also be given.

A significance test of observed data, generally to determine whether the (unknown) means of two populations are different, usually winds up with a score that is referred to a table, such as a t-, normal-, or F-table. The P value is then obtained from the table.

Although confidence intervals offer appraisals of variability and uncertainty, in some studies, such as certain large epidemiologic and demographic studies, biases are often greater threats to the validity of inferences than ordinary random variability (expressed in the standard deviation). Coding or typing errors may exaggerate the number of deaths from a cause, nonresponse to treatment may be selective (those patients more ill being less likely to respond), and so on.

4. Discuss eligibility of experimental subjects.

Reasons for and methods of selecting patients or other study units should always be reported, and, if the selection is likely to matter, the reasons should be reported in detail. The full range of potentially eligible subjects, or the scope of the study, should be precisely stated in terms that readers can interpret. It is not enough to say that the natural history of a condition has been seen in "100 consecutive patients." How do these patients compare with what is already known about the condition in terms of age, sex, and other factors? Are patients from an area or population that might be special? Are patients from an "unselected" series with an initial diagnosis, or do they include referral patients (weighted with less serious or more serious problems)? In comparing outcomes for patients who underwent surgery to outcomes for patients treated medically, were the groups in similar physical condition initially? What about probable cases not proved? Many other questions will arise in specific instances. Sometimes information is obvious (for example, if the investigator studied patients from one hospital because that is where he or she practices). Other questions about scope need answers. (Why begin on 1 January 1983? Why include only patients admitted through the emergency room?) Authors should try to imagine themselves as readers who know nothing about the study.

Although every statistically sound study has such scope criteria to determine the population sampled by the investigator, many also have more detailed eligibility criteria. Medical examples include the possible exclusion of patients outside a specified age range, those previously treated, and those who refuse randomization or are too ill to answer questions.

Which criteria are used to establish scope and which are used to establish eligibility may be uncertain, although both must be reported. Scope criteria push study boundaries outward, toward the full range of patients or other study units that might be considered as subjects, whereas eligibility rules narrow the scope by removing units that cannot be studied, that may give unreliable results, that are likely to be atypical (for example, the extremes of age), that cannot be studied for ethical reasons (for example, pregnant women in some drug studies), or that are otherwise not appropriate for individual study.

The first goal of reporting selection of subjects for the investigation is to state both scope and eligibility so that another knowledgeable investigator, facing the same group of patients or other study units, would make nearly the same decisions about including patients in the study.

The second goal is to provide readers with a solid link between the patients or cases studied and the population for which inferences will be made. Both scope and eligibility constraints can introduce substantial bias when results are generalized to other subjects, and readers need enough information to make their own assessment of this potential. Thus, reasons for each eligibility criterion should be stated. The two critical elements in setting the base for generalization are first to document each exclusion under the eligibility criteria with the reasons for that exclusion; and second, to present an accounting (often in a table) of the difference between patients falling within the scope of the study and those actually studied. The article should also say how patients excluded for more than one reason are handled; common approaches are to show specific combinations or to use a priority sequence. Such information helps the reader better understand how the study group is related to the population it came from, and also helps to assure that all omissions are accounted for. It should be so stated if no subject was ineligible for more than one reason.

Another critical element in reporting is to say how and when the scope and eligibility criteria were devised. Were scope and eligibility criteria set forth in a written protocol before work was started? Did they evolve during the course of the study? Were some eligibility criteria added at the end to deal with some problems not foreseen? For example, a written protocol might call for the study of "all" patients, but if only 5 percent of patients were female (or male), they might be set aside at this point — especially if they are thought to differ from male patients in ways relevant to the subject of the study.

5. *Give details about randomization.*

The reporting of randomization needs special attention for two reasons. First, some authors incorrectly use *random* as a synonym for haphazard. To prevent misunderstanding, tell readers how the randomization was done (coin toss, table of random numbers, cards in sealed envelopes, or some other method). Readers will then know that a random mechanism was in fact applied, and they can also judge the likelihood that it was subject to bias or abuse (such as peeking at cards). Second, randomization can enter in many ways. For example, a sample may be selected from a larger population at random, or study patients may be randomly allocated to treatments, or treated patients may be randomly given one or another test. Thus, it is not enough just to say that a study was randomized. The many possible roles of randomization can be dealt with by careful reporting to assure there is no ambiguity.

Even with randomization, imbalances occur, and these may need attention even if they do not call for special steps in the analysis. Stratification or matching may be used in combination with randomization to increase the similarity between the treated and control groups, and should be reported. Sometimes an assessment of the efficacy of stratification or matching in overcoming the imbalance is feasible; if so, it should be done and reported.

If the randomization was *blocked* (for example, by arranging that within each successive group of six patients, three are assigned to one treatment and three to another), reasons for blocking and the blocking factors should be given. Blocking should ordinarily affect statistical analysis, and authors should say how they used blocking in their analysis or why they did not.

6. *Describe the methods for, and success of, any blinding of observations.*

Blinding, sometimes called masking, is the concealment of certain information from patients or members of the research team during phases of a study. Blinding can be used to good effect to reduce bias, but because it can be applied in different ways, a research report should be explicit about who was blinded to what. An unadorned statement that a study was blind or double blind is rarely enough.

Patients may be blinded to treatment, or to the time that certain observations are made, or to preliminary findings regarding their progress. A decision to admit a patient to a study may be made blind to that patient's specific circumstances, and a decision that a patient randomized to treatment was not eligible may be made blind to the assigned treatment. The observer who classifies clinical outcomes may be blinded to the treatment, as may be the pathologist who interprets specimens or the technician who measures a chemical substance. These and other efforts to prevent bias by blinding should be reported in enough detail for readers to understand what was done.

The effectiveness of blinding should also be discussed in any situation where the person who is blinded may learn or guess the concealed information, such as by side effects that may accompany one treatment but not another. Such discoveries are particularly important for observations reported by patients themselves and for third-party observations of end points with a subjective component, such as level of patient activity.

A particularly critical aspect of blinding is whether the decision to admit a patient to a study was made before (or otherwise entirely and demonstrably independent of) any decision about choice of treatment to be used or offered. Where random allocation to treatments is used, the timing of randomization in relation to the decision to admit a patient should always be stated.

   7.   *Report treatment complications.*

Any intervention, or treatment, has some likelihood of causing unintended effects, whether the study is of a cell culture, a person, an ecologic community, or a hospital management system. Side effects may be good (quitting smoking reduces the risk of heart disease as well as the risk of lung cancer) or bad (drug toxicity). Side effects may be foreseen or unexpected. In most studies side effects will be of substantial interest to readers. Does a drug cause so much nausea that patients will not take it? If we stock an ecologic area with one species, what will happen to a predator? Does a new system for scheduling the purchase of hospital supplies at lower overall cost change the likelihood that some item will be exhausted before the replacement stock arrives?

Nearly every medical treatment carries some risk of complications — that is, of unintended adverse effects. Such effects should be sought at least as assiduously as beneficial effects, and they should be reported objectively and in detail. Treatment failure often gives the most useful information from a study. If no adverse effects can be found, the report should say so, with an explanation of what was done to find them.

   8.   *Give numbers of observations.*

The basic observational units should be clearly specified, along with any study features that might cause basic observations to be correlated. A study of acid rain might take samples of water from five different depths in each of seven different lakes — 35 measurements in all. But the relevant sample size for one or another purpose may be five (depths), or seven (lakes), or 35 (depths in different lakes). In a meta-analysis of such work[7] the whole study may count as only a single observation.

Similarly, a study in several institutions of rates of infection after surgery may be considered to have a sample size of three hospitals, 15 surgeons, 600 patients, or 3000 days of observation after surgery.

Choice of the basic unit of observation determines the proper method of analysis. This choice is critical, and may require an informed under-

standing of statistics as well as the subject matter. The analysis and reporting of correlated observations, such as the water samples and the infection
rates described above, raise difficult issues of statistical analysis that often
require expert statistical help.

A different kind of problem arises from ambiguity in reporting ratios,
proportions, and percents, where the denominator is often not specified
and may be unclear to readers.

Whatever the investigators adopt as their basic unit of observation, relationships to and possible correlations with other units must be discussed.
Such internal relationships can sometimes be used to strengthen an analysis (when a major source of difference is balanced or held constant), and
sometimes they weaken the analysis (by obscuring a critical limitation on
effective sample size). Complicated data structures need special attention
in study reporting, not just in study design, performance, and analysis.

9. Report losses to observation (such as dropouts from a clinical trial).

When the sample size for a table, graph, or text statement differs from
that for a study as a whole, the difference should be explained. If some
study units are omitted (for example, patients who did not return for
6-month follow-up), the reduced number should be reconciled with the
number eligible or expected by readers. Reporting of losses is often easiest
in tables, where entries such as "patients lost," "samples contaminated,"
"not eligible," or "not available" (for example, no 15-meter sample from a
lake with a maximum depth of 10 meters) can account for each study unit.

Loss of patients to follow-up, including losses or exclusions for noncompliance, should generally be discussed in depth because of the likelihood that patients lost are atypical in critical ways. Have patients not re-
turned for examination because they are well? Because they are still sick
and have sought other medical care? Because they are dead? Because they
do not wish to burden a physician with a bad outcome? Failure to discuss
both reasons for loss (or other termination of follow-up) and efforts to
trace lost patients are common and serious. Similarly, issues of noncompliance (reasons, as well as numbers) are often slighted by authors.

*10. References for study design and statistical methods should be to stand-
ard works (with pages stated) when possible rather than to papers where designs or methods were originally reported.*

An original paper can have great value for the methodologist, but often
does little to explain the method and its implications or the byways of cal-

culation or meaning that may have emerged since the method was first reported. Standard works such as textbooks or review papers will usually give a clearer exposition, put the method in a larger context, and give helpful examples. The notation will be the current standard, and the explanation will orient readers to the general use of the method rather than the specific and sometimes peculiar use first reported. For example, it would be hard to recognize Student's t-distribution in his original paper; indeed, t was not even mentioned. Exceptions to the general advice about using textbooks, review papers, or other standard works occur where the original exposition is best for communication and where it is the only one available.

11. Specify any general-use computer programs used.

General-purpose computer programs should be specified, along with the computer that ran them, because such programs are sometimes found to have errors.[8] Readers may also wish to know about these programs for their own use. In contrast, programs written for a specific task need not be documented, because readers should already be alert to the likelihood of errors in ad hoc or privately developed programs, and because they will not be able to use the same programs for their own work.

12. Put general descriptions of statistical methods in the Methods section. When data are summarized in the Results section, specify the statistical methods used to analyze them.

Where should statistical methods be described? There are good arguments for putting such material in one place, usually in the Methods section of a paper, but our preference[9] is generally to specify statistical methods at the places where their uses are first presented (see Chapter 1). Methods may differ slightly from one to another application within a given paper; and decisions about which results to report in full or which methods to use in exploring critical or unexpected findings generally depend on the data and earlier steps in the analysis. Keeping the specification of statistical methods close to their point of application will sometimes lead to more thought about choices and to better discussion of why a particular method was used in a particular way. Some editors, as well as some of our statistical colleagues, disagree, and authors should follow the instructions of the journal to which they submit their work. For example, after listing statistical methods used in the Methods section, Quigley et al.[10] report in their Results section about plasma inhibition levels, "This difference was significant by Fisher's exact test ($P<0.001$) for the pubertal girls. . . . By paired t-test, the girls who stayed in the same developmental group . . . had no significant change in plasma gonadotropin or estradiol levels between the initial and final assessments. Similarly, there was no significant difference in standard-deviation score for either plasma LH or FSH levels for the group of girls as a whole."

Statements such as "statistical methods included analysis of variance, factor analysis, and regression, as well as tests of significance," when divorced from the outcomes or reasons for their use, give the reader little help. On the other hand, if the only method was the use of chi-squared tests for 2×2 contingency tables, that fact might be sufficiently informative. Some general suggestions about reporting clinical trials have been discussed by Mosteller and associates.[11]

13. *Restrict tables and figures to those needed to explain the argument of the paper and to assess its support. Use graphs as an alternative to tables with many entries; do not duplicate data in graphs and tables.*

Authors have an understandable wish to tell readers everything they have learned or surmised from their data, but economy is much prized by scientific readers as well as editors. A basic point is that economy in writing and exposition gives an article its best chance of being read. Although many tables may help support the same basic point and might be appropriate in a monograph, an article generally requires only enough information to make its point — the mathematician's concept of "necessary and sufficient."

There are occasional exceptions. Sometimes a study generates data that have consequences beyond the article. For example, if information about certain biological or physical constants is obtained, it should be retained in the article. An author should inform the editor of this situation in a cover letter. Sometimes such data need to be preserved, but not in the article itself; many journals have some plan for the preservation and documentation of unpublished supporting material. Such plans are often mentioned in a journal's instructions to authors.

Whether tables or graphs better present material is sometimes a vexing question. Some readers go blind when faced with a table of numbers; others have no idea how to read graphs; unfortunately, these groups are not mutually exclusive, and some users of statistical data need to see quantitative findings in text. Overall there is a general failure to tolerate or understand the problems of any group that does not include oneself. Most of what we know about tables and graphs comes from the personal experiences of a few scholars, and little scientific information has been gathered on these subjects. Cleveland[12] has begun some scientific studies of what information can be communicated with graphs (for example, many people read bar charts better than pie charts). Tufte[13] has a beautiful book on the art of graphics.

In the field of tabular presentation, even less scientific investigation has been done, but there seems to be much value in some rules proposed by Ehrenberg[14]:

> Give marginal (row and column) averages to provide a visual focus. Order the rows and columns of the table by the marginal averages or some other measure of size or other logical order (keeping to the same order if there are many similar tables). Put figures to be compared into columns rather than rows (with larger numbers on top if possible). Round to two effective (significant) digits. Use layout to guide the eye and facilitate comparisons. In the text give brief summaries to lead the reader in the main patterns and exceptions.

To show the effect of Ehrenberg's rules, we chose Table 1 from the National Halothane Study[15] showing death rates observed in a pool of four high death rate operations (exploratory laparotomy, craniotomy, heart with pump, large bowel). Our primary interest is to compare death rates

Table 1. Death Rates in Proportions for High Death Rate Operations by Anesthetic Risk Levels.[15]

Anesthetic Risk Code	Halothane	Nitrous Oxide	Cyclopropane	Ether	Other
Unknown	0.11369	0.08682	0.08147	0.06148	0.09957
Risk 1	0.02454	0.02452	0.01634	0.01355	0.03358
Risk 2	0.05471	0.06893	0.04941	0.03812	0.05859
Risk 3	0.12471	0.16599	0.18187	0.11453	0.15306
Risk 4	0.15892	0.23140	0.18582	0.17919	0.35531
Risk 5	0.04665	0.06759	0.05725	0.04898	0.07606
Risk 6	0.22143	0.12996	0.17615	0.16008	0.17741
Risk 7	0.44164	0.43689	0.36689	0.62121	0.43348

among anesthetic groups with only a secondary interest in effect of physical status as measured by the code for anesthetic risk. It offers control to improve anesthetic comparisons.

Table 1 is obviously "busy" with five-digit numbers. In Table 2 we have rounded these to two decimals which lead to two significant figures in more than half the cells. Sometimes one has to have several different numbers of significant digits to preserve the decimal point. We converted to percentages instead of rates and thus eliminated both decimals and leading zeros.

Because of our interest in anesthetics we put anesthetics into rows, ordered according to a weighted average of death percentages for that anesthetic from highest to lowest. We have not included the weights.

The averages for anesthetic risk led us to rearrange considerably the risk columns according to a weighted average death rate because even though Risk 1 has the best and Risk 2 the second best prognosis, Risk 5 is nearly as low as Risk 2 because it represents Risk 2 patients with an emergency. Similarly Risk 6 is a mixture of Risk 3 and Risk 4 patients with an emergency. The Unknown Risk patients take a place between Moderate Complications with an emergency (Risk 5) and severe complications (Risk 3). In Table 2 entries generally rise from left to right and rise from bottom to top.

The text summarizing the table might read as follows, "The table shows that the overall weighted percentage of deaths in these high death rate operations is 9.3 percent. When the risk levels are arranged according to the average death rate for the risk, each anesthetic leads to a nearly monotonic increase in death rate according to risk, the order being Codes 1, 2, 5, Unknown, 3, 6, 4, and 7. The last five groups have sharply higher death rates with Risk 7 (moribund) giving 42.4 percent.

"Looking down columns, the death rates do not change much from one anesthetic group to another, though Other (a collection of many anesthetics) has the highest rate, not surprising because the anesthetist is probably avoiding the standard anesthetics because of some complication in the patient. Ether has the lowest rate, but from other data we know it is seldom used at some institutions and rarely in these high-risk operations.

"In the four low-risk code (1, 2, 5, Unknown) patients, Halothane, Cyclopropane, and Nitrous Oxide have very similar rates for these high death rate operations." The rearrangements and the simplifications have

Table 2. Death Rates in Percentages for High Death Rate Operations by Anesthetics Versus Anesthetic Risk Levels.[15]

Anesthetic Group	Anesthetic Risk Code								Weighted Average
	1	2	5	Un-known	3	6	4	7	
Other	3	6	8	10	15	18	36	43	11.7
Nitrous oxide	2	7	7	9	17	13	23	44	10.3
Cyclo-propane	2	5	6	8	18	18	19	37	9.8
Halothane	2	5	5	11	12	22	16	44	8.7
Ether	1	4	5	6	11	16	18	62	6.1
Weighted average	2.2	5.5	5.7	9.6	14.6	17.4	20.6	42.4	9.3

led to a much more understandable table than the original. Although these recommendations need not be slavishly followed, they can improve the readers' understanding. There is further discussion of presenting tables in Chapter 20.

14. Avoid nontechnical uses of technical terms in statistics, such as "random" (which implies a randomizing device), "normal," "significant," "correlation," and "sample."

Many words in statistics, and in mathematics more generally, come from everyday language and yet have specialized meanings. When statistical reporting is an important part of a paper, the author should not use statistical terms in their everyday meanings.

The family of *normal* (or Gaussian) distributions refers to a collection of probability distributions described by a specific formula. The distribution of *usual* or *average* values of some quantity found in practice is rarely normal in the statistical sense, even when the data have a generally bell-shaped distribution. *Normal* also has many other mathematical meanings, such as a line perpendicular to a plane. When we mix these meanings with the meaning of normal for a patient without disease, we have the makings of considerable confusion.

Significance and related words are used in statistics, and in scientific writing generally, to refer to the outcome of a formal test of a statistical hypothesis or test of significance (essentially the same thing). *Significant* means that the outcome of such a test fell outside a chosen, predetermined region. Careful statisticians and other scientists often distinguish between statistical and medical or social significance. For example, a large enough sample might show statistically significant differences in averages on the order of one tenth of a degree in average body temperature of groups of humans. Such a difference might be regarded as of no biological or medical significance. In the other direction, a dietary program that reduces weight by an average of 5 kg might be regarded as important to health, and yet this finding may not be well established, as expressed by statistical significance. Although the 5 kg is important, the data do not support a firm conclusion that a difference has actually been achieved.

Association is a usefully vague word to express a relation between two or more variables without any implication of either causation or its absence from one variable to another. *Correlation*, a more technical term, refers to a specific way to measure association, and should not be used in writing about statistical findings except in referring to that measure.

Sample usually refers to an observation or a collection of observations gathered in a well-defined way. To describe a sample as having been *drawn at random* means that a randomizing device has been used to make the choice, not that some haphazard event has created the sample, such as the use of an unstructured set of patient referrals to create the investigator's control group.

15. *Define statistical terms, abbreviations, and most symbols.*

Although many statistical terms such as mean, median, and standard deviation of the observations have clear, widely adopted definitions, different fields of endeavor often use the same symbols for different entities. Authors have extra difficulty when they need to distinguish between the true value of a quantity (a parameter such as a population mean, often symbolized by the Greek letter μ) and a sample mean (often written as $\bar{x}$).

We usually take for granted the mathematical symbols $=$, $+$, $-$, and $/$, as well as the usual symbols for inequalities (greater than or less than); we do the same for powers such as x^3, and for the trigonometric and logarithmic abbreviations such as sin, cos, tan, and log, although it is well to report what base the logarithms are using. Typography for ordinary multiplication differs, but is rarely a problem. Generally, symbols such as r for the correlation coefficient should be defined, as should n or N for the sample size, even though these are widely used.

Terms like *reliability* and *validity* are much more difficult, and they should always be defined when they are used in a statistical sense.

One difficulty with an expression such as $a \pm b$, even when a is a sample mean, is that b has many possibilities. (Some journals prefer the notation $a(b)$, but the ambiguities remain unchanged.) The author may use b for the observed sample standard deviation of individual measurements, or the standard error of the mean, or twice the standard error of the mean, or even the interquartile range, depending on the situation. The commonest ambiguity is not knowing whether b represents the standard deviation of individual observations or the standard error of the statistic designated by a. And no single choice is best in all situations. If the measure of variability is used only to test the size of its associated statistic, as for example in a P value that tests whether a correlation coefficient differs from zero, then use the standard error. If the measure of variability needs to be combined with other such measures, the standard deviation of single observations is often more useful.

The same difficulty occurs with technical terms. A danger is that a special local language will become so ingrained in a particular research organization that its practitioners find it difficult to understand that their use of words is not widespread. Nearly every laboratory has special words that need to be defined or eliminated in reports of findings.

When one or two observations, terms, or symbols are not defined, readers may be able to struggle along. When several remain uncertain, readers may have to give up because the possibilities are too numerous.

A well-established convention is that mathematical symbols should be printed in italics.[16-18] This practice has many advantages, including the reduction of ambiguity when the same character is commonly used to designate both a physical quantity and a mathematical or statistical quantity. In typescripts, an underline is generally used to indicate that a character is to be printed in italics, and authors may need to give special instructions to editors or printers if underlines are used for other purposes, such as to designate a mathematical vector (which might be printed both underlined and in italics).

We are indebted to Marcia Angell, Alexia Anctzak, M. Anthony Ashworth, John J. Bartko, Thomas Chalmers, Eli Chernin, Victor Cohn, David Hoaglin, Susan Horn, Edward J. Huth, Deborah A. Lambert, Kathleen N. Lohr, Thomas Louis, Thomas R. O'Brien, Kenneth Rothman, Stephen B. Thacker, Anne A. Scitovsky, Barbara Starfield, Wallace K. Waterfall, Jonathan Weiner, Alfred Yankauer, Cleo Youtz, and two reviewers, for their thoughtful and helpful advice, as well as the Methods Panel of the Institute of Medicine's Council for Health Care Technology.

Grant support: in part by grant GAHS-8414 from the Rockefeller Foundation, grant S8406311 from the Josiah Macy Jr. Foundation.

REFERENCES

1. International Committee of Medical Journal Editors. Uniform requirements for manuscripts submitted to biomedical journals. N Engl J Med 1991; 324:424-8.

2. International Committee of Medical Journal Editors. Uniform requirements for manuscripts submitted to biomedical journals. Ann Intern Med 1988; 108:258-65.

3. DerSimonian R, Charette LJ, McPeek B, Mosteller F. Reporting on methods in clinical trials. N Engl J Med 1982; 306:1332-7. [Chapter 17 of this book.]

4. Emerson JD, Colditz GA. Use of statistical analysis in the *New England Journal of Medicine.* N Engl J Med 1983; 309:709-13. [Chapter 3 of this book.]

5. Simon R. Confidence intervals for reporting results of clinical trials. Ann Intern Med 1986; 105:429-35.

6. Rothman KJ. Significance questing. Ann Intern Med 1986; 105:445-7.

7. Louis TA, Fineberg H, Mosteller F. Findings for public health from meta-analyses. Annu Rev Public Health 1985; 6:1-20.

8. Emerson JD, Moses L. A note on the Wilcoxon-Mann-Whitney test for 2 × k ordered tables. Biometrics 1985; 41:303-9.

9. Moses LE. Statistical concepts fundamental to investigations. N Engl J Med 1985; 312: 890-7. [Chapter 1 of this book.]

10. Quigley C, Cowell C, Jimenez M, et al. Normal or early development of puberty despite gonadal damage in children treated for acute lymphoblastic leukemia. N Engl J Med 1989; 321:143-51.

11. Mosteller F, Gilbert JP, McPeek B. Reporting standards and research strategies for controlled trials: agenda for the editor. Controlled Clinical Trials 1980; 1:37-58.

12. Cleveland WS. The elements of graphing data. Monterey, Calif.: Wadsworth Advanced Books and Software, 1985.

13. Tufte ER. The visual display of quantitative information. Cheshire, Conn.: Graphic Press, 1983.

14. Ehrenberg AC. The problem of numeracy. Am Statistician 1981; 35:67-71.

15. Bunker JP, Forrest WH Jr, Mosteller F, Vandam LD. The national halothane study. Bethesda, Md.: National Institutes of Health, National Institute of General Sciences, 1969.

16. Huth EJ. Mathematics and statistics. In: Huth EJ. Medical style & format: an international manual for authors, editors, and publishers. Philadelphia: ISI Press, 1987:170-6.

17. International Organization for Standardization. Units of measurement. 2nd ed. (ISO standards handbook 2). Geneva: International Organization for Standardization, 1982.

18. International Organization for Standardization. Statistical methods: handbook on international standards for statistical methods. (ISO standards handbook 3). Geneva: International Organization for Standardization; 1979:287-8.

ADDITIONAL REFERENCES

The following references may be helpful to readers who want to pursue these topics further.

1. Cochran WG. Sampling techniques. 3rd ed. New York: John Wiley, 1977.

2. Cohn V. News and numbers: a guide to reporting statistical claims and controversies in health and other fields. Ames, Iowa: Iowa State University Press, 1988.

3. Colton T. Statistics in medicine. Boston: Little, Brown and Co., 1974.

4. Committee for Evaluating Medical Technologies in Clinical Use. Assessing medical technologies. Washington, DC: National Academy Press, 1985.

5. Fleiss JL. Statistical Methods for Rates and Proportions. 2nd ed. New York: John Wiley, 1981.

6. Gardner MJ, Maclure M, Campbell MJ. Use of check lists in assessing the statistical content of medical studies. Br Med J 1986; 292:810-2.

7. Gore S, Altman DG. Statistics in practice. London: Taylor and Francis, 1982.

8. Ingelfinger JA, Mosteller F, Thibodeau LA, Ware JH. Biostatistics in clinical medicine. 2nd ed. New York: Macmillan, 1987.

9. Meinert CL. Clinical trials: designs, conduct, and analysis. New York: Oxford University Press, 1986.

10. Miké V, Stanley KE, eds. Statistics in medical research: methods and issues, with applications in cancer research. New York: John Wiley, 1982.

11. Cochran WG. Planning and analysis of observational studies. New York: John Wiley, 1983.

12. Shapiro SH, Louis TA, eds. Clinical trials: issues and approaches. New York: Marcel Dekker, 1983.

13. Sackett DL, Haynes RB, Tugwell P. Clinical epidemiology: a basic science for clinical medicine. Boston: Little, Brown & Co., 1985.

14. Snedecor GW, Cochran WG. Statistical methods. 8th ed. Ames, Iowa: Iowa State University Press, 1989.

17

~

REPORTING ON METHODS IN CLINICAL TRIALS

Rebecca DerSimonian, D.Sc., L. Joseph Charette, Ph.D.,
Bucknam McPeek, M.D., and Frederick Mosteller, Ph.D.

ABSTRACT   A clinical trial cannot be adequately interpreted without information about the methods used in the design of the study and the analysis of the results. To determine the frequency of reporting what we consider 11 important aspects of design and analysis, we surveyed all 67 clinical trials published in the *New England Journal of Medicine,* the *Lancet,* and the *British Medical Journal* from July through December 1979 and in the *Journal of the American Medical Association* from July 1979 through June 1980. Of the 11 items in the 67 trials published in all four journals, 56 percent were clearly reported, 10 percent were ambiguously mentioned, and 34 percent were not reported at all.

At least 80 percent of the 67 trials reported information about statistical analyses, statistical methods used, and random allocation of subjects, yet only 19 percent reported the method of randomization. Loss to follow-up was discussed in 79 percent of the articles, treatment complications in 64 percent, and admission of subjects before allocation in 57 percent, but eligibility criteria for admission to the trial appeared in only 37 percent. Although information about whether patients were blind to treatment was given in 55 percent, information about whether there was blind assessment of outcome was reported in only 30 percent. The statistical power of the trial to detect treatment effects was discussed in only 12 percent of the articles.

The clinical trials published in the *New England Journal of Medicine* reported 71 percent of the 11 items, those in the *Journal of the American Medical Association* 63 percent, those in the *British Medical Journal* 52 percent, and those in the *Lancet* 46 percent. These rates are significantly different ($P<0.001$). We recommend that editors improve the reporting of clinical trials by giving authors a list of the important items to be reported.

To evaluate published reports of clinical trials, readers need specific information about the methods used.[1-3] However, such reports frequently omit important features of design and analysis.[4-6] Mosteller and his co-workers[7] found that investigators reported only about 25 percent of five methodologic items in three samples drawn from an extensive survey of controlled trials.[8] The methodologic items sought included such basic matters as whether and how patients were randomized, what statistical method was used to analyze the data, the statistical power of the study, and the degree of blindness.

In this chapter we describe our studies of the reporting of clinical trials in four general medical journals: the *New England Journal of Medicine*, the *Lancet*, the *Journal of the American Medical Association*, and the *British Medical Journal*. We developed a survey method that would be appropriate for trials in a variety of fields of medicine, and we determined the frequency of reporting 11 methodologic items in 67 clinical trials from the four surveyed journals.

We recorded whether a particular methodologic aspect of a clinical trial was mentioned, not whether a particular method was used. If the author said that randomization was not carried out, that was coded positively as a report on the item, as was a statement that randomization had been carried out.

METHODS

Each issue of the four journals from July through December 1979 was searched for reports of comparative clinical trials. All the reports of this type that we found for this period were included in the study. So that the number of trials from each journal would be more nearly equal, the *Journal of the American Medical Association* was searched for an additional six months (through June 1980). We found and included a total of 13 clinical trials in the *New England Journal of Medicine*, 21 in the *Lancet*, 14 in the *Journal of the American Medical Association*, and 19 in the *British Medical Journal*.

We then surveyed the selected trials to determine how frequently certain aspects of design and analysis were reported. We selected 11 specific items on the basis of their importance to a reader in determining how much confidence should be placed in the authors' conclusions, their applicability across a variety of medical specialty areas, and their ability to be detected by the scientifically literate general medical reader. We chose the following items: (1) *eligibility criteria* (information explaining the criteria for admission of patients to the trial); (2) *admission before allocation* (information used to determine whether eligibility criteria were applied before knowledge of the specific treatment assignment had been obtained); (3) *random allocation* (information about random allocation to treatment); (4) *method of randomization* (information about the mechanism used to generate the random assignment); (5) *patients' blindness to treatment* (information about whether patients knew which treatment they were receiving); (6) *blind assessment of outcome* (information about whether the person assessing the outcome knew which treatment had been given); (7) *treatment complications* (information describing the presence or absence of side effects or complications after treatment); (8) *loss to follow-up* (information about the numbers of patients lost to follow-up and the reasons they were lost); (9) *statistical analyses* (analyses going beyond the computation of means, percentages, or standard deviations); (10) *statistical methods* (the names of the specific tests, techniques, or computer programs used for statistical analyses); and (11) *power* (information describing the determination of sample size or the size of detectable differences).

A total of 17 readers participated in the study. Two readers independently surveyed each paper for each item to determine whether the item was reported, not reported, or unclear. If an item was clearly not applicable to a particular study, it was regarded as reported. For example, if the patient clearly could not have been blind to treatment, this was regarded as reported. On their survey forms, readers indicated the location in the text of the data for each item found, to facilitate the adjudication of differences between the readers by a third reader.

The readers were university faculty members and graduate students with scientific backgrounds from departments of biostatistics, surgery, anesthesia, and health policy and management. (Their names appear in the acknowledgment.) We provided the readers with a substantial set of written instructions with examples. Over a period of a year, the Surgery Group of the Harvard Faculty Seminar on Health and Medical Practices developed these instructions in order to improve uniformity between readers by settling many of the inevitable ambiguities in advance.

The third reader assigned each item one of the following codes: R for "reported" if information about its presence or its inapplicability to the study was found or clearly implied, O for "omitted" if the information was not found, and ? if there was inadequate information (partial or ambiguous) about the item. The outcome R was initially always possible because when an item was not applicable, we awarded an R. The third reader based the final code on the opinions of the first two readers when they agreed and on further study of the paper to resolve disagreements.

Since we had three categories (R, ?, and O) adding up to 100 percent, a method using barycentric coordinates seemed appropriate for plotting. This approach permits three dimensions summing to a fixed total to be

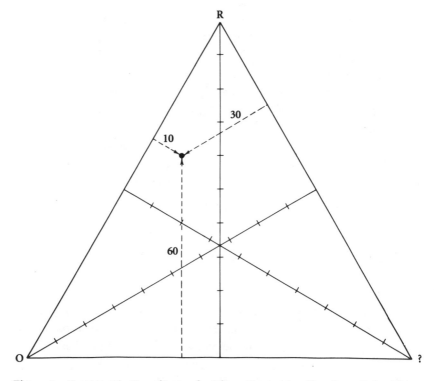

Figure 1. Barycentric Coordinates for Three Categories, Showing a Point Plotted When the Percentages of Articles with the Item Reported (R), Unclear (?), and Omitted (O) Are 60, 10, and 30, Respectively.

plotted in a plane, so that all coordinates are displayed both overtly and comparably, whereas if only two coordinates are plotted, one is suppressed and the others are overemphasized. The method uses an equilateral triangle with three coordinate axes, each running from the center of a base to the opposite vertex. At the base, the value of the coordinate is 0, and at the vertex it is 100. It is convenient to label the vertices with the codes R, ?, and O, as in Figure 1.

RELIABILITY

Three items produced a relatively high rate of disagreement between the initial two readers: eligibility criteria, 42 percent disagreement; admission before allocation, 49 percent; and blind assessment of outcome, 33 percent. All others had less than 20 percent disagreement. When the paired codes were R/O, adjudication was usually straightforward, because the adjudicator had the help of the location of the information. The flat disagreement usually meant that one coder had missed something. Removing the R/O code pairs changed the rate of disagreement over admission before allocation from 49 percent to 18 percent, thus implying that much of the coding problem in this item was in finding the information. The corresponding action for eligibility criteria and blind assessment of outcome only slightly reduced the rates of disagreement (to 37 percent and to 27 percent, respectively). These disagreements rested more on ambiguity of the report.

RESULTS

Table 1 summarizes the percentages of articles reporting each item for each journal and for all journals combined, as well as the overall mean scores of the four journals. At least 80 percent of the 67 articles report information about randomization, statistical analysis, and statistical method. Even though information about the use of random allocation is reported in 84 percent of the articles, only 19 percent report the method of randomization. Treatment complications are reported in 64 percent of the articles, eligibility criteria in 37 percent, and power in only 12 percent.

Figure 2 shows barycentric-coordinate plots for the 11 items in each of the four journals. The plots give the visual impression that reporting on the items has similar patterns in the *Lancet* and the *British Medical Journal*, whereas the

Table 1. Percentage of Articles Reporting Each Item in Each Journal.*

Journal	No. of Articles	Eligibility Criteria			Admission before Allocation			Random Allocation			Method of Randomization			Patients' Blindness to Treatment			Blind Assessment of Outcome		
		R	?	O	R	?	O	R	?	O	R	?	O	R	?	O	R	?	O
NEJM	13	77	23	0	85	8	8	100	0	0	15	23	62	62	0	38	46	23	31
JAMA	14	36	50	14	64	7	29	71	0	29	43	0	57	71	21	7	43	36	21
BMJ	19	21	58	21	63	5	32	95	0	5	16	11	74	37	21	42	26	21	53
Lancet	21	29	48	24	29	14	57	71	0	29	10	0	90	57	5	38	14	29	57
Total	67	37	46	16	57	9	34	84	0	16	19	7	73	55	12	33	30	27	43

Journal	No. of Articles	Treatment Complications			Loss to Follow-up			Statistical Analyses			Statistical Methods			Power			Mean, All Items		
		R	?	O	R	?	O	R	?	O	R	?	O	R	?	O	R	?	O
NEJM	13	92	0	8	100	0	0	92	0	8	92	0	8	15	0	85	71	7	23
JAMA	14	71	0	29	93	7	0	79	0	21	86	0	14	36	0	64	63	11	26
BMJ	19	58	5	37	74	11	16	100	0	0	79	5	16	5	5	89	52	13	35
Lancet	21	48	5	48	62	5	33	95	0	5	86	0	14	0	5	95	46	10	45
Total	67	64	3	33	79	7	15	93	0	8	85	1	13	12	3	85	56	10	34

* NEJM denotes the *New England Journal of Medicine*, JAMA the *Journal of the American Medical Association*, BMJ the *British Medical Journal*, R item reported, ? item unclear, and O item omitted.

New England Journal of Medicine and the *Journal of the American Medical Association* differ both from the two British journals and from one another.

In terms of the numbers of R codes, we found the order to be *New England Journal of Medicine, Journal of the American Medical Association, British Medical Journal,* and *Lancet.* The *New England Journal of Medicine* reported over 75 percent on seven items while reporting under 25 percent on two items. The *Journal of the American Medical Association* reported over

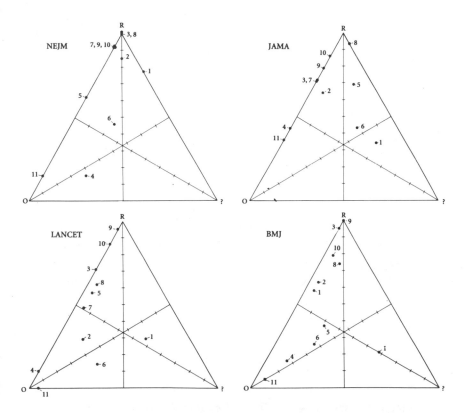

Figure 2. Barycentric-Coordinate Plots for the Percentages of the 11 Items Reported in Each of the Four Journals.
Figures indicate the items listed in Methods; NEJM denotes the *New England Journal of Medicine,* JAMA the *Journal of the American Medical Association,* BMJ the *British Medical Journal,* R reported, ? unclear, and O omitted. Data are from Table 1.

Table 2. Marginal Distribution of Numbers of Items Coded in 67 Papers.*

	No. of Items											
	0	1	2	3	4	5	6	7	8	9	10	11
	percentage of papers											
Reported (R)	0	0	3	10	9	13	19	13	21	9	0	1
Unclear (?)	25	45	21	7	1	0	0	0	0	0	0	0
Omitted (O)	1	12	22	13	24	4	9	6	7	0	0	0

*The figures across the top line of the table refer to the number of items. The other figures represent the percentage of papers given each of the three codes for the indicated number of items. See text for further details. Totals do not always add to 100 percent because of rounding.

75 percent on three items and under 25 percent on none. The *British Medical Journal* reported over 75 percent on three items and under 25 percent on three. The *Lancet* reported over 75 percent on two items and under 25 percent on three.

Table 2 shows the distributions of R, ?, and O codes by papers. It shows that just one paper received 11 Rs. At another extreme, two papers (3 percent) received only two Rs. The remaining papers were nearly evenly distributed, ranging from three to nine Rs. In round numbers, the average paper was assigned six Rs, one ?, and four Os. Thus, authors reported on about 10 percent of the items in an ambiguous manner and omitted reports on a third of them.

DISCUSSION

We have selected 11 important aspects of design and analysis and investigated how often they are reported in samples of articles from leading general medical journals — two from the United States and two from the United Kingdom. Several concerns influenced our selection of items for study. One was importance. We sought basic methodologic aspects that readers need to appraise the strengths of reported clinical trials. We wanted to survey the reporting standards of therapeutic trials in a variety of medical and surgical specialties. For this reason, we chose basic aspects of methodology that are applicable across medical fields — aspects that could be identified by broadly educated scientists, not just by experts in the various special fields.

In assessing a controlled clinical trial, a reader asks who the subjects were and how they were selected for admission to the trial. We surveyed each article to find the eligibility criteria used. We usually found some information about eligibility criteria, but frequently it was inadequate. A paper might say who the subjects were but might fail to be clear on whether all such subjects or only some unspecified subsets had been included. If the selection criteria are not clearly stated, a reader is uncertain about who the subjects were and how they were selected. It is difficult to generalize the findings of such a trial to groups other than the subjects themselves. We looked for information to determine whether the eligibility criteria were applied without knowledge of the specific treatment to be assigned. Here, again, we had difficulty finding this information. Unless admission to the trial is settled before knowledge of the assigned treatment group is obtained, there may be bias in the selection of eligible subjects.

Since randomization is a major bias-reducing technique used in controlled clinical trials, we sought specific information about random treatment assignment or about whether such assignment was possible in each trial. An author's assurance of random treatment assignment is not convincing unless the method used to generate the random assignment is discussed. Some methods of random allocation are sure and straightforward. Others may be effective but are uncertain in actual use, and others may appear to be random but actually have serious weaknesses. In scientific usage, random does not mean haphazard. To assure readers that the randomization was done appropriately, the method should be described. Not only must the randomization be done well, but it is best if its execution can be checked later if questions arise, as they frequently do. Flipping coins, tossing dice, or drawing cards may tempt investigators to interfere with the process, and such procedures also cannot be checked. Randomization by the use of alternate cases, odd and even birthdays, or hospital record numbers has similar weaknesses.[7]

Random numbers from one of the published random-number tables or pseudo-random numbers generated by a well-studied computer method offer good sources of randomization. Ideally, after being entered into the trial, the patient is allocated to treatment from a central source. Carefully prepared sealed envelopes can work, but only if they are completely opaque even when held up to a strong light, and if investigators do not break the seals until entrance has been confirmed.

Another bias-reducing technique common to clinical trials is blinding. Many papers that we surveyed reported that the therapy given was concealed from the patient or the physician or both (double blinding). However, most reports stop after using the term "blind" or "double-blind" and leave the reader uncertain of exactly what has been concealed from whom.

Three parties may be involved in a clinical study, and each may be subject to strong hopes and prejudices about the trial. First of all, patients have their own concerns: a desire to get well, perhaps a desire to please the physicians, and a strong self-interest that the treatment will work. Secondly, the physician who gives or orders the treatment naturally hopes for success. Thirdly, someone must evaluate the patient's response to treatment. Sometimes the evaluator is the treating physician; at times, a third party determines responses. Obviously, evaluators who are aware of the treatment given may have their own prejudices and hopes, and these may bias the findings, despite conscious efforts to be fair.

Like randomization, the methods of achieving blindness should be reported to give the reader important information for judging the adequacy of a trial's defenses against bias. In some trials, not all degrees of blindness can be easily provided, but this aspect requires careful consideration and reporting by the authors. Although many reports mention double blinding, one often cannot tell whether all three parties (patient, treating physician, and evaluator) are thereby blinded. Reports should be specific on this point. In this survey we considered only the reporting of blinding of patients and evaluators.

Physicians know that treatments sometimes produce effects other than those intended. To determine the usefulness of a treatment, readers need to assess the nature and frequency of these side effects and their implications for patient care. A report should describe an active search for side effects or complications and discuss those that are found.

Clinical trials study the effects of a deliberate intervention. Readers want to know what happened to all the subjects treated. Sometimes investigators cannot collect outcome data from all subjects because some of them die, move away, decline to continue to participate, or become lost from the study group for other reasons. Especially in trials with long-term follow-up, large trials, and trials with complicated protocols, some data are likely to be missing. When dropouts are properly reported, the reader can often assess the effect of missing data on the trial's conclusions.

The uncertainty associated with small sample sizes usually requires statistical inference to evaluate the effects. The reader needs to know what statistical methods were used and how. When the author merely tucks in such expressions as "$P<0.05$" without identifying the statistical test, readers cannot satisfy themselves that the methods were appropriate.

Finally, information about the power of the trial to detect differences is required to give some idea of the adequacy of the plan for the investigation.[6] Although confidence limits summarize the uncertainty of the outcome, a brief discussion of power gives the reader a better feel for the strength of the conclusion. Small sample size frequently leads to trials whose power to detect differences among treatment groups is weak. Freiman et al.[6] (Chapter 19 of this book) have stated that in 50 percent of papers reporting no significant differences between therapies, a 50 percent improvement in performance would not have been found. If no statistically significant difference is found, the reader who knows that the study was strong enough to have a good chance of detecting a clinically important difference can conclude that the matter is fairly well settled. When power is not discussed, the reader has a right to suspect that the study was not large enough to detect important differences. Judging from the frequency with which clinical trials failed to mention power, most investigators seem not to have realized their obligation to report on this item.

Emerson, McPeek, and Mosteller[9] reviewed the responses to the same 11 items for 84 surgical trials appearing from July 1981 through June 1982 in six leading general surgical journals: the *American Journal of Surgery*, the *Annals of Surgery*, the *Archives of Surgery*, the *British Journal of Surgery*, *Surgery*, and *Surgery, Gynecology and Obstetrics*. Table 3 shows the average percentages of R, ?, and O for the two studies. The percentages are similar in the two studies, though the surgery study reported 13 percent higher on patients' blindness to treatment, 15 percent higher on treatment complications, and 20 percent lower on statistical methods used.

Kelen et al.[10] evaluated the reporting of the 11 items in 45 interventional clinical trials appearing in three acute care journals from January 1980 through June 1983. They found a 40 percent overall level of reporting.

Jacobsen et al.[11] assessed 3 of the 11 items in 42 randomized controlled experiments on humans published in nursing research journals from 1980 through 1984: they found 5 percent reporting on power, 14 percent on method of randomization, and 45 percent on loss to follow-up compared to our respective 12, 19, and 79 percent.

Although less directly comparable, findings from perinatal medicine[12] and from infectious disease research[13] yield similar conclusions.

We found substantial differences among four general medical journals in the frequency with which they reported individual items and in the mean frequency with which they reported all 11 items. Table 1 shows the score of each journal for each of the 11 items, as well as the mean scores of the four journals. Of course, taking an average across the 11 items gives them equal weight. Others may prefer different weights for the various items. However, small changes in weights would change such an average relatively little and thus would be unlikely to change the relative standings of the four journals. Despite our reporting on 11 measures for each article, the number of articles studied from each journal was relatively small, and therefore the differences in mean scores among the journals needed careful statistical evaluation.

To examine the variation among the four journals, we assigned an R the score of 1 and any other outcome a score of 0. We then carried out a one-way analysis of variance on total scores (running from 0 to 11) for the papers, and we found $F = 7.9$ (with 3 and 63 degrees of freedom) — a significance level of less than 0.001. The journals differ in their performance, with an estimated standard deviation of 1.15 for the true mean score across journals. This compares with the estimated average standard devia-

Table 3.   Comparison of Results from the Current Study to Those in a Similar Study of Surgical Investigations Using the Same 11 Items.

	Percentages of Articles Reporting		
	R	?	O
67 Clinical trials, this study	56	10	34
84 Surgical trials[9]	59	5	36

tion of residuals of 1.75 within journals. Thus, scores among journals do not vary as much as scores for papers within journals.

To assess the variation in the response to the 11 items, which runs from 12 percent for power to 93 percent for statistical analysis, we again scored R as 1 and other outcomes as 0 and applied Cochran's Q test[14] for matched samples. This yielded a chi-square (with 10 degrees of freedom) of 224.5, which has a significance level comparable to 16 standard deviations out in a standard normal distribution. Thus, the items differ highly in reporting, as expected. The important point is not that the rates differ, but that they average 56 percent instead of nearly 100 percent.

Editors have the power to control what is published. Although not all may agree on our specific list of items, editors could greatly improve the reporting of clinical trials by providing authors with a list of items that they expected to be strictly reported. Since authors usually have all the information required to report on all the items that we studied, 100 percent reporting could be readily achieved.

Several authors have proposed a set of methodologic guidelines to aid investigators and editors in the preparation and review of manuscripts.[15-17] The guidelines address basic issues in the design, conduct, and analysis of clinical trials. More recently, Bailar and Mosteller described in detail the guidelines for presenting statistical aspects of scientific research as outlined in the 1988 edition of the Uniform Requirements for Manuscripts Submitted to Biomedical Journals.[18,19]

Although the reporting that we discuss is important and basic to an appreciation of the strength of an investigation, we know that a badly executed study may report all items and that a sound investigation may fail to report some of these strengthening measures. Of course, investigators often have good reason for weaknesses in design, but the reasons for weaknesses in reporting can be few.

We are indebted to the Surgery Group of the Harvard Faculty Seminar (John Bailar, Benjamin Barnes, Anne Bigelow, Roger Day, Gregg Dinse, John Emerson, Mary Ettling, Karen Falkner, John Gilbert [deceased], Katherine Godfrey, John Hedley-Whyte, Ellen Hertzmark, Harmon Jordan, Philip Lavori, Robert Lew, Alan Lisbon, Thomas Louis, Kay Patterson, Marcia Polansky, John Raker [deceased], James Rosenberger, Stanley Shapiro, Michael Stoto, Jane Teas, and Grace Wyshak) for their cooperation in developing and testing the questionnaire and instructions, serving as primary readers, and giving advice in this research, and to William Goffman for advice and bibliographic assistance.

REFERENCES

1. Chilton NW, Barbano JP. Guidelines for reporting clinical trials. J Periodont Res 1974; 9:Suppl 14:207-8.

2. O'Fallon JR, Dubey SD, Salsburg DS, Edmonson JH, Soffer A, Colton T. Should there be statistical guidelines for medical research papers? Biometrics 1978; 34:687-95.

3. Levenstein MJ, Bishop YMM. Analysis and reporting as causes of controversies. In: Rosenoer VM, Rothschild M, eds. Controversies in clinical care. New York: Spectrum, 1981:1-23.

4. Reiffenstein RJ, Schiltroth AJ, Todd DM. Current standards in reported drug trials. Can Med Assoc J 1968; 99:1134-5.

5. Mosteller F. Problems of omission in communications. Clin Pharmacol Ther 1979; 25:761-4.

6. Freiman JA, Chalmers TC, Smith H Jr, Kuebler RR. The importance of beta, the type II error, and sample size in the design and interpretation of the randomized controlled trial: survey of 71 "negative" trials. N Engl J Med 1978; 299:690-4. [Chapter 19 of this book.]

7. Mosteller F, Gilbert JP, McPeek B. Reporting standards and research strategies for controlled trials. Contr Clin Trials 1980; 1:37-58.

8. Staquet MJ, ed. Randomized trials in cancer: a critical review by sites. New York: Raven Press, 1978.

9. Emerson JD, McPeek B, Mosteller F. Reporting clinical trials in general surgical journals. Surgery 1984; 95:572-9.

10. Kelen GD, Brown CG, Moser M, Ashton J, Rund DA. Reporting methodology protocols in three acute care journals. Ann Emerg Med 1985; 14:880-4.

11. Jacobsen BS, Meininger JC. Randomized experiments in nursing: the quality of reporting. Nurs Res 1986; 35:379-82.

12. Tyson JE, Furzan JA, Reisch JS, Mize SG. An evaluation of the quality of therapeutic studies in perinatal medicine. J Pediatr 1983; 102:10-3.

13. MacArthur RD, Jackson GG. An evaluation of the use of statistical methodology in the *Journal of Infectious Diseases.* J Infect Dis 1984; 149:349-54.

14. Cochran WG. The comparison of percentages in matched samples. Biometrika 1950; 37:256-66.

15. Zelen M. Guidelines for publishing papers on cancer clinical trials: responsibilities of editors and authors. J Clin Oncol 1983; 1:164-9.

16. Simon R, Wittes RE. Methodologic guidelines for reports of clinical trials. Cancer Treat Rep 1985; 69:1-3.

17. Gardner MJ, Machin D, Campbell MJ. Use of check lists in assessing the statistical content of medical studies. Br Med J 1986; 292:810-2.

18. Bailar JC III, Mosteller F. Guidelines for statistical reporting in articles for medical journals. Amplifications and explanations. Ann Intern Med 1988; 108:266-73. [Chapter 16 of this book.]

19. International Committee of Medical Journal Editors. Uniform requirements for manuscripts submitted to biomedical journals. Ann Intern Med 1988; 108:258-65.

18

~

STATISTICAL CONSULTATION IN CLINICAL
RESEARCH: A TWO-WAY STREET*

Lincoln E. Moses, Ph.D., and Thomas A. Louis, Ph.D.

ABSTRACT   The results of clinical research often rest on statistical interpretation of numerical data. Thus, effective collaboration between clinician and statistician can be crucial. Interaction in the planning phases of a project can identify tractable scientific and statistical problems that will need attention and can help avoid intractable ones. The central requirement for successful collaboration is clear, broad, specific, two-way communication on both scientific issues and research roles.

C linical research often depends for its conclusions on the correct design and performance of studies to collect numerical data and on the subsequent interpretation of the results. Statistical issues arise in all three of these areas, and they must be correctly addressed, or the usefulness of the whole investigation can be threatened.

Statistical issues may or may not be straightforward; the clinician may or may not be statistically experienced and sophisticated. If the complexity of the statistical problem is greater than the clinician's readiness to deal with it, then recourse to biostatistical consultation is likely to advance the investigation. Some statistical problems may not be readily apparent and may therefore be discovered only through such a consultation.

The kind of transaction between the clinician and the statistical consultant that we are discussing is scientific, not primarily technical. This chapter aims to indicate — to both parties — ways to improve the prospects of

---

*Portions of this chapter have appeared elsewhere (Moses L, Louis TA. Statistical consulting in clinical research: the two-way street. Stat Med 1984; 3:1-5) and are included here with the permission of John Wiley.

successful consultation. Boen and Zahn[1] and Hand and Everitt[2] discuss aspects of statistical consultation, and Baskerville[3] proposes a plan for training consulting statisticians. All three works are bountiful sources of examples of the consultative process and of additional references.

This chapter focuses neither on the consultative interaction that takes only five minutes nor on the mature relationship between a clinical and a statistical investigator that has developed through joint work on several projects. Rather, we address situations in which the integrity of the research effort may depend on sound statistical design, analysis, and interpretation, all of which may require a joint effort extending at least over several weeks. The merit of such a cooperative undertaking rests largely on the participants' success in dealing with two classes of activities, which we call scientific interaction and coordination of affairs.

SCIENTIFIC INTERACTION

The consultation may start with a question that seems limited, such as "Would you give me a reference to the Mann–Whitney test?" or "What computer package can do a logistic regression on our computer?" Or the question may seem (and be) broad, as in the following examples: "We are preparing a research proposal. Would you look over the statistics part? May we name you as a consultant?" or "A journal has just rejected this paper because of the statistics. Will you help me prepare it for resubmission?" However the transaction may begin, its sound progress depends crucially on one thing: the clinician and the statistical consultant must ultimately be dealing with the same problem. At the beginning the scientific problem is unknown to the consultant. Before he or she can be helpful, the statistician must correctly understand the investigation — its purposes, motivating questions, materials, techniques, and measurements; if he or she offers advice based on a misapprehension of these features, the advice may be wrong and useless or even damaging. Thus, a clear, mutual understanding of the problem is the single most important element in the consultation.

Step-by-step communication is necessary for success. As noted above, the statistician must first gain a correct understanding of the substantive problem. When a point seems to be understood, he or she should check it by saying, "Now, let me tell *you* — the clinician — about this point and see if I have it right." This step may have to be repeated. Similarly, the clinician needs to understand the essential statistical features of the project; the investigator can best check his or her comprehension of a concept by explaining it to a colleague or to the statistician.

Each partner should have high expectations of the other. Statistical thinking is in a large measure scientific thinking, once it is understood. Clarity from the statistician is a proper expectation. Similarly, the statistician may reasonably expect that the clinician can make clear most information that he or she understands well. The iterative style of communication helps to ensure that these high expectations will be met. Of course, jargon is an unacceptable encumbrance in the consultation. The statistician should not use it and should not accept it from the clinician.

A broad definition of the problem can lead more quickly to a correct understanding of a highly specific concept than would a narrowly focused treatment of that concept alone. This principle applies to both clinical and statistical issues, with two main consequences. First, each collaborator should be prepared to do some reading in the other's field. Second, each partner should be wary of attempting to protect the other by not mentioning topics that may be relevant, merely to avoid what may be perceived as needless complication. Although in everyday life communication may be eased by the omission of details that seem inessential or difficult to explain, that practice can lead to trouble in scientific collaboration.

Both participants' understanding of the problem needs to be specific. The worth of a study may turn on details. For example, the difference between a sound randomized clinical trial and a collection of anecdotal material depends on the answers to questions like these: How were the patients for this investigation chosen? How was it decided which ones would receive which treatment? Did the person assessing post-treatment status know which treatment the patient had received? Did the patient know?

Specific knowledge about the measurement process can also be essential. Measurements of the same phenomenon may vary from day to day; to assess that possibility in a study, it is necessary to know which observations were made on which days, with which piece of apparatus, and by which technician or interviewer. Some measurements are actually composites of others, and the variability of the composite will depend on the variability of its components. If the statistical consultant is involved in planning the investigation, his or her understanding the details may permit a more effective and efficient experimental design.

Broad, specific, and iterative communication is, therefore, essential to the consultation. But additional measures can also help the collaboration to succeed. For instance, it can be useful at an appropriate stage for one of the partners to attend a meeting of the other's colleagues, one addressed to the subject of the investigation. The clinician may invite the statistician to staff meetings or may be invited to a seminar of statistical staff and students. From such meetings, new ideas and helpful criticism often emerge.

Frequently, the collaboration is advanced by observation. The statistician may more quickly gain a fuller understanding of the study by seeing the apparatus in use, watching the diagnostic procedure, and looking at the recording processes. Similarly, the clinician who undertakes to follow the data-editing steps and statistical calculations may gain insight from doing so.

Written communication may advance the work notably. Questions, requests, information, and tentative proposals all tend to gain specificity in written form. Memoranda are not subject to memory decay and can be discussed with knowledgeable colleagues of the recipient (and the sender). Memoranda may be used advantageously to record the current stage in the participants' thinking, the nature of an issue that urgently demands resolution, or the details of a proposal that will require consultative work. It pays to be aware of situations in which a memorandum may be the preferred method of communication.

Thorough communication may take time, but shortcuts must be avoided. This recommendation parallels good medical practice, which calls for taking a medical history before prescribing treatment. The parallel reaches further; it may be that the analysis (or treatment) will not ultimately be changed by the fuller understanding (or history). In that case, the payoff is the confidence that the right course has been taken. Boen and Zahn[1] have a similar view of the role of consultation.

COORDINATION OF AFFAIRS

When two or more people share a task, certain difficulties may crop up as the work goes forward. Forethought can eliminate many such problems, especially when willing participants have joined together in a mutually at-tractive project. Problems are likely to be fewer and smaller when partners have worked together before. In a collaboration between clinician and stat-istician, there may be benefit in asking certain questions in advance and agreeing on the answers. Issues that merit such consideration are the schedule, resources, decisions about acknowledgment versus coauthorship, and the use of data.

SCHEDULE

What is the schedule for the project? Are there deadlines, and can they be met? Most clinicians and many statisticians grossly underestimate the time necessary to complete the project and overestimate the time available. Data must be collected before they can be analyzed and before a report can be written, and each step takes time.

If it can be foreseen that the work will proceed in stages, the partici-pants should decide how much time should be allowed for each stage and who is responsible for completing it. For example, data need to be prepared for computer entry, entered, and checked for validity before analysis can begin. Generally, the clinician will have to participate in the first and third of these activities, especially by providing feedback on acceptable ranges for data and by checking on numbers that are out of range.

RESOURCES

Are the resources (budget, computer time, and personnel) adequate, or are some changes needed? Many long-term projects require a statisti-cal collaborator, not simply an occasional consultant. A person who is trained to use statistical computing packages should be identified early so that the analysis recommended by the statistician can be carried out. Occasionally, standard computing packages cannot perform appropriate analyses for a particular investigation, and special-purpose routines must be used or new programs must be written. However, statistical packages for microcomputers now provide all standard and many specialized analyses and graphics. Therefore, statistical input sharply focuses on what to do and what to make of results, enterprises requiring a large dose of statistical input. In addition, the statistician should ensure that the requisite hardware, software, and organizational structures are available to produce valid and timely data and analyses.

If financial issues arise in relation to the work, then clarity about them is crucial. Are there to be charges for the statistical consultation? How will the services of programmers be funded? Computing costs? Key punching? Secretarial support?

ACKNOWLEDGMENT VERSUS COAUTHORSHIP

Sooner or later a decision about authorship of the study will need to be made. High-quality statistical input in the design and analysis phases of a project can be as important a scientific contribution as that provided by the medical team. Early discussion of authorship may lead either to an agreement to defer the decision until the participants' relative contributions can be assessed or to a tentative decision subject to change.

If coauthorship does not seem appropriate, then acknowledgment is in order. The statistician should be acknowledged for his or her advice — if it is taken — thus establishing responsibility as well as credit. Acknowledgments that are too broad or otherwise inaccurate can be unfair; it follows that any acknowledgment should be reviewed and accepted by the statistician.

USE OF THE DATA

It may be wise to agree in advance on arrangements concerning possible future uses of the study's data and statistical analyses. Either party may someday wish to use the material in other articles or in textbooks. What steps should be taken if such an occasion arises?

The message of this discussion of potentially troublesome issues is not that there are many ways to get into difficulty but that it is well to consider early, and to agree about, the essential steps in completing the work. The collaborators should talk about the logistics of the project, resolve any immediate problems quickly, and be aware of potential problems not yet settled. It can be helpful to record the results of these discussions in a joint memorandum that may bring to light some previously unnoticed misunderstandings. We see disadvantages to this method, too; sometimes the air is chilled by the utterance "Let's put that in writing." But the possibility of preparing a memorandum about logistics deserves explicit consideration.

THE BENEFITS OF STATISTICAL CONSULTATION

We began this chapter by observing that clinical investigations often present problems that demand statistical treatment — correct statistical treatment. Of course, collaboration with a biostatistician will usually help with such problems. Interaction between clinician and statistician in the planning phases of a project can identify tractable scientific and statistical problems that will need attention and can help avoid intractable ones. After the data are in, it can be too late for statistical attention. As a colleague, Helena Kraemer (personal communication), has remarked,

> If consultation is at the *Post Hoc* stage, it may be that objectives cannot be accomplished (sampling bias, poor design, etc.). It is the statistician's responsibility to state this frankly. We cannot do magic, and we can't participate in cover-ups. It is as well that researchers know our limitations in advance. This is a particular problem when the first consultation takes place after a research paper is rejected for publication because of poor methodology. Not much one can do!

A study that has been planned in the light of statistical considerations is less likely to have statistical flaws. In addition, it can occasionally be much more cost effective, saving time, money, or both. Sometimes the scope of a study can be broadened at little or no cost by making use of familiar patterns of statistical design in experiments. Admittedly, however, a good study often costs more than a poor one. Finally, intellectual benefits often accrue to either collaborator, or both, and to their students and co-workers.

We are indebted to the Harvard study group, Helena Kraemer, Ph.D., Peter Gregory, M.D., and Byron William Brown, Jr., Ph.D., for their feedback and encouragement.

REFERENCES

1. Boen JR, Zahn DA. The human side of statistical consulting. Belmont, Calif.: Lifetime Learning Publications, 1982.

2. Hand DJ, Everitt BS, eds. The statistical consultant in action. Cambridge: Cambridge University Press, 1987.

3. Baskerville JC. A systematic study of the consulting literature as an integral part of applied training in statistics. Am Stat 1981; 35:121-3.

19

~

THE IMPORTANCE OF BETA, THE TYPE II ERROR, AND SAMPLE SIZE IN THE DESIGN AND INTERPRETATION OF THE RANDOMIZED CONTROLLED TRIAL

SURVEY OF TWO SETS OF "NEGATIVE" TRIALS

Jennie A. Freiman, M.D., Thomas C. Chalmers, M.D., Harry Smith, Jr., Ph.D., and Roy R. Kuebler, Ph.D. *

ABSTRACT Seventy-one "negative" randomized controlled trials were reexamined to determine whether the investigators had studied large enough samples to give a high probability (>0.90) of detecting a 25 percent and a 50 percent therapeutic improvement in the response. Sixty-seven of the trials had a greater than 10 percent risk of missing a true 25 percent therapeutic improvement, and with the same risk, 50 of the trials could have missed a 50 percent improvement. Estimates of 90 percent confidence intervals for the true improvement in each trial showed that in 57 of these "negative" trials, a potential 25 percent improvement was possible, and 34 of the trials showed a potential 50 percent improvement. Many of the therapies labeled as "no different from control" in trials using inadequate samples have not received a fair test. Concern for the probability of missing an important therapeutic improvement because of small sample sizes deserves more attention in the planning of clinical trials.

A follow-up of this study 10 years later, this time including 65 "negative" randomized controlled trials from 1988, revealed no essential improvement in recognition of the importance of beta.

*Deceased.

Over 40 years have elapsed since the publication of the first clinical trials employing randomization.[1] Since then the randomized controlled trial has gradually become accepted as the most effective way of determining the relative efficacy and toxicity of a new therapy. A persistent objection to the trial has been its inability to document the efficacy of some therapies considered by their advocates to be clinically most useful. This inability could result from insensitivity, or it could be that the initial enthusiasm for the therapy was based on inadequate or biased trials.

The purpose of the present study was to examine a number of "negative" trials ($P > 0.05$) published in the literature to determine whether the therapy under investigation had been afforded all due process in its trial. Appropriate statistical analyses were based on the data presented by the authors. It was found that most studies included too few patients to provide reasonable assurance that a clinically meaningful difference (i.e., therapeutic effect) would not be missed. Published reports gave few details of the prior planning, so that it is unclear to what extent the hazards of insufficient trial size were taken into account.

Proper methodology for randomized controlled trials[2,3] requires careful planning before the trial begins. This planning must consider what the size of the trial ought to be for an important clinical therapeutic effect to be detected if it exists. The choice of size involves the statistical concepts of Type I and II errors, the probability of these errors (often referred to as alpha [α] and beta [β], respectively), and the size of a clinically important effect of therapy (a predetermined value of delta [Δ]).

To facilitate an understanding of the deficiencies of many negative trials, a description of the technical aspects of the proper choice of the size of a clinical trial is presented below.

THE PROBLEM

In a typical randomized controlled trial the aim is to estimate the true response rate, P_t, for a treatment under study and compare it with the estimate of the true response rate, P_c, for a competing control therapy, whether it is an existing standard treatment or a placebo. The treatment group of patients yields the observed response rate, $\hat{P}_t$, which is the estimate of P_t, and the control group of patients produces the observed response rate, $\hat{P}_c$, which is the estimate of P_c. The observed difference, $\hat{P}_c - \hat{P}_t$, is then an estimate of the effectiveness of the treatment.

But even if P_t and P_c are truly equal, non-zero observed differences $(\hat{P}_c - \hat{P}_t)$ will occur by the workings of chance in the samples of patients. Various sizes of observed difference will occur with various probabilities. When the difference actually observed has a very small probability of being exceeded by chance, we make the "possible but not probable" argument and reject the notion that $P_c = P_t$. Of course, we have to have a rule for how small that "small" probability must be. This is the famous "level of significance," α.

The procedure used is the classic test of a hypothesis. We set up the null hypothesis (H_0) that $P_c - P_t = 0$. After we obtain $\hat{P}_c - \hat{P}_t$ we can figure out the probability that a difference at least as large in magnitude as that actually observed would occur if H_0 were true. This is the well-known P value. If then $P \leq \alpha$ we reject H_0.

When we reject H_0, of course, we are running a risk of making an erroneous decision. The two rates P_t and P_c may indeed be equal, so that our rejection decision is an error. This false-positive error (made by rejecting H_0 when in fact it is true) is called the Type I error. By using our criterion $P \leq \alpha$ for rejection rule, we have fixed the probability of a Type I error at α, and thus our control of the size of that risk.

Our test procedure, however, runs another risk. If $\hat{P}_c - \hat{P}_t$ turns out so that $P > \alpha$, we shall decide, "Do not reject H_0," meaning the observed difference is not statistically significant. Clearly, this may be an error too. The true difference, $P_c - P_t$, may be non-zero, and yet chance delivers an observed difference, $\hat{P}_c - \hat{P}_t$, that is not large enough to fit our rule for rejecting H_0. We then make the error of not rejecting H_0 when in fact H_0 is false. This false-negative error is called the Type II error, and its probability is designated β.

The probability of a Type II error is not one single value like the probability of a Type I error. The probability of a Type I error is calculated on the basis that H_0 is correct — that is, on the basis of $P_c - P_t = 0$. The probability of a Type II error is based on the situation in which H_0 is false — that is, $P_c - P_t \neq 0$. But if $P_c - P_t$ is not zero, there is an infinity of values that such a difference (say Δ) can have. And for each value of Δ there is a probability of a Type II error when we run our test of H_0. These values make up an entire curve of β as a function of Δ. This function is often called the operating characteristic curve of the test. It has a large, distinguished record in industry, where it has been a central feature of the quality control that rules on acceptance or rejection of inspected lots of manufactured products.

Figure 1 is an example of an operating characteristic curve for the kinds of randomized controlled trials that we are considering. It is based on one of the trials analyzed in this review.[4] In this case the control group numbers $n_c = 64$, the treatment group is of size $n_t = 66$, $\alpha = 0.05$, and the end point is case fatality. It takes the response rate of controls as base-line figures P_c ($= 0.297$, or a mortality rate of 29.7 percent) and plots the value of β against values of $\Delta = P_c - P_t = 0.297 - P_t$. We can enter the graph on the horizontal scale with any value of Δ that interests us, and read off on the vertical scale the value of β — the probability of not rejecting H_0: $\Delta = 0$. Note that when we enter the graph with $\Delta = 0$, we read $\beta = 0.95$ — similar to our choice of $\alpha = 0.05$ as the probability of rejecting H_0 when it is true ($\Delta = 0$).

In Figure 1 we have shown two examples of Δ that are often of clinical interest: a 25 percent reduction in mortality from the control mortality rate, P_c, and a 50 percent reduction from P_c (respectively, $\Delta = 25$ percent of $0.297 = 0.074$, and $\Delta = 50$ percent of $0.297 = 0.149$). These examples show that the randomized controlled trial would have a 77 percent likelihood of ruling in favor of $\Delta = 0$ when Δ actually is 0.074 (a 25 percent reduction in mortality) and a 42 percent likelihood of saying there is no difference when a 50 percent reduction in mortality is true.

If the β values for the Δs in which the investigator has a special interest are unsatisfactorily high, his or her recourse is to increase the size of the treatment and control groups. Tables for minimum sample size to satisfy a variety of predetermined values of α, β, and Δ are available for a reasonable variety of $\hat{P}_c$ values. Such tables take into account the random nature of both $\hat{P}_c$ and $\hat{P}_t$, and give "sample size per group" since the most efficient test procedure has equal sample sizes for treatment and control. They usually consider $1 - \beta$ rather than β, since $1 - \beta$ expresses the probability of avoiding the Type II error — that is to say, the probability of detecting the difference, Δ. In this sense one uses the term "power" (of the test), which is defined as $1 - \beta$.

For the example considered in Figure 1, let us ask what sample sizes would be required to maintain α at 0.05 but reduce β to 0.10 (instead of 0.42) when Δ is a 50 percent reduction from P_c. A minimum sample-size table[5] gives the answer: 202 in each group, or about three times the original size of the reputedly negative trial.

Such sample sizes in prior planning are often larger than what the investigator has in mind. When the size required is beyond practicable limits,

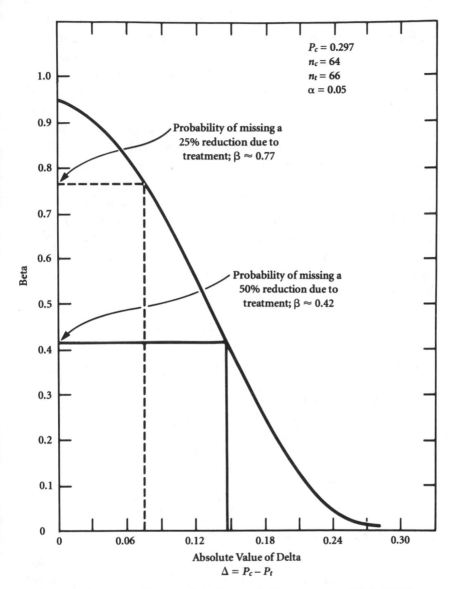

Figure 1.   Operating Characteristic Curve of a Representative Clinical Trial.

something has to give: either admitting higher α or β risk (or both) or settling for a larger Δ required for discrimination, or both. For instance, the sample sizes given in the example will allow a test with $\alpha = 0.10$ and power $1 - \beta = 0.75$ of detecting a reduction of 0.20 in P_c (i.e., a reduction of two thirds in the rate). It does seem important to face the harsh alternatives before the trial is embarked on. Sometimes, it will appear clear that the available resources offer virtually no chance of attaining acceptable definite results. On the other hand, one might argue that too small studies should still be performed when the opportunity presents itself, because later combinations may be meaningful.

One further question should be considered in the initial planning of a trial. For ethical reasons, contingency plans should be made for examining the data at intervals. Clearly, it is not justified to continue assigning sick patients to a therapy that appears highly likely to be less effective than originally perceived, even though the numbers studied at the time of looking at the data are appreciably lower than those required by the original trial plan. Of greater pertinence to this paper on negative trials, however, is the study in which the standard or placebo therapy is winning the trial at these interval looks but not at the $\alpha = 0.05$ significance level. Again, the rules for stopping such a trial should be considered at the planning stage.

Details behind determination of sample size are given by Fleiss[5] and in the extensive work of Feinstein on "clinical biostatistics,"[6] in which a particularly instructive chapter is entitled "Sample Size and the Other Side of Statistical Significance."

STUDY OF PUBLISHED NEGATIVE TRIALS

CRITERIA AND CHARACTERISTICS

We encountered 110 negative randomized controlled trials in a collection of over 300 positive and negative studies compiled over 10 years. Initially, they were considered negative if the authors concluded that the therapy under examination was not superior to the control therapy because the observed difference between the responses to the experimental and control therapy was not statistically significant ($P>0.05$). In other words, the null hypothesis could not be rejected at the 5 percent level of significance. The

110 negative trials were reduced to 71 (see Table 1) by means of the following rigid criteria imposed to ensure negative trials (a list of the original 71
articles is available from T.C. Chalmers).

First of all, the paper had to state explicitly that the comparison of two
treatments did not reach statistical significance ($P>0.05$) — i.e., statements
like "No significant reduction in mortality," "Increased mortality rate in
control group is not significant at the 5 percent level," and "This difference
is not statistically significant" had to appear in the paper.

Secondly, an admissible randomized controlled trial also had to be "not
significant at the 10 percent level" (two-tailed test).

Thirdly, all trials had to have discrete end points for measurement of
effectiveness — i.e., the measure of efficacy had to be one of three kinds: a
decrease in mortality, a decrease in the number of patients with complica-

Table 1.   Characteristics of the 71 Admissible Negative Trials.

Characteristic	No. of Trials
Source of trial	
Lancet	19
New England Journal of Medicine	11
Journal of the American Medical Association	6
Other	35
Year of Publication	
1975–77	26
1970–74	29
1965–69	12
1960–64	4
Category of disease	
Cardiovascular	27
Gastrointestinal	9
Other	35
Type of controls	
Placebo	25
Standard treatment	22
Other	24
End point	
Mortality	41
Complications or lack of improvement	30

tions, or a reduction in the number of patients with no improvement. Thus, all trials were binomial in structure, assessing definable "success–failure."

Finally, there had to be enough patients in the control group and in the treatment group so that the expected value in all four categories was greater than 5.

The characteristics of the 71 admissible randomized controlled trials are shown in Table 1. The 71 negative trials came from 20 different journals, but principally from the *Lancet,* the *New England Journal of Medicine* , and the *Journal of the American Medical Association.* Although the papers were published over a span of years, 1960–1977, the distribution by year was heavily skewed to the later years.

For the repeat analysis, a random sample was obtained of 1100 randomized controlled trials systematically added to a database by weekly perusal of *Current Contents.* Both the title page of each journal and the index of each *Current Contents* issue were reviewed for randomized trials. We reviewed 277 of the 1100 abstracts to find 65 papers which fulfilled the criteria for selection outlined above.

METHODS OF ANALYSIS

For each admissible negative randomized controlled trial the following data were recorded: observed response rates of the experimental group ($\hat{P}_t$) and the control group ($\hat{P}_c$), and the difference ($\hat{P}_c - \hat{P}_t$); sample size of the treated group, n_t; sample size of the control group, n_c; level of significance used in the paper (α, the probability of a Type I error — i.e., the probability of falsely concluding that a difference exists); size of the difference in the response rates that is clinically important to detect (since this point was rarely mentioned in the papers under consideration, it has been arbitrarily assumed that a reduction of 25 and 50 percent in the observed control rate* would be common choices for "clinically important" differences); and power of the statistical test of significance ($1 - \beta$) — i.e., the probability of detecting the true clinically important difference with the recorded trial size.

---

*All calculations are based on one of three different unfavorable control rates: mortality rate, complication rate, or no-improvement rate.

For purposes of statistical analysis, we calculated the operating characteristic curve, as explained above, for each of the original 71 negative clinical trials, using the numbers studied and the observed proportion for the
control treatment, $\hat{P}_c$, as the null hypothesis true value.

To determine $100(1 - \alpha)$ percent confidence intervals, a 90 percent
confidence interval for the true difference between control and treatment percentages was calculated for each trial. This assessment is helpful in interpreting a study since it gives information in addition to the
reject-or-nonreject procedure of the hypothesis test: it gives a bracket
within which the investigator can have high confidence that the true
difference does lie. Whenever the bracket includes zero, we cannot reject
the null hypothesis that asserts that zero difference. But the exact location and width of the confidence interval suggest a good deal about the
direction in which the truth lies and the adequacy of the sample size for
pinning it down.

RESULTS

ANALYSIS OF THE PROBABILITY β — PLANNING STAGE

We calculated the level of β for each paper, employing the data listed
above, assuming the level of statistical significance to be $\alpha = 0.05$ for a one-
tailed test, and assuming two levels of important clinical difference between control and treatment, $\Delta_1 = 0.25\hat{P}_c$ and $\Delta_2 = 0.50\hat{P}_c$.

The distribution of the βs for the two cases in the original series (1 and
2) is shown in Table 2. For a reduction of 25 percent in the control rate, 50
percent of the trials had βs in excess of 74 percent. Only four of the trials
(5.63 percent) were large enough to ensure a $\beta \leq 0.10$, the usually accepted
standard for clinical trials. The picture is slightly better if a 50 percent reduction in the end-point percentage is considered clinically important.
Fifty percent of the trials had βs in excess of 40 percent. Thirty percent had
$\beta \leq 0.10$ (i.e., power ≥ 0.90).

The data for the later series in Table 3 revealed minimal improvement in
the cumulative percentage of patients with βs of 20 or below, 12 percent
versus 7 percent for 25 percent reduction and the same for 50 percent reduction. The substantial proportion of studies for the randomly selected

Table 2. Distribution of the βs (Original Series, 1960–1977).

β (%)	Case 1 (25% Reduction in P_c)			Case 2 (50% Reduction in P_c)		
	Frequency	Cumulative Frequency	%	Frequency	Cumulative Frequency	%
0–10	4	4	5.63	21	21	29.58
11–20	1	5	7.04	1	22	30.99
21–30	2	7	9.89	4	26	36.62
31–40	2	9	12.68	9	35	49.30
41–50	5	14	19.72	5	40	56.34
51–60	7	21	29.58	4	44	61.97
61–70	2	23	32.39	8	52	73.24
71–80	16	39	54.93	9	61	85.92
81–90	25	64	90.14	9	70	98.59
91–100	7	71	100.00	1	71	100.00

series, as in the original series, was undersized as indicated by the calculations of β ($P<0.001$).

Thus, either these trials were almost uniformly undersized in the planning stage or the expected reduction in the end-point percentage due to the treatment under consideration was very much in excess of a 50 percent reduction, and the reduction did not take place.

CONFIDENCE INTERVALS FOR THE TRUE PERCENT REDUCTION

The 90 percent confidence intervals for all 71 randomized controlled trials in the original series are plotted in Figure 2. Twenty-five (35.2 percent) of the trials reported less mortality, etc., in the control group than in the treated group. In the remaining 46 (64.8 percent) the treated group showed a reduction in mortality, etc., when compared with the control group. All the confidence intervals included zero — i.e., no difference was still possible. In 57 (80 percent), the intervals included a 25 percent reduction in end point, and in 34 (49 percent) a 50 percent reduction.

Table 3. Distribution of the βs (Second Series, 1988).

β (%)	Case 1 (25% Reduction in P_c)			Case 2 (50% Reduction in P_c)		
	Frequency	Cumulative Frequency	%	Frequency	Cumulative Frequency	%
0–10	4	4	6.15	21	21	28.77
11–20	4	8	12.30	5	26	35.62
21–30	2	10	15.38	2	28	38.36
31–40	0	10	15.38	3	31	42.47
41–50	1	11	16.92	9	40	54.79
51–60	7	18	27.69	5	45	61.64
61–70	5	23	35.38	8	53	72.60
71–80	13	36	55.38	2	55	75.34
81–90	14	50	76.92	7	62	95.38
91–100	15	65	100.00	3	65	100.00

COMPARISON OF THE β CALCULATIONS FOR THE TRIALS WITH THE OBSERVED CONFIDENCE LIMITS FOR THE TRUE PERCENT REDUCTION OBTAINED FROM EACH TRIAL

Table 4 shows the relation between pretrial values of β for detecting 25 percent and 50 percent true reductions in the end point for each trial and the observed 90 percent confidence-interval estimates of the true percent reduction for each trial. Of the 67 trials whose βs for missing a 25 percent reduction in response rate were larger than 10 percent, the confidence intervals of 57 (85 percent) included a potential 25 percent reduction in the end-point percentage — e.g., a reduction in mortality, etc., of 25 percent.

Similarly, 34 (68 percent) of the 50 trials whose β was greater than 10 percent for missing a 50 percent reduction in response rate had interval estimates that included a potential true 50 percent reduction.

Stated in another way, only 15 percent and 32 percent of the trials that had not studied sufficient patients for a power of 0.90 to detect reductions of 25 and 50 percent, respectively, had justification in stopping supported by the fact that the 90 percent confidence interval of the estimated true

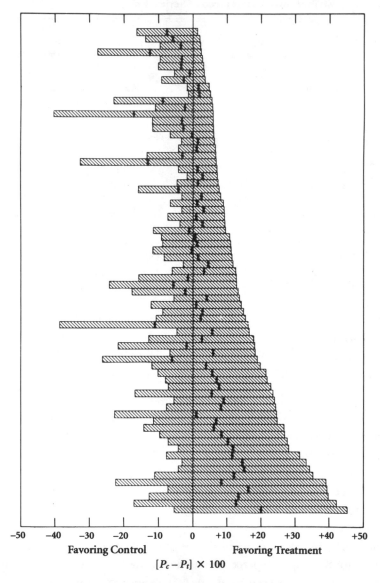

Figure 2.   Ninety Percent Confidence Limits for the True Percentage Difference
for the 71 Trials (Original Series, 1960–1977).
The vertical bar at the center of each interval indicates the observed value,
$\hat{P}_c - \hat{P}_t$, for each trial.

Table 4. Relation between Pretrial Estimates of β and Inclusion of 25 Percent and 50 Percent Reductions within 90 Percent Confidence Intervals for the True Difference.

$\beta > 10\%$	90% Confidence Interval, Including					
	25% Reduction			50% Reduction		
	yes	no	totals	yes	no	totals
Yes	57	10	67	34	16	50
No	0	4	4	0	21	21
Totals	57	14	71	34	37	71

difference did not include those reductions. The data were similar for the second study (see Figure 3).

DISCUSSION

The failure to attain a level of statistical significance does not necessarily mean that the two treatments being compared are identical. This reservation holds especially if a small sample size confers insufficient power for the statistical test employed. Analysis of both series (136 negative randomized controlled trials) shows that investigators often work with sample sizes too small to offer a reasonable chance of successfully rejecting the null hypothesis in favor of the treatment when differences are as large as 25 percent and 50 percent.

Both investigators and clinicians should be concerned about being able to detect an important clinical difference. All investigators are limited by both resources and time. If they are to run a trial from which meaningful, reliable conclusions may be drawn, they should be aware of the limitations of the trial. An important limitation of any trial is inherent in the statistical method employed. Thus, in planning a randomized controlled trial, the investigators must estimate the null position (H_0) and decide what difference they are interested in detecting. Then if they are not adept at statistics, they may consult a statistician to determine the sample size needed to demonstrate such a difference if it exists, within generally acceptable error limits — for example $\alpha = 5$ percent, $\beta = 10$ percent. Alternatively, if the investiga-

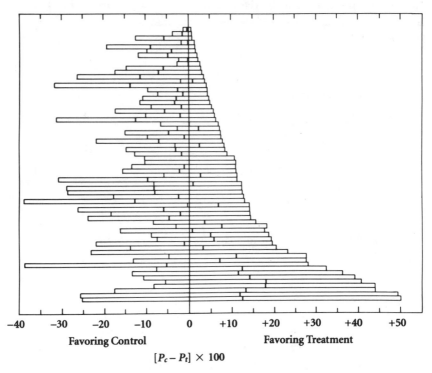

-40   -30   -20   -10   0   +10   +20   +30   +40   +50

Favoring Control   Favoring Treatment

$$[P_c - P_t] \times 100$$

Figure 3.   Ninety Percent Confidence Limits for the True Percentage Difference
for the 65 New Trials (Second Series, 1988).
The vertical bar at the center of each interval indicates the observed value,
$\hat{P}_c - \hat{P}_t$, for each trial.

tors decide what difference they would view as clinically important and can
guess how many patients they expect to enter the trial, a statistician can easily
determine with what probability such a difference is likely to be found if it exists (power of the test) as well as what chance exists that real differences of
other magnitudes will be missed (β error). In either case, an operating characteristic curve should be constructed before the trial is undertaken. Of course,
different investigators will disagree about what a clinically significant difference would be for a given trial. They will also disagree on the risk they are willing to take of missing a meaningful effect of a new regimen.

Since β is specifically conditioned upon the control value, α probability,
and sample size, it will change as any of these three variables changes. The

α level is generally constant and the control value may be gleaned from the available literature. However, since the investigators can rarely predict ahead of time exactly how many patients will enter and complete the trial, their preliminary estimate cannot be exact. Nonetheless, the value of calculating the probability of a Type II error before a trial is run lies in the fact that the investigators can consider various permutations of the number of patients they will enlist, on the one hand, and the difference they would be interested in, on the other, to determine whether their trial is feasible and worthwhile, before they go ahead and conduct it.

One change in the statistical handling of clinical trial data has become so prominent since the original publication[7] of this article in 1978 that it is important to dampen the suggestion that trials should not be undertaken if sufficient patients are not available to satisfy the goals of a β of 0.10 and an α of 0.05. The techniques of meta-analysis have been so well developed that the combination of undersized trials may offer important information which would not be obtained if the trials had not been undertaken and reported.[8] Although many argue that one or two large trials will be much more valid than many small ones, the increased heterogeneity of multiple small trials may make the overall conclusion more applicable to the practice of medicine. Under ideal circumstances, both very large trials with a fixed protocol and multiple small trials with varying protocols are useful.

Ethical considerations will also enter into the decision whether to continue a trial until an acceptable β has been achieved when the trends indicate that a clinically meaningful result is unlikely. For instance, some of these 136 trials may have been stopped before an acceptable sample size (as might have been calculated before the start of the trial) was achieved because the trend favored the control treatment. Although not mentioned by the authors and possibly not performed, calculation of the 90 percent confidence limits on the true difference between treatment and control would have indicated the extent to which a true difference favoring the experimental treatment was missed.

It is conceivable that β and the confidence intervals of the difference were considered in the 71 original negative papers, but the problem was mentioned only rarely. Only one of the original 71 papers stated that α and β were considered before the start of the study. In only 18 papers was a trend recognized in the discussion, whereas in 14 the need for a larger sample size was mentioned. In none of the trials with a trend favoring the control group were the confidence limits mentioned and the decision to stop stated to be based on them. In most studies the lack of a difference significant at the 5 percent level was taken to mean that no clinically meaningful difference existed.

In the 1988 series, 8 percent of the papers considered α and β before the start of the studies and 11 percent calculated either a post-hoc β or an estimate of the number of patients necessary to establish the observed difference as statistically significant. One of the 65 trials from the 1988 series referred to the original paper in this series in discussing the small number of patients in the trial.

This review has not dealt with the reasons for stopping the studied trials. The purpose has been to indicate the potential weakness in small negative clinical trials. This defect of many clinical trials was pointed out by Schoolman et al.[9]

The almost total lack of discussion of α and β in the introductions of the papers and the minimal discussion of the implications of the small numbers indicate a need for greater education of investigators, reviewers, and editors in the techniques of the clinical trial.

The minimal improvement in the 10-year period indicates very modest impact of the original paper, although it has been cited 429 times in the medical literature in the 12 years since its publication. The mild increase in the recognition of the importance of beta and the larger number of trials with a trend favoring the control group may be a reflection of the selection of the papers at random from a published series rather than the selection from papers that had been collected in a non-random manner as in the first series.

The conclusion is inescapable that many of the therapies discarded as ineffective after inconclusive negative trials may still have a clinically important effect. All is not lost, however, if one believes in the validity of combining similar studies of reasonable quality to achieve an answer that may not be obtained by any one study alone.

REFERENCES

1. Medical Research Council. Streptomycin treatment of pulmonary tuberculosis. Br Med J 1948; 2:769-82.

2. Peto R, Pike M, Armitage P, et al. Design and analysis of randomized clinical trials requiring prolonged observation of each patient. Br J Cancer 1976; 34:585-612, 1977; 35:1-39.

3. Burdette WJ, Gehan EA. Planning and analysis of clinical studies. Springfield, Ill.: Charles C Thomas, 1970.

4. Clausen J, Felsky M, Jørgensen FS, et al. Absence of prophylactic effect of propranolol in myocardial infarction. Lancet 1966; 2:920-4.

5. Fleiss JL. Statistical methods for rates and proportions. New York: John Wiley, 1973:176-94.

6. Feinstein AR. Clinical biostatistics. St. Louis: CV Mosby, 1977:320-34.

7. Freiman JA, Chalmers TC, Smith H Jr, Kuebler RR. The importance of beta, the type II error and sample size in the design and interpretation of the randomized control trial. Survey of 71 "negative" trials. N Engl J Med 1978; 299:690-4.

8. Sacks HS, Berrier J, Reitman D, Pagano D, Chalmers TC. Meta-analyses of randomized control trials: an update. [Chapter 23 of this book.]

9. Schoolman HM, Becktel JM, Best WR, et al. Statistics in medical research: principles versus practices. J Lab Clin Med 1968; 71:357-67.

20

WRITING ABOUT NUMBERS

Frederick Mosteller, Ph.D.

In writing about numbers, as in other tasks in scientific writing, there are no absolute rules, and good practice depends on what sort of document is being prepared. The goals of trying to write the truth and communicating well with the reader may be at odds with considerations of length and the interests of the audience, as well as with ground rules of journals or editors. This chapter provides advice, with examples, on writing about numbers in the biomedical literature.

Because authors need to decide whether to provide numerical data primarily in text or in tables, this chapter discusses first the allocation of numbers. Then it gives advice on issues that arise more often in the text, makes a few suggestions about numbers in tables, and ends with some remarks about symbols.

NUMBERS IN TABLES OR TEXT?

Among the problems facing authors is how to allocate numbers to tables and text. In the *New England Journal of Medicine*, "[d]ata presented in tables should in general not be duplicated in the text or figures."[1] Many other journals have a similar rule because space is too expensive to allow numbers to be presented twice. In writing reports and textbooks authors may have more room to maneuver.

Readers may differ in their attitudes toward numbers. Some, almost like sponges, can sop up numbers from tables and interpret them readily. Others like to have the numbers in the text, and still others go snow-blind when collections of numbers appear in one place. They like numbers to be explained one at a time and to be few and far between. Thus, the distribution of numbers should be influenced by the audience's preferences and customs as well as by the requirements of documentation.

Whatever the reasons for writing, include the message of the table in the text. Even the spongelike readers may not get the message the author wants to convey from reading the table, because tables often have several messages.

To help the reader understand a table, the text often needs to include an explanation of the source or meaning of one of the numbers. This explanation may be especially necessary when the rows and columns have headings that are severely abbreviated. (The abbreviations should be explained in footnotes to the table.)

NUMBERS IN THE TEXT

Although some manuals of style[2-6] go into detail about handling numbers in the text, the rules have many exceptions. Some manuals are oriented more toward the humanities or journalistic writing than toward scientific or technical writing. Again, ease of reading and clarity are goals that should override style sheets, though editors can often cleverly revise to meet style-sheet rules. Some specific issues are taken up below.

USING WORDS OR NUMERALS

Some journals and other sources of advice to writers have a rule that numbers smaller than 10 should be written out in words and larger ones should be given in Arabic numerals. Or they may recommend writing out isolated two-digit numbers as well. Such rules are satisfactory when the numbers are unimportant in themselves or when nothing is to be gained by following other rules. Such a rule by itself does not, however, recognize the various possibilities.

For example, to say "Three physicians met before the operation" can be satisfactory when the exact number does not matter. It is sometimes simpler to give a concrete number than a vague one. In the statement above, "A few physicians met before the operation" is equally informative if we do not care whether two or seven met, but only that a small group did. When accuracy is possible, it is usually preferable to vagueness. The reader who gets the impression that the author cannot keep track of or count small numbers may conclude that the author is equally incapable of dealing with larger numbers.

If a number is the first word in a sentence, it still seems good practice to write it out in words, though some authors writing about mathematics now start sentences with numerals or even symbols.[7] Although the *Lancet* often begins sentences with Arabic numerals, the practice has not become widespread. When other rules come into play, as discussed below, it is better to recast the sentence to avoid having the number at the beginning.

Usually, numbers are better written as Arabic than Roman numerals in scientific work. Sometimes Roman numerals can be used as ordinals (a number representing the position, or order, as 5th) or labels, as in outlines, page numbers in a preface, and labels for categories, when the number of instances is small. For actual quantification, such as a date, Roman numerals should be avoided, since their use may create mistakes. For a nonmedical example, the disputed authorship of 10 of the 12 *Federalist Papers* in question would be resolved if we knew that Hamilton had made a mistake in recording an "X" in a lone Roman numeral.[8] If he had, then his note about authorship would have assigned 10 papers to Madison that Hamilton otherwise seemed to be claiming for himself. The other two disputes would be explained because they came late in the series, when Hamilton wrote almost all the papers. Beyond this, many people read Roman numerals inaccurately.

EXHAUSTION AND CHECKING

The simple use of categories for sorting leads to frequencies, and the author reassures the reader by using the principle of exhaustion so that all cases are accounted for.

Consider the sentence "Of the 84 patients with myocardial infarction, 22 had subendocardial infarction, and 62 transmural infarction, with 78 discharged alive — a mortality in the hospital of 7 percent."[9] This example illustrates the exhaustion principle, and we are comforted to see that 22 + 62 = 84. It also illustrates the difficulty that arises when figures for patients alive and dead are not given at the end of the sentence. Told that 78 patients discharged alive leads to a mortality of 7 percent, the reader has to subtract 78 from 84 before taking a quotient to check the 7 percent. This difficulty might have been avoided with ". . . 78 discharged alive and 6 dead, for a hospital mortality of 7 percent." The example is easy to follow because all the numbers are in numeric form and therefore they stand out. Putting the total first makes it easy to spot.

LARGE NUMBERS AND PRECISION

Although scientific journals encourage precise writing, numbers with many distinct digits can lose readers in details when what may be needed primarily is a grasp of the magnitude. Rounding tends to emphasize the order of magnitude, as does the use of scientific notation. For example, "Nearly 1 million patients were admitted to this class of hospitals in 1980" may be preferable to "This class of hospitals admitted 969,537 patients in 1980." Unfortunately, the author may need to include the actual number in the article, as well as to make sure the reader understands that the number is about a million. In such a case, it is necessary to work in the round number in discussing the magnitude. More generally, the author must decide whether to emphasize precision, magnitude, or both.

NUMBERS CLOSE TOGETHER

Putting unrelated numbers side by side confuses the reader, at least temporarily. For example, the sentence "In 1980, 969,537 patients were admitted to this class of hospitals" dazzles the reader because the numbers are run together. Perhaps worse examples are the sentences "For $375, 125 women were vaccinated" and "This group of patients with leukemia had an average white-cell count of 257, 112 lymphocytes and 145 other types."

PARALLELISM AND SIMILARITY

When two sequences have paired items, maintaining the same order in both keeps the matching straight. Also, when numbers come from the same series, they should be written in the same manner.

In describing parallel or similar groups, maintain order and similarity of statement to keep the reader with you. For example, "Among the 30 patients receiving treatment A, 8 contracted pneumonia and 1 of these died; and among the 28 patients receiving treatment B, 4 contracted pneumonia and 2 of these died" is better than "Among the 30 patients receiving treatment A, 8 contracted pneumonia and 1 died, and also 2 died who received treatment B, of whom there were 28 in all, 4 suffering from pneumonia."

Be sure that numbers to be compared are presented in the same way — words or numerals. Either, and even both, can be useful in special circumstances, but keep related numbers alike. For example, write "Among the 78 patients, 14 had fever and 3 had jaundice," not "Seventy-eight patients had 14 cases of fever and three of jaundice." The *Lancet*'s device of allowing an Arabic numeral to lead off a sentence would produce parallel treatment here if we also replaced "three" by "3."

When numbers are important for their size and are to be compared, Arabic numerals should be preferred to words. For example, "Among 78 patients, 8 contracted pneumonia" is preferable to "Among 78 patients, eight contracted pneumonia" and also to "Among seventy-eight patients, 8 contracted pneumonia." The first version suggests that 78 and 8 are to be compared and, without saying so, that about 10 percent contracted pneumonia.

Complications arise when two sets of numbers must be treated simultaneously. Often the numbers in one set act as labels, while in the other set the size of the numbers counts. In this circumstance, it may be useful to present one set in numerals and the other in words. For example, either "Three groups of patients included 15 in the one-dose group, 12 in the two-dose group, and 27 in the three-dose group" or "The one-, two-, and three-dose groups included 15, 12, and 27 patients, respectively" is better than "In the study, 15 received 1 dose, 12 received 2, and 27 received 3."

NUMBERS IN TABLES

To discuss numbers in tables, we need the concept of significant figures. In the phrase "significant figures," the word "significant" refers to the degree of accuracy the number seems to give. The context is usually that of measurement.

The numbers (a) 23,000, (b) 230, (c) 2.3, and (d) 0.0023 all have two significant digits or figures — namely, 2 and 3. In scientific notation, they would be written as follows: (a) 2.3×10^4, (b) 2.3×10^2, (c) 2.3, and (d) 2.3×10^{-3}. In scientific and medical writing, zeros following the last non-zero digit are not ordinarily regarded as significant digits, unless something is specified about accuracy, and in a decimal fraction the opening zeros to the right of the decimal point do not count as significant figures. The exception is that in scientific notation it is understood that all digits in a coefficient are significant. Without scientific notation, one may need to emphasize the accuracy of a number such as 300 if exactly 300 is meant.

The number of significant figures gives a hint of the accuracy of the number. For example, 98.2° has three significant digits and might be regarded as correct to within 0.05°. (One should not count heavily on this level of accuracy.) If, however, measurements were taken only to the nearest 10°, a report of 98.2° might mislead the reader about the accuracy of the number, and a one-significant-digit report of 100° might well be regarded as correct to within half a degree. Therefore, in these ambiguous circumstances, the author should tell what degree of accuracy is intended, as nearly as possible.

In scientific notation, 2.3×10^4 equals 23,000, and writing the coefficient as 2.3 suggests that the coefficient is correct to within 0.05, whereas 2.30×10^4 suggests that the coefficient 2.30 is correct to within 0.005. In Table 1, the first number in the right-hand column has this interpretation.

Tables can have two main purposes: to record and preserve the numbers for later reference, or to communicate a message for immediate comprehension and use. If the data are especially valuable for their accuracy or extent, then the purpose may be to preserve the numbers. Tables of physical constants, normal laboratory values, or population censuses, for example, have this feature, and so a great number of significant digits may be given for the benefit of a future user. In such cases, we keep as many digits as the data afford or the user may wish. Sometimes the recording ensures that the exact number can be provided to the reader, as in an unpublished appendix to a paper. The *New England Journal of Medicine* handles this situation by depositing extensive tables of important data with the National Auxiliary Publications Service and providing a footnote to the text. This service makes microfiche or photocopies of such tables available for a moderate charge to those who request them. Thus, to preserve numbers because of their accuracy or extent, direct publication in an article or report is not always needed.

Instead of recording or preserving numbers, many scientific tables communicate messages that the authors wish to deliver. For this purpose, the rule should be to use as few digits as will still deliver the message, because the fewer the digits, the more comprehensible the numbers. Readers who

Table 1.   An Illustration of Scientific Notation: Species and Quantity of Predominant Isolates from Vaginal Secretions of Referred Patients with Nonspecific Vaginitis.*

Species	No. of Positive Cultures (%) †	Mean Viable Count in Positive Cultures
Gardnerella vaginalis	17 (100)	1.60×10^8‡
Bacteroides		
B. *capillosus*	8 (47)	3.67×10^7
B. *bivius*	4 (24)	2.28×10^7
B. *disiens*	3 (18)	5.67×10^7

* Adapted from Spiegel at al.[10] (partial table).

† Data on 17 referred patients.

‡ Data on 12 patients only.

are daunted by one six-digit number may find whole tables of them incomprehensible, but most people get something out of comparing one- or two-digit numbers.

The discussion above emphasizes understanding in analyzing numbers within an article. When the numbers are to be used again in secondary analyses, a high degree of accuracy in the primary numbers can be very useful. Consequently, statistics on improvements in therapy, such as percentage survival, average length of hospital stay, and test statistics such as t and χ^2, should be given accurately (usually to three significant figures), and significance levels should, when possible, be given to two significant figures, rather than merely reported as $P<0.05$ or $P>0.05$, for example.

A good deal of folklore and personal experience suggest that when numbers are to be compared, they are better understood when lined up vertically instead of horizontally. Table 2 illustrates all the principles just mentioned. Note that when there are fewer digits in a column, the numbers are easier to compare. Consider whether eliminating more digits would improve understanding at the cost of precision. Note also that reading across is difficult, not only because the numbers are farther apart, but also because the eye is unsure that it is on the appropriate line. Blank lines can aid the eye in reading across, but they weaken the vertical comparisons.

Table 2.   An Illustration of the Greater Ease of Comparing Vertical Rather than Horizontal Numbers and of Comparing Numbers with Few Digits Rather than Many: Measurement of Antibody Radioactivity in Resected Tumors and Adjacent Normal Tissues.*

Patient No., Tumor Site	nCi ^{131}I in Tumor†	Tumor Weight g	nCi ^{131}I/g of Tumor	nCi ^{131}I/g of NC Mucosa	nCi ^{131}I/g of NC Serosa	nCi ^{131}I/g of Fat	$\dfrac{^{131}\text{I/g of tumor‡}}{^{131}\text{I/g of NC Mucosa}}$
8, cecum	1954	67	29.2	8.4	—	5.5	3.5
10, right colon	2092	60	34.9	7.1	—	7.3	4.9
11, sigmoid	1831	67	27.3	5.2	3.9	3.3	5.3
15, right colon	592	30	19.8	7.4	4.1	2.4	2.7
17, sigmoid	561	11	51.0	13.0	10.6	5.8	3.9
20, cecum	1432	28	51.1	9.8	7.2	5.7	5.2
25, right colon	1150	36	31.7	20.9	13.2	8.4	1.5
26, sigmoid	253	16	15.8	9.0	4.9	2.0	1.8

*   Adapted from Mach et al.[11] Samples consisted of 5 to 20 fragments of 1 to 2 g of tumor or adjacent normal tissues, dissected normal colon mucosa (NC mucosa), external part of bowel wall (NC serosa), or peripheral fatty tissue (fat) and were counted in a gamma-scintillation counter. Nanocuries (nCi) were corrected for the physical decay of ^{131}I during the period between injection and counting.

†   Total radioactivity recovered in the tumor. To convert curies to becquerels, multiply by 3.7×10^{10}.

‡ ^{131}I radioactivity per gram of tumor divided by ^{131}I per gram of normal colon mucosa.

Although using fewer digits promotes understanding, it is also true that many digits may be needed for internal comparisons. (Calculation is a different matter, and it is not treated here.)

In Table 3, Section A (on hospital billings) shows attractive uniformity, but it leans more toward preserving the data than toward making them comprehensible. As Section B illustrates, the same table with the decimals dropped (or rounded off, if preferred) is more readable. When one is reducing the number of digits for ease of comprehension, it does not matter much whether the later digits are dropped or rounded off. Dropping them may be more convenient. When rounding, a good rule is to round off to the nearest number, or if rounding a 5, to the nearest even number. Round 95.1 to 95, but 95.5 to 96.

Two main points are made in the original text about Section A.[12] First, the costs for the various percentage groups of patients are skewed so that the 10 percent with the highest costs pay from 42 to 47 percent of the total. Second, the hospitals had similar figures. These points might be made more precisely by taking the median or mean out of each row and displaying the differences from the median or mean (residuals).

Section C displays the median for each row from Section B and shows the differences from that number. For example, from the first row of Section B, we can see that the median of the six numbers is 29 because it is both the third- and the fourth-largest number. We write 29 at the left as in Section C, and then we replace the observations in the row by their residuals from 29 ($31 - 29 = 2, 29 - 29 = 0, 23 - 29 = -6$, and so forth). Then we repeat the process for the other rows. Section C shows the similarities among hospitals A, D, and E and the dissimilarities of the others, while giving a good summary figure for each row. If row means, rather than medians, are preferred, residuals could be based on those figures. Section C could take the place of both Section A and Section B.

Like the totals in the text, the totals in tables should check. Unless individuals can belong to more than one category, percentages should add up to nearly 100. Except when one is dealing with two categories, because of rounding, the percentages will often not add up to exactly 100. Keeping more decimal places does not solve this problem.[13] It is customary to add a footnote stating, "The percentages do not add up to 100 because of rounding errors." When an author gives, without comment, many columns of percentages based on frequencies, each adding up to exactly 100, the

Table 3.   Formats for Tables of Homogeneous Numbers: Hospital Billing Distribution (One Year).*

High-Cost Patients (%)†	Percentage of Total Hospital Billings Accruing to Corresponding Percentage of Patients					
Section A	Hospital A	Hospital B1	Hospital B2	Hospital C	Hospital D	Hospital E
5	31.7	29.8	23.0	27.6	31.4	29.9
10	45.2	45.1	36.1	41.9	46.8	45.5
20	63.9	63.0	59.6	59.1	66.0	64.0
30	74.6	76.3	75.1	72.0	75.6	73.8
40	82.6	86.3	83.9	78.5	82.6	80.3
50	87.6	92.1	90.1	85.1	87.6	85.9
Section B (Section A with decimals dropped)						
5	31	29	23	27	31	29
10	45	45	36	41	46	45
20	63	63	59	59	66	64
30	74	76	75	72	75	73
40	82	86	83	78	82	80
50	87	92	90	85	87	85
Section C (Each level and residual by hospitals)						
	Median	Residuals				
5	29	2	0	−6	−2	2	0
10	45	0	0	−9	−4	1	0
20	63	0	0	−4	−4	3	1
30	74.5	0	2	0	−2	0	−2
40	82	0	4	1	−4	0	−2
50	87	0	5	3	−2	0	−2

* Adapted from Zook and Moore.[12] Billings were summed over all hospitalizations for the same disease in one year. Distributions were weighted by the inverse of the 1976 hospitalization frequency to obtain a random sample of patients.

† Patients were ranked on the basis of charges for one year, on a scale ranging from the highest 5 percent to the highest 50 percent.

reader has reason to suppose that the author has "fudged" the numbers a bit in a manner that has not been described (except for such special sample sizes as 2, 4, 5, 10, and multiples of 5, where two-digit percentages always do add up to 100). I recommend for scientific writing that the numbers not be fudged. In nonscientific writing, the failure of the percentages to add up to 100 may be a distraction that prevents the reader from paying attention to the main points of the discussion, and so the recommendation does not apply. For intellectual honesty, a footnote might state that the percentages have been adjusted to add up to 100. There is further discussion of presenting tables in Chapter 16; see especially the National Halothane Study example.

SYMBOLS

Unfortunately we cannot arrange to have a single set of symbols to cover all occasions, and different symbols customarily mean different things in different contexts. Still, for general use in statistical writing, the lists in Table 4 may be helpful. Whatever notation is chosen, conventional or not, the definitions of any letters should be specified. The reader should not have to guess that n is the sample size or that π is the usual 3.14. . . . .

In the physical sciences and in mathematics, it is an international convention to use italic letters for variables such as p for probability, t for time, and x for the ubiquitous unknown of algebra. Also roman letters are used for the abbreviations for functions such as cos for cosine or log for logarithm or ln for natural logarithm (logarithm to the base e).

Some symbols that cause relatively little trouble are =, meaning "equals" or sometimes "which equals"; $\neq$, meaning "does not equal"; $\doteq$ or $\approx$, meaning "approximately equal"; and $\equiv$, usually meaning "equal by definition." Thus, if we define the circumference (C) of a circle in terms of the radius (r), we could say "$C = 2\pi r$, where $\pi = 3.14$. . . ."

More trouble comes from the inequality signs, $>$ and $<$. These symbols mean "greater than" and "less than," respectively. For example, $3>2$ and $-10<2$. In the correct sequence $0.2<P<0.5$, the interpretation is that P exceeds 0.2 and is exceeded by 0.5, and therefore it is in the interval from 0.2 to 0.5. Some writers think these symbols are a frame and write incorrectly that $0.2<P>0.5$. If this expression means anything, it means that P exceeds 0.2 and that it also exceeds 0.5, and so the latter part of the statement would have conveyed all the information. In text, it is preferable to write out expressions like greater than instead of using > as an abbreviation, though parenthetical remarks such as ($r>0.87$) can be used.

In some physics works <x> has been used to mean the average value of x. It is more usual to indicate averages by putting a bar over a symbol, so that $\bar{x}$ is the average of the xs and $\overline{\log x}$ is the average of the logarithms of the xs.

Because of typesetting costs, editors ordinarily prefer "knocked-down" fractions to those that are "built up" — i.e., they prefer a/b to $\frac{a}{b}$. In knocking down a built-up fraction, the danger is that the algebra will go wrong. For example,

$$\frac{a+b}{c} \neq a + b/c = a + \frac{b}{c}.$$

Instead,

$$\frac{a+b}{c} = (a+b)/c,$$

Table 4. Symbols Used in Statistical Writing.

ENGLISH ALPHABET

a, b, c, d	often stand for constants, expecially c because it is the first letter of "constant"; also, b is often used for coefficients in a regression equation.
e	may be reserved for the base of the natural logarithms, 2.71828 . . . ; also used for errors.
f	used both for mathematical function and for frequency.
F	special statistic used in the analysis of variance; also often a cumulative distribution function.
g, h	a mathematical function.
i, j	often used for an integer, but sometimes for an imaginary number.
k	an integer or constant (first letter of German word "Konstant").
l, o	often avoided because of confusion with 1 and 0 (one and zero).
m	an integer.
n, N	an integer, especially a sample size (n) or a population size (N).
p, P	a probability.
q	a probability, usually the complement ($q = 1 - p$) of some probability, p.

Table 4. (Continued)

r, s, t, T	integers; each also has a common technical meaning: r, a sample correlation coefficient; s, a sample standard deviation; t, a special statistic called Student's t, after the pseudonym of William S. Gosset (t is also often used to indicate time); T, a generalized t-statistic for higher dimensions called Hotelling's T.
u, v	variables.
w	often used for weighting.
x, y, z	usually variables; z is sometimes a variable with zero mean and a unit standard deviation.

GREEK ALPHABET

α	the significance level of a statistical test, also called the probability of a Type I error.
β	the power of a statistical test is $1 - \beta$, and β is the probability of accepting the null hypothesis when it is false (called a Type II error); also a coefficient in a regression equation.
δ, Δ	a difference.
ϵ	usually a small number.
θ	a parameter to be estimated.
κ	a statistical measure of association in a contingency table (table of counts).
λ	often the mean of a Poisson distribution.
μ	a population mean.
ν	a frequency, or a raw moment.
π, Π	$3.14 \ldots$, a true probability, or proportion; $\prod_{i=1}^{n}$ indicates a product.
σ	a population standard deviation.
Σ	a summation operator.
τ	a measure of association.
ϕ, Φ	a mathematical function, sometimes the Gaussian (normal) density function and distribution function.
ρ	a population correlation coefficient.
χ	when squared, a statistic measuring goodness of fit, among other things.
ψ	a mathematical function.

and so knocking down requires care and knowledge and often more effort than inserting a solidus, /, also called a "shilling mark."

A related complication applies to the use of the square-root symbol $\sqrt{}$. In $\sqrt{a+b}$, the horizontal bar or vinculum is actually a parenthesis that groups $a + b$. Without it, $\sqrt{a} + b$ might mean $b + \sqrt{a}$ or it might mean $\sqrt{a+b}$. Nevertheless, the square-root sign without the vinculum is often used in mathematical writing, even though it may be unclear how far the radical extends. For example a coefficient in the probability density function of the Gaussian distribution involves $\sqrt{2\pi\sigma^2}$, which could be written $\sqrt{2\pi}\,\sigma$ or as $\sigma\sqrt{2\pi}$. Some authors write instead $\sqrt{}2\pi\sigma$, which is satisfactory for those who know that the radical applies only to 2π, but not for those unfamiliar with the formula. Apparently, the radical without the vinculum is like words for Humpty Dumpty in *Through the Looking Glass*: it means what the author intends it to mean.

One way out of this difficulty, though not an attractive one, is to use fractional exponents. In the examples above, $\sqrt{a+b}$ could be replaced by $(a+b)^{1/2}$; $\sqrt{2\pi\sigma^2}$ could be replaced by $(2\pi\sigma^2)^{1/2}$, by $(2\pi)^{1/2}\sigma$, or even by $\sigma(2\pi)^{1/2}$.

Blyth[14] uses parentheses instead of the vinculum while employing the radical; for example, he uses $\sqrt{}(2\pi(n-\alpha))$. This usage seems clear to me, though I would be uneasy about $\sqrt{}(2\pi)\sigma^2$. Chaundy et al.[15] give an excellent discussion of the root sign as well as of many issues in printing mathematical formulas. Two additional useful sources of advice are the American Medical Association's style manual[16] and Huth's chapter[17] on numbers and related matters.

I am indebted to John Bailar, David Dorer, Karen Falkner, Katherine Godfrey, Colin Goodall, David Hoaglin, Edward Huth, Lois Kellerman, Philip Lavori, Robert Lew, Lillian Lin, Thomas Louis, Marjorie Olson, Kay Patterson, Marcia Polansky, and Kerr White for their helpful discussion and comments on the manuscript.

REFERENCES

1. Information for authors. N Engl J Med 1980; 303:60.

2. Chicago manual of style. 13th ed. Chicago: University of Chicago Press, 1982.

3. Campbell WG, Ballou SV. Form and style: theses, reports, term papers. 4th ed. Boston: Houghton Mifflin, 1974.

4. A manual for authors of mathematical papers. Providence, R.I.: American Mathematical Society, 1973.

5. Style manual. Revised ed. Washington, D.C.: Government Printing Office, January 1973.

6. Bureau of the Census manual of tabular presentation. Washington, D.C.: Government Printing Office, 1949.

7. Cheng C-S. Optimality of some weighing and 2^n fractional factorial designs. Ann Stat 1980; 8:436-46.

8. Mosteller F, Wallace DL. Inference and disputed authorship: the Federalist. Reading, Mass.: Addison-Wesley, 1964.

9. Schroeder JS, Lamb IH, Hu M. Do patients in whom myocardial infarction has been ruled out have a better prognosis after hospitalization than those surviving infarction? N Engl J Med 1980; 303:1-5.

10. Spiegel CA, Amsel R, Eschenback D, Schoenknecht F, Holmes KK. Anaerobic bacteria in nonspecific vaginitis. N Engl J Med 1980; 303:601-7.

11. Mach J-P, Carrel S, Forni M, Ritschard J, Donath A, Alberto P. Tumor localization of radiolabeled antibodies against carcinoembryonic antigen in patients with carcinoma: a critical evaluation. N Engl J Med 1980; 303:5-10.

12. Zook CJ, Moore FD. High-cost users of medical care. N Engl J Med 1980; 302:996-1002.

13. Mosteller F, Youtz C, Zahn D. The distribution of sums of rounded percentages. Demography 1967; 4:850-8.

14. Blyth CR. Expected absolute error of the usual estimator of the binomial parameter. Am Stat 1980; 34:155-7.

15. Chaundy TW, Barrett PR, Batey C. The printing of mathematics: aids for authors and editors and rules for compositors and readers at the University Press. London: Oxford University Press, 1954:76-7.

16. American Medical Association. Manual of style. 8th ed. Baltimore: Williams & Wilkins, 1989.

17. Huth EJ. Medical style and format. Philadelphia: ISI Press, 1987:153-61.

SECTION V

Reviews and Meta-Studies

21

~

MEDICAL TECHNOLOGY ASSESSMENT

John C. Bailar III, M.D., Ph.D., and Frederick Mosteller, Ph.D.

ABSTRACT   Medical technology involves a wide range of actions to screen for, diagnose, treat, or prevent disease or improve the working of the health care system. Assessment of medical technology combines information about safety and efficacy with social values, costs, side effects, acceptability, and legal issues to reach conclusions about the value of the technology under study. The methods used to get information include randomized controlled trials, epidemiologic studies, postmarketing surveillance, cost-benefit and cost-effectiveness analysis, modeling, and decision analyses. Each of the proposed uses of a new technology generates special problems in assessment.

Industry, government agencies, private foundations, professional societies, and independent physicians carry out assessments, but, except for the Food and Drug Administration which has legal responsibility for approval of drugs and some devices, no organization is in charge of medical technology assessment in the United States. Therefore many areas of medicine and surgery are evaluated poorly or not at all.

Patients, physicians, industry, and payers all have an important stake in assessment because the flow of new technologies exceeds our ability to appraise them, and yet many of them give little or no benefit and some even do harm. The Agency for Health Care Policy and Research has recently launched a program of outcomes research with special attention to controversial technologies.

Unless we invest carefully in new studies of safety, cost, and effectiveness and learn to assimilate the results of our clinical experience, we will often fail to make good choices from the buffet of new medical technologies.

Such things as a drug, a device or instrument, or an operation may constitute a medical technology; a technology that has been little studied and little used is for most practitioners a new technology, whatever the length of time since it was first proposed. We will discuss the definition of a medical technology below, but, by almost any definition, many new ones are introduced each year. Similarly, many become less commonly used or even disappear. The reasons for the rise and fall of technologies are numerous and complex. They include new information about benefits and risks, and the emergence of competing technologies that may be regarded as better, as well as costs, technical feasibility, acceptability to patients and medical personnel, and available skills and facilities.

Relman[1] refers to the "third revolution in health care, which is the era of assessment and accountability." He notes that "To control costs, without arbitrarily reducing access to care or lowering the quality of care, we will have to know a lot more about the safety, appropriateness, and effectiveness of drugs, tests, and procedures, and the way care is provided by our medical institutions and professional personnel."

Because in the end practitioners must choose treatments for the benefit of their patients, they want to be aware of benefits and dangers, and often of costs. But new technologies are coming on so rapidly that the pace of invention outruns the pace of evaluation and assessment. For example, a new surgical technique may be widely adopted before long-term complications are generally recognized. A new technology often generates revenue for the institution or practitioners using it, and, in the case of drugs and devices, for the manufacturers as well. It therefore increases the cost of medical care, sometimes by a wide margin. The appeal of the technology may lead to inappropriate proliferation of facilities, as appears to have happened with the newer imaging technologies. The need for long-term assessment of costs, risks, and benefits, all in the broadest sense, has never been greater.

Rational decisions about the use or abandonment of a technology must depend critically on an evaluation of that technology; i.e., on an assessment of its effects under actual or expected conditions of use. In this chapter we will examine some of the scientific bases for decisions to introduce or abandon a medical technology. Many Original Articles in the *New England Journal of Medicine* and, we believe, many similar articles in other medical journals bear on some technology. Our discussion is written especially for readers who, for their own decision making, must try to understand the implications of articles on technology assessment. Assessment involves nearly every statistical technique and method in general use and has caused the creation of many special techniques just for these purposes.

WHAT IS A MEDICAL TECHNOLOGY?

A medical technology is expressed in actions, and we use the term to cover some well-defined part of the range of actions intended to screen for, diagnose, treat or prevent disease, promote good health, reduce costs, facilitate the delivery of services, or otherwise improve the workings of health-related activities. A drug or a surgical procedure may be a technology, but so may techniques of physiotherapy, diagnostic imaging, methods of helping patients to stop smoking, the organization of intensive care units, methods of medical record keeping, and hospital supply procedures. For the purposes of this chapter, we exclude techniques of research, e.g., the randomized controlled trial or other research methods discussed elsewhere in this book. They are used in technology assessment, but we do not count them as medical technologies.

An important consideration is the range of activities that is to be considered a single technology. Is some new combination of old drugs a new technology? What about a new dosage schedule? At what point do the successive steps from a radical mastectomy to lumpectomy of a breast cancer constitute a new technology? Or when does the evolution of a new policy on immunization define a new technology? Such questions can be important in some contexts (reimbursement and medical liability come to mind), and they are central to the precise definition of research questions in scientific work. We give them little attention here, but only because "technologies" coalesce and split from paper to paper, depending on the specific object of the research, and because a technology is generally defined by the authors of each paper in a way that is adequately explicit for their own purposes.

A technology usually becomes more narrowly defined as research increases and understanding about it develops, but there are exceptions where the idea of a technology is greatly broadened again, even to the whole of prevention and therapy for a major disease.[2] Such variations are not often a matter for concern, as long as each research report is understood on its own terms. However, they can be serious when several reports — or even a whole field — are reviewed, especially with a method called meta-analysis. In the latter instance variations in a technology from one report to another can hamper efforts to analyze the data (see Chapters 22 and 23).

WHAT IS MEDICAL TECHNOLOGY ASSESSMENT?

"Medical technology assessment" is the term we use to designate a conscious, formal attempt to examine all relevant evidence needed to determine the optimal use, if any, of a medical technology. Assessments nearly always compare two or more technologies (sometimes including "no action") to provide some guidance to practitioners about clinical choices. Because medical technology assessments examine the whole of the evidence, they rarely rely on a single study and often have features of a review. However, they differ from many reviews in their tight focus on clinical decision making and in their primary intent to improve practice rather than to summarize present information and guide new research, although these activities are sometimes side benefits of assessments.

We make a distinction between a medical technology assessment and research studies to support such an assessment. For example, a randomized controlled trial may compare the clinical effects of some new drug to those of a standard, but a full technology assessment of that drug would also look at its effects in other settings and with other populations, its costs, unwanted side effects, acceptability to patients, and perhaps many other things. Thus not every study of a medical technology qualifies as a "medical technology assessment" in our use of the term. Fuchs and Garber[3] have recently compared what they called the "new" technology assessment, which includes costs and social issues, with the "old," which concentrates on efficacy and safety. Our usage is close to their "new" technology assessment.

To determine the frequency of reporting such assessments in a major medical journal we examined Volume 323 (second half of 1990) of the *New England Journal of Medicine*. The volume contained 124 Original Articles, of which we categorized 69 (56 percent) as reporting full technology assessments or data critical to technology assessments, usually the latter. Among these, 54 dealt with treatment and 15 with diagnosis. The studies of treatment included 30 reports on the efficacy of one treatment, 7 that compared two or more treatments, and 8 that focused on the way a treatment is used, such as dose level or the best age for administration. Among the remainder, 4 reported on side effects, 3 on feasibility of use, 1 on the prevalence of use, and 1 on a cost-benefit analysis.

Although a "full" medical technology assessment might be difficult to define because complicated technologies bring unexpected questions, safety and efficacy are base-line issues. Beyond this, costs, ethics, laws and regulations, and social implications are important. For example, in vitro fertilization raises issues in each of these categories, and medical practitioners, ethicists, and the courts will be busy for years sorting out the consequences. Such matters as the resources required to carry out a technology, including, for example, the possible need to create a new profession, are part of a full assessment. Other questions center on needs for the technology and the probable extent and circumstances of its use.

Lack of assessment is not just a matter of inadequate review of the literature; often the needed research has not been done, so that major decisions must be made by both the policy-maker and the individual practitioner in the absence of full understanding. While little has been done to develop a national program of medical technology assessment, the need is increasingly recognized. Relman[4] has proposed setting aside a few tenths of one percent of the national health budget for such assessment work; this small proportion would still be enough to develop a very substantial body of needed information.

OUTCOMES RESEARCH AT AHCPR

In 1989, the Agency for Health Care Policy and Research (AHCPR) began a substantial program of research on the outcomes of medical intervention. One important part of this program is its access to the immense amount of information collected through the Medicare program, including the billing information collected by the Health Care Financing Administration and other organizations. Billing information often tells about current medical practices and the changes in treatments.

This effort was stimulated in part by early efforts of Wennberg,[5] and Wennberg and Gittelsohn,[6] who found that the rates of some medical procedures varied considerably from place to place in ways that could not be easily explained by variations in disease incidence. For example, in 13 similar catchment areas of Vermont in 1969 through 1971 the age-adjusted rates for tonsillectomy and adenoidectomy varied from 4 to 41 operations per 1000 per year (among those under 26 years of age). Prostatectomy varied from 11 to 38 per 10,000 per year, and hysterectomy from 20 to 60 per 10,000. Such variations were also found for other surgical procedures in Vermont. These variations suggest that physicians do not agree about appropriate use of technology — in this case, certain surgical procedures — and that research to improve assessments of these technologies might improve their application.

In addition, Fowler et al.[7] collaborated with urologists in Maine to find out what happened to patients who had had prostate surgery. Among operated patients, four percent later reported persistent incontinence and five percent impotence. The number who died within three months after operation had not been determined previously and was found to be substantially larger than had been assumed. These and related findings suggested that broader and more systematic study of outcomes might again improve the application of medical technologies.

The use of financial claims data to tell what treatments or other medical technologies perform best under what circumstances is a matter of substantial research within the Patient Outcome Research Teams (PORTs) that the AHCPR has established for the procedures listed in Table 1. The diseases and procedures have been chosen for study on the basis of their importance to patients, their costs, and the perceived uncertainty about their outcomes. Each PORT will study its disease or procedure for several years. Initially, the AHCPR plans to use existing data about usual treatments

Table 1. PORTs and Guideline Panels Established as of 1991 by AHCPR.

PORTs[8]	Guideline Panels[9]
Acute myocardial infarction	Benign prostatic hypertrophy (BPH)
Benign prostatic hypertrophy (BPH) and localized prostate cancer	Cataracts in otherwise healthy eyes
Biliary tract disease	Depression treated in outpatient community-based settings
Cataracts	Pressure sore (management)
Cesarean section (obstetrical decision making in labor and delivery)	Sickle cell anemia
Chronic ischemic heart disease	Urinary incontinence
Diabetes	
Hip fracture and osteoarthritis	
Low back pain	
Pneumonia	
Total knee replacement	

without launching new clinical trials, but PORTs do use meta-analytic results from trials and other studies to guide their work.

AHCPR has also set up Guideline Panels (see Table 1) in areas parallel to the PORTs to establish standards for clinical practice. For example, one panel deals with pain. It uses meta-analyses of the literature together with special medical knowledge to help summarize, interpret, and develop guidelines for practice; as new findings are published, the guidelines will be reviewed and revised. Its first priority is postoperative pain; it will then turn to chronic pain, especially that from cancer.

APPROACHES TO MEDICAL TECHNOLOGY ASSESSMENT

Selection of methods for the evaluation of medical technology will depend on whether the technology involves prevention, screening, diagnosis, or treatment.

Diagnostic tests can be used to illustrate many of the ideas of medical technology assessment, and we start with them. A short section then discusses the rather simple extensions to screening procedures. We then turn to treatment of established disease, which brings in some important new ideas, and finally to prevention.

DIAGNOSTIC TESTS

In its simplest form, the assessment of a diagnostic technology involves two dichotomies: disease present or absent, and diagnostic test result positive or negative. The four combinations are conveniently recorded in the 2×2 table of Table 2, where a, b, c, d are numbers of test results ob-

Table 2. Schematic Outcomes of a Diagnostic Test.

Test Result	Disease		Total
	Present	Absent	
Positive	a	c	$a + c$
Negative	b	d	$b + d$
Total	$a + b$	$c + d$	n

served and $n = a + b + c + d$ is the total number of test results examined. Two of these counts, a and d, correspond to correct test results (true positive and true negative, respectively), while c is the number of false positive results and b counts the false negatives. Because the counts are highly dependent on the sample size n, it is customary to express them as rates. For example, $c/(a + c)$, the proportion of positive test results that are false positive, is called the false-positive rate. Its complement $a/(a + c)$, the true-positive rate, has a special name, predictive value positive. Similarly, $b/(b + d)$ is the false-negative rate and $d/(b + d)$ is the predictive value negative.

One would like to make both of the predictive values (positive and negative) as large as possible, but, in practice, efforts to increase one often have the opposite effect on the other. A classic example is blood pressure. A diagnostic approach that uses a confirmed diastolic pressure of 95 mm Hg to separate hypertensive from normotensive patients may be deemed to miss too many patients who have the disease. Lowering the figure to 90 mm Hg will miss fewer hypertensive subjects, but may greatly increase the false-positive rate (and reduce predictive value positive). The usual approach is to use some cut-off figure that allows both false positives and false negatives, but generally with more of the former because false positives tend to be less serious medically than false negatives. (For example, further workup may readily expose a false positive; or the consequences of unneeded treatment may be less serious than the consequences of missed treatment.) The optimum balance of false positives and false negatives is beyond the scope of this chapter, but is explored in Weinstein et al.,[10] Tosteson et al.,[11] Doubilet et al.,[12] and Phelps and Mushlin.[13]

The 2 × 2 table in Table 2 is fine for many purposes, but numeric results for the predictive values are highly dependent on the proportions of subjects who do and do not have the disease (the prior probabilities), and hence on the specific population studied. For example, doubling the number of persons who have the disease (and not changing the number who do not) doubles the value of b, and may almost double the false-negative rate. Thus, application of the test in a new setting, where the proportion of subjects who are ill is different or not known, can produce results that are hard to interpret for individual patients. A way around this is to take proportions in the other direction: $a/(a + b)$ is called the *sensitivity* of the test, and $d/(c + d)$ is its *specificity*. These are not dependent on the proportion of subjects who are ill; for example, doubling that proportion doubles both a

and *b* so that the sensitivity is unchanged. We would like to have both sensitivity and specificity as large as possible, but two problems remain. One is that, as with the predictive values, attempts to increase one generally decrease the other, and for the same reasons. The other is that sensitivity and specificity still cannot be usefully interpreted for individual patients without some knowledge of the prior probabilities that subjects do or do not have the disease. When a disease is very rare, even when both sensitivity and specificity are high — for example, 99 percent — the false positives will be much more frequent than the true positives.

In practice, each of these four measures — predictive values positive and negative, sensitivity, and specificity — conveys important information, but in general they are not enough to determine the appropriate role and use of a diagnostic test. As noted, interpretation of test results depends on the probability (before testing) that a subject is ill, and application of the test may also involve considerations of remedial action if results are positive, cost, time required to obtain test results, quality control in a community setting, and acceptability to patients (pain, time, nature and size of specimens required). A full technology assessment must consider all of these.

SCREENING

Technology assessment of screening tests (for persons who are asymptomatic but may have early disease or disease precursors) differs in some ways from that of diagnostic tests (for persons in whom there is a specific indication of possible illness). Most aspects of assessing diagnostic tests carry over to screening, but the differences can be important. One is that for screening tests the proportion of truly affected persons is likely to be small, so that many or most positive results are false positive. This is not necessarily serious if the screening test procedure is embedded in a broader program for further study of each initially positive finding; then the evaluation should focus on the whole process rather than the initial results. For diagnostic tests, in contrast, many or most patients have a medical problem that demands investigation; thus more weight may be given to such things as diagnostic precision and accuracy, and less to acceptability to patients.

A second difference is that, with screening tests, questions about how and how much long-term outcomes are changed are likely to be prominent. Earlier detection of disease is not helpful unless earlier intervention is helpful. This is sometimes true (e.g., hypertension), but testing for early, asymptomatic glaucoma has been widely abandoned because it is not clear that outcome is changed. Similarly, current questions about mammography of women under 50 center on whether their survival can be improved, though this is not an issue for older women. Screening for cancer of the urinary bladder can detect early disease, and the disease can be successfully treated by surgery, but only at the cost of cystectomy for many patients whose disease would never progress and thus who would not ultimately benefit from the operation.

A third critical difference between technology assessments of screening tests and diagnostic tests is the emphasis on cost. A program to test millions of people to find a small percentage who have early disease or its precursors cannot command the financial resources per subject that are available to support diagnostic testing, where yields of positive results are higher and patients already have conditions that demand accurate diagnosis and relief. There may also be substantial differences in how the sequence of steps in the medical investigation are arranged, and these differences are fair game for technology assessment, including procedures for the recruitment and scheduling of subjects, methods of quality control, record keeping, and follow-up.

THERAPY

Technology assessments of therapy differ in important ways from assessments of both diagnostic and screening tests. We again start with a 2×2 table (Table 3), but its nature and use have changed.

Here, the only important percentages run horizontally, and interest centers on the comparison of cure rates: $a/(a + b)$ vs. $c/(c + d)$. Chapters 4, 5, and 6 deal in some depth with three different approaches to the generation of

Table 3. Schematic Table for Outcomes of New vs. Standard Therapy.

Treatment	Outcome		Total
	Cured	Not Cured	
New	a	b	$a + b$
Standard	c	d	$c + d$

data of this form, and we will not repeat those discussions. Instead, we will focus on a few aspects that assume special relevance in technology assessment. As with diagnostic tests, outcomes often come in degrees (not just cured or not cured), and special attention must be given to how a treatment works in the broad spectrum of practice as well as in a research environment.

Therapeutic technologies often carry some risk, and this risk can be justified if it is necessary to prevent a greater harm from failure to treat or from treatment by a less effective means. However, the risks of treatment are often hard to study, and they may not get prominent attention in the first reports on a new therapeutic technology. Thus, a technology assessment may be incomplete because complications have not been identified and reported in the same depth as the intended beneficial outcomes. Further, honest but enthusiastic investigators may unconsciously bias their selection of patients, application of support measures, or measures of outcome in ways that support a promising new treatment. Some of the biases can be controlled by such measures as randomization and blinding, but these are not always possible, and even a randomized controlled trial with adequate blinding can produce biased results if the various controls fail. The assessment of a therapy may be complicated by long-delayed critical outcomes (e.g., long-term survival after cancer therapy, or delayed side effects of treatment) or by the need to weigh multiple good and bad results.

Randomized Controlled Trials

Randomized controlled trials protect against biases — not only those that have been thought of (such as age, sex, and severity of disease), but also against many that may not have been considered. Thus randomized controlled trials are often regarded as the strongest approach to the comparison of clinical therapies. This does not mean that they are the only method of learning about comparisons, nor that such trials are always well carried out, but that randomized trials have the advantage of promising initial equivalence of the groups being compared, a criterion that other methods tend to fail in advance. For example, when we compare the performance of a current group of patients with that of patients treated in the past, we must always hope that not much has changed between times. When nonrandomized comparisons are made between groups, some of whom are treated in different ways, there is always the risk that the choice of treatment was dictated by considerations of a patient's condition and that the comparison may not be a fair one.

We can, of course, recite a litany of ways for clinical trials to go wrong, but for every way a randomized controlled trial can fail, other methods can have similar problems. Some kinds of investigations do have benefits a randomized controlled trial cannot offer. For example, if we need to know how often certain treatments are used, a sample survey or other method of collecting descriptive information would be preferred.

In developed countries, much of medical care focuses on comfort and convenience and relief of morbidity rather than on lifesaving, and therefore assessment needs to be appropriate to these goals. Through the past two decades, clinical investigators have done much to create methods of assessing quality of life. By now the number of scales is large enough that standard usage is a greater need than new ways of making the assessments. We can expect such methods to be used increasingly in future clinical trials even though they are not routinely used in 1991. For an introduction see Mosteller and Falotico-Taylor[14] and the references cited therein.

Any investigation takes time, and, before it is mounted, thought should be given to the rate of innovations in the field. When innovations are coming thick and fast, a treatment may be obsolete before it is evaluated. Against this, however, must be weighed the fact that the results of evaluations not begun now will not be available later.

PREVENTION

Technology assessment of actions meant to prevent illness differs from that of treatment in ways that resemble the differences between assessment of diagnostic and screening technologies. One critical difference is that prevention is often highly decentralized to a broad range of health professionals and even to much or all of the general public. Examples of preventive technologies include smoking cessation (and subtechnologies to help smokers attain that end), dietary change to reduce the intake of salt or cholesterol and saturated fat, regular exercise, ingestion of aspirin to prevent myocardial infarction, and dietary supplements to prevent cancer. Some such measures have been widely adopted despite the lack of sound assessment or even in the face of evidence that they do not work (e.g., vitamin supplements in persons without evidence of deficiency). Others (e.g., reduction of dietary cholesterol and saturated fat) have been subjected to technology assessment, although the assessment itself may lack a critical element of scientific validity and public credibility.

Another way in which technology assessments of preventive measures often differ from those of other interventions is in the nature of the data. Because most serious diseases are rather uncommon on a day-to-day or year-to-year basis, studies of preventive measures must be large, and most must be long-term. This means, in turn, that they are expensive, and that few will be done. The technology assessor may have only one major field study to work with (e.g., the field trial of the Salk vaccine), or even none (e.g., the efforts to reduce risk of breast cancer by reducing dietary fat). Thus the body of evidence available for assessing a prevention technology may be limited, so that uncertainty about findings may be greater than for assessing a treatment technology.

A third difference is that the longer duration of most field trials of preventive measures (up to 15 years for the proposed Women's Health trial[15] of dietary change and breast cancer) creates numerous difficulties. One is the sheer delay — results are not available when they are needed. Another is the likelihood of unintended switches. Research subjects assigned to the intervention may quit, even if they remain under active follow-up, while those designated as controls may adopt the preventive step on their own, as seems to have happened in the Multiple Risk Factor Intervention Trial[16] (MRFIT) of steps to reduce cardiac disease. Because subjects who change their own management may differ in critical ways from those who do not, and also because such noncompliance may be expected in the public at large if the technology is recommended for broad use, it is customary (and proper) to analyze the data according to intention to treat—that is, according to the treatment assigned, regardless of how well the subject adhered to the prescribed regimen. Analysis by intention to treat prevents some kinds of bias and may provide more realistic assessments, but increases sample size requirements, sometimes dramatically.

A further problem with the long time cycle in studies of preventive technologies is that by the time results are available, the technology itself may have changed enough to have diminished the value of the field trial.

SOURCES FOR MEDICAL TECHNOLOGY ASSESSMENT

In the case of drugs and some devices, the Food and Drug Administration (FDA) has responsibility for deciding on the basis of evidence brought by a sponsor (generally the manufacturer) whether the new drug or device may be prescribed for a specific indication. Once approved, a drug can be used by physicians in whatever manner they think wise, even if it has not been tested for its safety and efficacy in the new situation.

One pressure for assessment, therefore, comes from industry in the area of drugs and some devices. But there is little interest in the assessment of some technologies because they do not appear especially profitable to industry and hence lack sponsors. Large areas of medicine can go unassessed because of lack of interest or funding, even though the treatments may be important and costly or may lead to severe consequences for patients. Nevertheless clinical investigators and others want to know what treatments are preferable, so they often carry out comparative evaluations of technologies.

Many medical technology assessments are published in the periodical literature of medicine, and can be identified and retrieved by the usual methods; Medline, Medlars, and the Cumulative Index Medicus of the National Library of Medicine can be especially helpful. Up-to-date textbooks and monographs may also include many assessments. Sometimes, however, other sources are needed, perhaps to learn more about a specific topic, to identify unpublished reports, or to study the methods used in preparing technology assessments. Useful information is often available on request from medical specialty organizations, manufacturers of drugs and devices, and registries of specific medical diseases or conditions (such as cancer registries or orthopedic implant registries). Particularly helpful is a two-volume set[17,18] from the National Academy of Sciences. Experienced medical librarians, and often research investigators, can search through such sources to find reports about specific technologies or specific medical problems.

METHODS CONTRIBUTING TO MEDICAL TECHNOLOGY ASSESSMENT

Many methods exist to collect and interpret information relevant to a medical technology assessment. The oldest and most common is the literature review: one or several persons collect what they believe is the corpus of relevant published material (sometimes supplemented with unpublished materials), interpret it, and offer summary judgments and conclusions. A thoughtful and detailed review by a knowledgeable person can provide much insight and valuable advice, but its inherent subjectivity and lack of structure are sometimes problems. Thus many persons and groups have turned to more standardized methods, often using groups of experts, to reach conclusions they can agree on in the light of information from the literature, and the results of presentations and feedback from the other experts. The best known of these are the consensus conferences sponsored by the National Institutes of Health, Office of Medical Applications of Research[18, p205] (OMAR) and the Delphi method.[17, p130-1] The FDA and other regulatory agencies commonly do technology assessments in the course of their review of requests for approval of new products or new uses of older products, and these may be made available. The package insert, familiar to practicing physicians, presents in brief form the findings from a detailed medical technology assessment performed by the FDA.

An unusually systematic type of literature review, supplemented by discussion with expert persons and groups, is carried out by the Office of Health Technology Assessment at the request of the Health Care Financing Administration.[19] In addition the Office of Technology Assessment has produced many medical technology assessments in response to requests from congressional committees.

A special set of methods called meta-analysis combines quantitative information from many studies of the same procedure to reach quantitative conclusions; these methods are becoming widely used in studies of health-related technologies. Chapters 22 and 23 of this book present an overview of the strengths and scope of meta-analysis. Meta-analyses, like other investigations that contribute to medical technology assessments, commonly use several distinct kinds of data resources. In addition to randomized controlled trials, some of the most important are discussed below.

REGISTRIES

Special organizations collect and maintain data concerning the occurrence of a few specific diseases or procedures and their outcomes. Examples are the Framingham Heart Study, the International Bone Marrow Transplant Registry, and the cancer registry for the United States called the Surveillance, Epidemiology, and End Results (SEER) program that records selected data items regarding all cancers in about 11 percent of the U.S. population. These data provide information about risk factors and can assist in the technology assessment of treatments. A 1991 collection of papers reviews the contributions of registries, including those mentioned above, to medical technology assessment.[20]

EPIDEMIOLOGIC STUDIES

When experiments are not available or are unlikely to be carried out, case–control studies and cohort studies are used to compare cases and controls. For example, the proportion exposed among cases with the disease can be compared with the proportion exposed among controls who do not have the disease. Special methods exist to appraise risk factors and associations; for example, by comparing exposure for cases with those for noncases in apparently comparable populations, or by following cohorts through time. The associations found may indicate causes.

POSTMARKETING SURVEILLANCE

Because most randomized trials can have comparatively few patients, rare adverse events may not be observed during the testing. Later, when the technology is widely used, surveillance of adverse events by the manufacturer, the FDA, or the Centers for Disease Control can alert the medical profession to possible problems.

COST-BENEFIT ANALYSIS AND COST-EFFECTIVENESS ANALYSIS

Various forms of analysis relate costs and benefits of treatment. A count of lives saved may not by itself be much of a measure of benefit. If the lives are saved only for brief periods, or if the patients continue indefinitely in a coma without hope of further recovery, many people would regard the benefits as small or, in the latter circumstance, even negative. Consequently, economists have suggested measuring the dollar cost per year of life saved because this gives an opportunity to compare benefits from different health expenditures. Or if there is reason to be concerned about the quality of the life saved, then the concept of "quality-adjusted life years" is sometimes invoked. Sometimes years saved before some cutoff age, such as 65, are counted and those beyond are not. Needless to say such definitions can be controversial. When benefits cannot be translated well to dollars, sometimes other assessments can be made, such as time spent outside of institutions, percent of time in good health, or years in productive work.

MODELING

Mathematical or computational methods are sometimes used to study the performance of processes over time when it is not easy to see what the outcomes of a procedure will produce.

DECISION ANALYSIS

In complicated problems with multiple choices and with multiple outcomes representing different benefits and losses, decision analysis offers ways to organize information and evaluate strategies. As with the other methods we have discussed, the findings feed into a total assessment, but do not automatically produce a decision. A decision analysis can be part of a cost-benefit analysis.

These and other methods for assessing medical technologies have been discussed at length in two publications of the National Academy of Sciences.[17,18]

CONCLUSION

Medical technology assessment has always been important for individual practice and the progress of medical science, but until recently it has generally been done informally, incompletely, as an additional burden on investigators and policy-makers — and not very well. The recent marked increase in the concepts, methods, and practice of medical technology assessment has already led to substantial improvements in the selection and application of technologies, and these improvements will surely continue. But we do need many more assessments. As Relman[4] says, "Unless we make a major investment in new trials of the safety, cost, and relative effectiveness of the drugs, tests, procedures, and operations we now employ so generously and unless we systematically collect and analyze the results of our clinical experience, we will never know enough to make discriminating choices."

REFERENCES

1. Relman AS. Assessment and accountability: the third revolution in medical care. N Engl J Med 1988; 319:1220-2.

2. Bailar JC, Smith EM. Progress against cancer? N Engl J Med 1986; 314:1226-32.

3. Fuchs VR, Garber AM. Sounding board: the new technology assessment. N Engl J Med 1990; 323:673-7.

4. Relman AS. Reforming the health care system. N Engl J Med 1990; 323:991-2.

5. Wennberg JE. Small area analysis and the medical care outcome problem. In: Sechrect L, Perrin E, Bunker J, eds. Research methodology: strengthening causal interpretations of nonexperimental data. Washington, D.C.: Department of Health and Human Services, Public Health Service, Agency for Health Care Policy and Research, 1990.

6. Wennberg J, Gittelsohn A. Small area variation in health care delivery. Science 1973; 182:1102-8.

7. Fowler FJ Jr, Wennberg JE, Timothy RP, Barry MJ, Mulley AG Jr, Hanley D. Symptom status and quality of life following prostatectomy. JAMA 1988; 259:3018-22.

8. Agency for Health Care Policy and Research. Medical Treatment Effectiveness Program (MEDTEP): active research projects as of September 30, 1990. AHCPR program note. Rockville, Md.: Department of Health and Human Services, 1990. (OM 91-0517)

9. Agency for Health Care Policy and Research. Clinical guideline development. AHCRP program note. Rockville, Md.: Department of Health and Human Services, 1990. (OM 90-0086)

10. The use of diagnostic information to revise probabilities. In: Weinstein MC, Fineberg HV, Elstein AS, et al. Clinical decision analysis. Philadelphia: W.B. Saunders, 1980:75-130.

11. Tosteson AN, Rosenthal DI, Melton LJ 3rd, Weinstein MC. Cost effectiveness of screening perimenopausal white women for osteoporosis: bone densitometry and hormone replacement therapy. Ann Intern Med 1990; 113:594-603.

12. Doubilet P, McNeil BJ, Weinstein MC. The decision concerning coronary angiography in patients with chest pain. A cost-effectiveness analysis. Med Decis Making 1985; 5:293-309.

13. Phelps CE, Mushlin AI. Focusing technology assessment using medical decision theory. Med Decis Making 1988; 8:279-89.

14. Mosteller F, Falotico-Taylor J, eds. Quality of life and technology assessment. Washington, D.C.: National Academy Press, 1989.

15. Marshall E. News and comment. Science 1990; 250:1503-4.

16. Multiple Risk Factor Intervention Trial Research Group. Multiple risk factor intervention trial: risk factor changes and mortality results. JAMA 1982; 248:1465-77.

17. Institute of Medicine, Committee for Evaluating Medical Technologies in Clinical Use. Assessing medical technologies. Washington, D.C.: National Academy Press, 1985.

18. Goodman C, ed. Medical technology assessment directory: a pilot reference to organizations, assessments, and information resources. Washington, D.C.: National Academy Press, 1988.

19. Lasch K, Maltz A, Mosteller F, Tosteson T. A protocol approach to assessing medical technologies. Int J Tech Assess Hlth Care 1987; 3:103-22.

20. Burdick E, McPherson MA, Mosteller F, eds. The contribution of medical registries to technology assessment. Int J Tech Assess Hlth Care 1991; 7:123-99.

22

~

COMBINING RESULTS FROM INDEPENDENT INVESTIGATIONS

META-ANALYSIS IN CLINICAL RESEARCH

Katherine Taylor Halvorsen, D.Sc., Elisabeth Burdick, M.S.,
Graham A. Colditz, M.D., D.P.H., Howard S. Frazier, M.D.,
and Frederick Mosteller, Ph.D.

ABSTRACT   Meta-analysis refers to a collection of methods for combining quantitative information from several sources to give a summary statistic together with its uncertainties. The results strengthen our knowledge beyond that contributed even by multiple single studies and may guide diagnosis and treatment of patients and point toward future research.

We illustrate the variety of topics, methods, and statistical techniques through closer scrutiny of a few meta-analyses published during 1989. The many statistical methods for combining data take different forms, and they are based upon different assumptions. We explain the importance, and review the reporting, of eight attributes of meta-analysis: methods of searching, eligibility criteria, number of articles, outcome variables, study design, results used for combining, homogeneity, and statistical methods. These items may help the reader evaluate a meta-analysis.

Meta-analysis is defined as the quantitative analysis of two or more independent studies to integrate the findings and describe features of the studies that contribute to variation in their results. Meta-analysis in medical research often uses the accumulated evidence about a treatment or procedure to provide guidance to clinicians and to suggest directions for future research.

The statistical techniques appropriate for meta-analysis provide both the clinician and the medical investigator with quantitative summaries of the results of several studies, usually evaluations of therapies or of diagnostic methods. The methods ordinarily give a combined estimate or test, to-gether with an estimate of its uncertainty, and thus offer a perspective on the degree of knowledge of the issue. Meta-analysis contributes to our be-lief in the presence or absence of a treatment effect and gives some notion of the size of the effect or of the diagnostic value of a procedure. When the variability of an effect across studies is large, we usually need to know why. Often the relation of some other variable such as age, severity of disease, or size of dose may explain much of the variability observed.

For several reasons, the results in a collection of studies of similar treat-ments differ to some degree. Some studies may be too small to demonstrate clearly a treatment effect that would be apparent in a larger study. The pa-tients in the studies may not be comparable, the treatments applied may differ from one study to another, and the skills of the care givers may differ. When a difference between treatments is small relative to the inherent vari-ability of the data, an occasional report may, by chance alone, show statisti-cal significance in the "wrong" direction. Combining the results of several studies through the techniques of meta-analysis can provide stronger evi-dence *for* or *against* a treatment effect than one can derive from any of the individual studies because it produces a more precise estimate of the effect (i.e., an estimate with smaller standard error or a narrower confidence interval).

MacMahon et al.[1] combined results from 14 randomized controlled trials of the effects of prophylactic lidocaine hydrochloride on early ven-tricular fibrillation and death in patients with suspected acute myocardial infarction. No single trial was compelling by itself because the trials were modest in size. During follow-up intervals of 1 to 4 hours in the trials of intramuscular lidocaine injection (6961 patients) and 24 to 48 hours in the trials of intravenous lidocaine infusion (2194 patients) a total of 103 cases of ventricular fibrillation and 137 deaths were recorded. Overall, allocation to lidocaine was associated with a 35 percent reduction in risk of ventricu-lar fibrillation (95 percent confidence interval: 3 to 56 percent reduction). There was, however, no evidence of beneficial effect on mortality. Indeed, the odds of early death were 38 percent greater among patients allocated to lidocaine (95 percent confidence interval: 2 percent reduction to 95 percent increase). Further follow-up and larger trials are required to determine the safety and therefore the usefulness of lidocaine in acute myocardial infarction.

In a second example, the Collaborative Group[2] presents a large meta-analysis of early reports of therapy for patients with breast cancer. The analysts combined data from 28 trials of tamoxifen citrate (16,513 women) and 40 trials of other cytotoxic chemotherapies (13,442 women). In all, some 8106 deaths were available for the analyses. Women over 50 who were treated with tamoxifen had a statistically significant 20 percent reduction in the annual odds of mortality during five years of follow-up. Women under 50 who were assigned to other cytotoxic chemotherapies experienced a 22 percent reduction in the annual odds of death through five years of follow-up, though women 50 or older had no clear evidence that other chemotherapy reduced mortality. Direct comparisons showed that combination chemotherapy was significantly more effective than single agent chemotherapy. Because of its size, this meta-analysis was able to show clearly that both tamoxifen and combination chemotherapy can reduce five-year mortality for women with breast cancer.

MEDICAL TOPICS OF META-ANALYSES REPORTED IN 1986–1989

The present study uses the meta-analyses selected by Sacks et al. (see Chapter 23); the analyses were published in the medical literature between 1986 and 1989. The criteria for inclusion on the list were that data from more than one study had to be combined and that at least one of the studies pooled had to be a randomized controlled trial. If we do not apply the randomization constraint, then in 1989 and 1990, more than 100 medical meta-analyses were published in each year. Sometimes we have no randomized trials but have epidemiologic studies such as case–control studies or cohort studies. These may still be combined in a meta-analysis, though with more caution about the potential for bias.

To show the spread of meta-analyses across clinical disciplines from 1986 through 1989, we classified our list of meta-analyses (except for a few that Sacks et al. added after their initial search) by area of application; 13 headings describe medical specialties or problems or interventions and one provides for methodological papers. Some papers fall into more than one category and the boundaries are not sharply drawn in order to indicate generally what fields have been getting attention. General internal medicine has by far the largest number of meta-analyses, with 33 in the four-year period. Cardiology follows with 15 meta-analyses in the same period, surgery has 10, and gastroenterology has 8 in those four years. Each of the remaining ten headings (intensive care, obstetrics/gynecology, neurology, oncology, pediatrics, psychology, rheumatology, infectious disease, nephrology, methods) have 2 to 6 meta-analyses in the four-year period.

VARIETY IN CONTENT AND METHODS USED IN META-ANALYSES

To illustrate the variety of both content and method, Table 1 lists 7 meta-analyses[3-9] published during 1989 that met the inclusion criteria described above. It gives the medical focus of each study, methods of search for articles to include in the meta-analysis, eligibility criteria for inclusion of studies, number of studies included, outcome measure used in the meta-analysis, statistical methods used for combining, and results obtained. For inclusion in Table 1, we chose studies that show the wide range of medical areas and have fuller reporting. For example, all studies in Table 1 listed their methods of search.

The methods of search included computer searches and hand searches of journals, as well as informal methods, such as letters to scholars working in the field. Some studies not discussed here did not report their methods of search.

Eligible studies varied from blinded, randomized, controlled trials to trials that had neither blinding nor randomization. Again, some meta-analysts did not report the types of studies included in their combination. The number of studies included in Table 1 varied from 4 to at least 22. The outcome measure depended on the problem studied. A variety of statistical methods were needed because of differences in the kinds of data and situations being combined. For example, in the analysis by Kairys[6] of steroid treatment of laryngotracheitis, or croup, the outcome measure is the proportion of patients on drug therapy with significant improvement in symptoms minus the proportion of patients on placebo with significant improvement. Total mortality is used as the outcome measure in several studies (Infant-Rivard,[5] O'Connor[7]) and relative risk and risk difference are used in others (Romero,[8] Schmader[9]).

In combining data from clinical outcomes, the main situations lead to continuous measurements (such as blood pressures) and counts (such as the number who died). Sometimes we transform the original data from studies so that their originally incompatible data can be combined and analyzed. In clinical medicine at present, the most common statistics used to combine data come from counts in 2 × 2 tables — typically an experimental group and a control group, each with counts of favorable and unfavorable outcomes. Three main statistics are used for summarizing these 2 × 2 tables: risk difference (e.g., difference in proportion of adverse events), risk ratio (e.g., ratio of proportions of adverse events), and odds ratio (e.g., ratio of odds of an event in the treatment group to the odds in the control group). The appropriate choice of statistics for averaging and reporting variation in the meta-analysis depends on which of these measures is to be used. The risk difference has the advantage of directly measuring the reduction in deaths or increases in cures, but it may not be best or even appropriate for a specific analysis.

A special form of transformation called *effect size* — a technical term not to be confused with the plain meaning of size of effect — combines differences and standard deviations of measurements to avoid the incommensurate units for some collections of studies. For example, different studies of outcome may use time to one of several different adverse events, yet in the meta-analysis the investigator wishes to combine them because in each instance longer times imply better outcomes.

Beyond the issues of types of data, investigators often distinguish between situations with *fixed effects* and those with *random effects*. Fixed effects arise when the studies included in the meta-analysis are regarded as the only members of the universe of concern — for example, a study of the three hospitals in town A. Random effects arise when the studies are a sample from a larger universe of possible studies, so we need to allow for variation from one sample of studies to another. A good analogy arises when several institutions study the results of applying the same treatment to their patients; if these are the only institutions that concern us, the fixed effect situation applies, but if results are to be generalized to a larger population of institutions, then random effects apply. Thus, the random effects model requires two levels of sampling — first of institutions and then of patients within each sampled institution — while the fixed effects model uses only the latter. These distinctions call for differing statistical methods in the meta-analysis.

Table 1. Characteristics of Seven Selected Meta-Analyses Published During 1989.

First Author	Medical Focus	Methods of Search	Inclusion Criteria	No. of Studies	Outcome Measure	Statistical Methods Used	Results Obtained
Clark[3]	Injectable gold salts in rheumatoid arthritis	MEDLINE computer search and hand search	RCT;* in English, comparing gold versus placebo	4	Any outcome reported in 2 or more of the 4 studies and using the same scale	Combined effect size, method of Hedges and Olkin[10]	Significant improvement in joint count (30%), grip strength (14%), functional capacity (13%), and hemoglobin concentration (5%) for treatment over placebo, but side effects were more frequent.
Goldstein[4]	Use of progestational agents in pregnancy	BRS† and NLM‡ computer searches, Oxford database of perinatal trials, Current Contents, bibliographies	High-risk pregnancies, RCT* of progestogen treatment during pregnancy, provides data on outcome of pregnancy	15	Counts of 5 possible outcomes of pregnancy	Modified Mantel–Haenszel technique (Yusuf et al.[12]) and random effects model of DerSimonian and Laird[13]	Authors conclude that progestogens should not be used outside of randomized trials at present because no statistically significant effect of the treatment could be detected.

Infante-Rivard[5]	Endoscopic variceal sclerotherapy in long-term management of variceal bleeding	NLM‡ computer search and bibliographies	RCT,* in English, published 1970–1987, comparing serial sclerotherapy to other treatments, most patients followed more than one year	7	Total mortality and mortality related to bleeding	Weighted average risk difference estimates, overall risk difference, Mantel–Haenszel[14] test of overall risk difference	Patients with esophageal varices that have bled may benefit from the inclusion of repeated sclerotherapy in their long-term management regimen. An overall risk difference of –0.15 indicated that sclerotherapy reduced the number of deaths by 25%.
Kairys[6]	Steroid treatment of laryngotracheitis, or croup	MEDLINE 1950 to 1987, bibliographies, hand search, and personal communications	RCT,* blind assessment, only difference between experimental and control groups was use of steroids; in English	9	Proportions improved 12 and 24 hours post treatment; proportion requiring intubation	Modified Mantel-Haenszel technique (Yusuf et al.[12]) for combining odds ratios	The use of steroids in children hospitalized with croup is associated with a significantly increased proportion of patients showing clinical improvement 12 and 24 hours post treatment (odds ratios 2.25 and 3.19, respectively), and a significantly reduced incidence of endotracheal intubation (odds ratio 0.21).

(Table continues)

Table 1. (Continued)

First Author	Medical Focus	Methods of Search	Inclusion Criteria	No. of Studies	Outcome Measure	Statistical Methods Used	Results Obtained
O'Connor[7]	Effects of cardiac rehabilitation programs with exercise	Computer search and personal communications	Published, RCT* of rehabilitation after MI[§] with "objective and reproducible entry criteria"	22	Total mortality, cardiovascular mortality, sudden death, fatal MI[§], nonfatal MI[§]	Modified Mantel-Haenszel technique (Yusuf et al.[12]) for combining odds ratios	After an average of 3 years of follow-up the odds ratio for total mortality (0.80), cardiovascular mortality (0.78), and fatal reinfarction (0.75) were significantly lower in the rehabilitation group than in the comparison group. The odds ratio for sudden death was significantly lower in the treatment group at 1 year (0.63) but not at 2 or 3 years.
Romero[8]	Relationship between asymptomatic bacteriuria and preterm delivery/ low birth weight	MEDLINE and bibliographies	RCT,* published 1966–1986, in English, clearly stated definitions of asymptomatic and "prematurity," data given in paper	19	Relative risk and risk difference	Typical relative risk (Rothman and Boice[15]), typical risk difference (Rothman[16])	Cohort studies showed that untreated asymptomatic bacteriuria during pregnancy significantly increased rates of low birth weight (relative risk 1.54) and preterm delivery (relative risk 2.00) as com-

| Schmader[9] | Current therapies (antiviral, corticosteroids) for post-herpetic neuralgia in the immunocompetent host | MEDLINE, Index Medicus, and bibliographies, published 1968–1987, in English | Controlled study, immunocompetent subjects, 1 month or more of follow-up, study not limited to ophthalmic herpes zoster | 21 | Difference in proportion with syndrome; syndrome defined as pain after skin has healed | Combined log odds ratio used to estimate the pooled odds ratio and its 95% confidence interval; Pearson correlation coefficient for test association | pared to non-bacteriuric patients. Antibiotic treatment significantly reduced the risk of low birth weight (relative risk, 0.56) as compared to placebo treatment. Currently there is no proven useful therapy for post-herpetic neuralgia. The benefits of acyclovir and corticosteroids are limited. A clear consensus definition of post-herpetic neuralgia is needed to improve future investigations. |

---

*   RCT: randomized, controlled, clinical trial
†   BRS: Bibliographic Retrieval Service
‡‡ NLM: National Library of Medicine
§   MI: myocardial infarction

As the statistical column in Table 1 indicates, many authors have devised statistics and tests for combining information from several studies.[10-17] Indeed, the name alone may be ambiguous because the same authors such as Cochran[11] or Mantel and Haenszel have created several tests. We need not go into such details here. Our main purpose is to give an idea of the need for such a multiplicity of methods.

REPORTING IN META-ANALYSIS

Readers need information about the design and analysis of a study to evaluate and interpret its results. Chalmers et al.[18] outlined a scoring system for evaluating the quality of meta-analyses (see Chapter 23), and Halvorsen[19] proposed a checklist of eight important aspects of design and analysis for use in assessing the reporting of meta-analyses.

In this chapter we introduce the clinician to topics to be considered when reading a meta-analysis. In Chapter 23, the authors explain the details of quality assessments currently being used for research in meta-analysis.

THE REPORTING STANDARD

We will evaluate the reporting of the following eight items: (1) methods of search for relevant articles; (2) eligibility criteria for articles; (3) number of articles included; (4) outcome variables used for combining results; (5) study design; (6) results used in combining (information indicating whether the meta-analysis used counts or other data from the individual studies such as a summary statistic — that is, a test statistic, P value, or estimate from each study); (7) statistical methods used in meta-analysis; and (8) homogeneity. When the outcomes of studies are homogeneous, about as alike as sampling variation would allow, the mean of the study outcomes summarizes them well, but if results vary substantially from study to study, then the extent of variability among studies needs attention and reporting.

DISCUSSION

The medical literature is vast; in 1991, of some 13,000 journals in English and many foreign language journals, the MEDLINE database covers only 2938. Clearly, it is impossible to review more than a small fraction of what is available; therefore, the methods used in a search may have a strong influence on the sample obtained. For this reason, item 1, methods of search, requires that authors describe how they conducted their search for studies. They should clearly describe the sources of research reports and should tell readers which types of journals and perhaps which other sources they included in their search and from what period the publications were drawn. If the investigators make a computerized search, they need to report the databases used (e.g., MEDLINE). Readers knowledgeable about the literature will then have a good idea whether the search has covered the areas most relevant to their interests.

Goldstein et al.[4] provide a good example of reporting the breadth of a search in their meta-analysis of randomized controlled trials of progestational agents in pregnancy. They tell the reader,

> A preliminary list of papers was provided by the *Oxford Database of Perinatal Trials* (Chalmers et al., 1986).[20] Papers were also searched for through the Bibliographic Retrieval Service (BRS) and the National Library of Medicine (NLM) computer systems based on medical subject heading key words (e.g., pregnancy, clinical trials, random allocation, double-blind method, progesterone, sex hormones), by routine searching of *Current Contents* (a weekly publication listing recently published articles by journal), and by searching the references of studies and review articles as they were obtained.

Clinicians also need to know the kinds of patients and treatments that produce the reported conclusions, because they want to know whether the results apply to their patients. This information is often implicit in the criteria used to select studies for meta-analysis. Item 2 asks whether authors reported specific information about the medical problems, patient characteristics, treatment methods, and features of study design they used to select studies for their sample. If the eligibility criteria are broad and include several possible diseases, kinds of patients, or treatments, then the authors should describe the sample they actually obtained.

Items 3 and 4 tell the reader how many results the authors combined, how they specified the outcome variable, and how that variable was measured. Meta-analyses based on larger numbers of studies are likely to be more reliable for given sample sizes within studies. How outcome is measured may be a key issue because differences in methods can produce substantial effects. Item 5, study design, should specify at least whether the investigators combined results from randomized trials, from nonrandomized trials, from observational studies, or from more than one of these types of studies. Other things being equal, randomized trials are regarded as being of higher quality.

Item 6 tells the reader what results from the individual studies were used to obtain the combined statistics. Readers need to know the specific statistical procedures used (item 7) to be able to interpret a combined significance test or estimate. Authors should provide sufficient information about their statistical analysis, including the names of widely available computing programs, so that readers can perform similar analyses.

Authors should also tell the reader whether they have considered the homogeneity of the results (item 8) and give their assessment of the homogeneity. Lack of homogeneity means that outcomes in new situations cannot be expected to be reliably predicted from the mean of the observed studies, and that study-to-study variability must also be taken into account.

REPORTING METHODOLOGICAL META-ANALYSES

The present chapter and that by Sacks et al. (Chapter 23) may be considered as belonging to another type of meta-analysis. The meta-analyses described in Table 1 address the same scientific question as the studies they use. Other methodological meta-analyses, including this chapter, address a different question. This second type of meta-analysis may use primary reports to make inferences about, for example, the quality of reporting (see Chapter 17), the quality of scientific research (see Chapter 23), the power of statistical tests used for analysis (see Chapter 19), the sources of funding, and other features of research. They could be regarded as "meta-meta-analyses."

Not all eight items discussed above are applicable to this second type of meta-analysis. We have reported methods of search for the meta-analyses analyzed here, and we have defined the criteria for inclusion in this study and reported the number of articles found. The remaining five items are not necessarily applicable because, as is the case in this paper (except in Table 1), this analysis does not seek to answer the research questions that the individual studies address, nor does it integrate the outcome measurements produced by the studies we have examined.

Preparation of this chapter was partly supported by Grant Number 05936 from the Agency for Health Care Policy Research and by the Alfred P. Sloan Foundation.

REFERENCES

 1. MacMahon S, Collins R, Peto R, Koster RW, Yusuf S. Effects of prophylactic lidocaine in suspected acute myocardial infarction. An overview of results from randomized controlled trials. JAMA 1988; 260:1910-6.

 2. Early Breast Cancer Trialists' Collaborative Group. Effects of adjuvant tamoxifen and of cytotoxic therapy on mortality in early breast cancer: an overview of 61 randomized trials among 28,986 women. N Engl J Med 1988; 319:1681-92.

 3. Clark P, Tugwell P, Bennett K, Bombardier C. Meta-analysis of injectable gold in rheumatoid arthritis. J Rheumatol 1989; 16:442-7.

 4. Goldstein P, Berrier J, Rosen S, Sacks HS, Chalmers TC. A meta-analysis of randomized control trials of progestational agents in pregnancy. Br J Obstet Gynaecol 1989; 96:265-74.

 5. Infante-Rivard C, Esnaola S, Villeneuve JP. Role of endoscopic variceal sclerotherapy in the long-term management of variceal bleeding: a meta-analysis. Gastroenterology 1989; 96:1087-92.

 6. Kairys SW, Olmstead EM, O'Connor GT. Steroid treatment of laryngotracheitis: a meta-analysis of the evidence from randomized trials. Pediatrics 1989; 83:683-93.

 7. O'Connor GT, Buring JE, Yusuf S, et al. An overview of randomized trials of rehabilitation with exercise after myocardial infarction. Circulation 1989; 80:234-44.

 8. Romero R, Oyarzun E, Mazor M, Sirtori M, Hobbins JC, Bracken M. Meta-analysis of the relationship between asymptomatic bacteriuria and preterm delivery/low birth weight. Obstet Gynecol 1989; 73:576-82.

 9. Schmader KE, Studenski S. Are current therapies useful for the prevention of post herpetic neuralgia? A critical analysis of the literature. J Gen Int Med 1989; 4:83-9.

10. Hedges LV, Olkin I. Statistical analysis and methodology. New York: Academic, 1984.

11. Cochran WG. The combination of estimates from different experiments. Biometrics 1954; 10:101-29.

12. Yusuf S, Peto R, Lewis J, Collins R, Sleight P. Beta blockade during and after myocardial infarction: an overview of the randomized trials. Prog Cardiovasc Dis 1985; 27:335-71.

13. DerSimonian R, Laird N. Meta-analysis in clinical trials. Controlled Clin Trials 1986; 7:177-88.

14. Mantel N, Haenszel W. Statistical aspects of the analysis of data from retrospective studies of disease. JNCI 1959; 22:719-48.

15. Rothman KJ, Boice JD. Epidemiologic analysis with a programmable calculator. NIH Pub #79-1949. Washington, D.C.: United States Department of Health, Education, and Welfare, 1979.

16. Rothman KJ. Modern epidemiology. Boston: Little, Brown, 1986.

17. Fleiss JL. Statistical methods for rates and proportions. New York: John Wiley, 1981:165-8.

18. Chalmers TC, Berrier J, Hewitt P, et al. Meta-analysis of randomized controlled trials as a method of estimating rare complications of non-steroidal anti-inflammatory drug therapy. Aliment Pharmacol Therap 1988; 2(Suppl 1):9-26.

19. Halvorsen KT. Combining results from independent investigations: meta-analysis in medical research. In: Bailar JC, Mosteller F, eds. Medical uses of statistics. Waltham, Mass.: NEJM Books, 1986:392-416.

20. Chalmers I, ed. Oxford database of perinatal trials. Oxford: Oxford University Press, 1986.

23

~

META-ANALYSES OF RANDOMIZED
CONTROL TRIALS

AN UPDATE OF THE QUALITY AND METHODOLOGY

Henry S. Sacks, Ph.D., M.D., Jayne Berrier, M.A.,
Dinah Reitman, M.P.S., Daniel Pagano, B.A.,
and Thomas C. Chalmers, M.D.

ABSTRACT   Meta-analysis attempts to analyze and combine the results of previous reports. We have updated our 1987 survey of 86 meta-analyses of randomized control trials in the English-language literature with an additional 78 reports. We have evaluated the quality of these meta-analyses using a scoring method we have devised that lists 23 items in six major areas — study design, combinability, control of bias, statistical analysis, sensitivity analysis, and application of results. Of the 23 individual items, the mean number satisfactorily addressed was 7.63 ± 2.84 (mean $\pm$ S.D.) for 40 papers published from 1955 to 1982, 6.80 ± 3.86 for 66 papers published from 1983 to 1986, and 11.91 ± 4.79 for 58 papers published from 1987 to 1990 ($F = 31.3$, $P < 0.001$).

We conclude that there has been a noticeable improvement over this short time, but there still exists an urgent need for improved methodology in searching the literature, evaluating the quality of trials, and synthesizing the results.

A number of papers have appeared in the medical literature that attempt to evaluate and combine the results of previous studies. Light and Smith[1] were among the first to propose pooling original data from various research studies. Glass was the first to refer to this type of research as "meta-analysis."[2] A dictionary defines "meta" as "more comprehensive, transcending — used with the name of a discipline to designate a new but related discipline designed to deal critically with the original one."[3] Thus, meta-analysis is a new discipline that critically reviews and statistically combines the results of previous research. A 1985 review of meta-analysis in the public health field emphasized its growing importance.[4]

The purposes of meta-analysis include the following: (1) to increase statistical power for primary end points and for subgroups; (2) to resolve uncertainty when reports disagree; (3) to improve estimates of size of effect; (4) to answer new questions not posed at the start of individual trials; and (5) to bring about improvements in the quality of the primary research. These functions are applicable to randomized control trials because, like other types of clinical research, they are often undersized.

This paper will update the survey we published in 1987,[5] describe some criteria to be considered by those who perform meta-analyses as well as those who utilize them, and suggest areas that need further work.

METHODS

The English language medical literature (from January 1966 through July 1990) was searched for papers that pooled the results of controlled clinical trials. Papers were found in *Current Contents* and by computer searches of the National Library of Medicine (NLM) and Bibliographic Retrieval Services, Inc. (BRS) databases by looking for reviews of specific subjects and using the terms "meta-analysis," "pooled or pooling," or "combined or combining" in title, abstract or, where available, full text. Other sources included references in papers found by the above methods and personal communication.

Our criteria for inclusion of papers in this analysis were that data from more than one study must be combined, and at least one of the studies pooled must be a randomized control trial (it turned out that there was always more than one). Meta-analyses of diagnostic evaluations and epidemiologic studies were not included, nor were those dealing only with methodology — the emphasis was on therapy.

Each paper was evaluated independently by two investigators using a scoring sheet which lists what we consider to be the important elements of a meta-analysis. The evaluators were blinded as to the name of the journal and the authors of the paper.

RESULTS AND DISCUSSION

From a total of 177 papers retrieved, 86 studies met the inclusion criteria in our first survey. Since then we have found an additional 78. When the same results were reported or updated in more than one publication, only the latest publication was cited but the evaluation was based upon a composite of all available information.

The earliest meta-analysis we found was an article on the powerful effects of the placebo by Henry K. Beecher, published in 1955.[6] Only 4 other meta-analyses of clinical trial data published before 1970 were found; 20 were published during the 1970s and 139 since 1980. Some of the increase may be due to better coding and retrieval. The meta-analyses found were published in over 50 different journals and 1 book. As shown in Table 1, only 10 journals have published more than 2 meta-analyses.

We believe that the important qualities of any meta-analysis can be divided into six major areas: study design, combinability, control of bias, sta-

Table 1.   Publication of Meta-Analyses of Randomized Control Trials by Journal and Date.

Journal	Number of Meta-Analyses			
	1955–1982	1983–1986	1987–1990	Total
American Journal of Medicine	2	6	6	14
Journal of the American Medical Association	1	4	7	12
Lancet	6	3	2	11
New England Journal of Medicine	3	4	1	8
British Medical Journal	2	2	2	6
Gastroenterology	4	0	1	5
American Psychologist	2	0	2	4
Australian and New Zealand Journal of Psychiatry	0	3	0	3
Annals of Internal Medicine	0	1	2	3
Cancer	1	1	1	3
Other journals	19	42	34	95
Totals	40	66	58	164

Table 2.   Comparison of Quality Features among Meta-Analyses.

	1955–1982 (n = 40) Adequate	1983–1986 (n = 66) Adequate	1987–1990 (n = 58) Adequate
	number (%)		
Prospective design			
Protocol	5 (13)	2 (3)	13 (22)
Literature search	10 (25)	24 (36)	40 (69)
List of trials analyzed	37 (93)	56 (85)	54 (93)
Log of rejected trials	4 (10)	11 (17)	24 (41)
Treatment assignment	38 (95)	17 (26)	46 (79)
Ranges of patients	13 (33)	12 (18)	36 (62)
Ranges of treatment	20 (50)	25 (38)	39 (67)
Ranges of diagnosis	18 (45)	21 (32)	34 (59)
Combinability			
Criteria	17 (43)	26 (39)	39 (67)
Measurement	5 (13)	17 (26)	27 (47)
Control of bias			
Selection bias	0	0	7 (12)
Data-extraction bias	0	0	7 (12)
Interobserver bias	0	7 (11)	11 (19)
Source of support	17 (43)	14 (21)	16 (28)
Statistical analysis			
Statistical method	22 (55)	40 (61)	45 (78)
Statistical errors	15 (38)	31 (47)	38 (66)
Confidence intervals	14 (35)	27 (41)	49 (84)
Subgroup analysis	28 (70)	39 (59)	45 (78)
Sensitivity analysis			
Quality assessment	9 (23)	10 (15)	15 (26)
Varying methods	6 (15)	12 (18)	25 (43)
Publication bias	3 (8)	11 (17)	24 (41)
Application of results			
Caveats	37 (93)	31 (47)	41 (71)
Economic impact	0	2 (3)	3 (5)

tistical analysis, sensitivity analysis, and problems of applicability. The scoring results are listed in Table 2 and a discussion of each major area follows below.

We found that, after an initial learning period, two evaluators agreed on over 90 percent of items scored; the majority of disagreements were due to oversights.

In the sections that follow, we have divided the papers into three time periods, selected to give roughly equal numbers in each: 1955 through 1982, 1983 through 1986, and 1987 through 1990.

STUDY DESIGN

In meta-analysis, like any other form of research, it is important to try to make the process as rigorous and well defined as possible.

PROTOCOL

As with any scientific endeavor, the questions to be answered, the criteria for inclusion in the study, and the methodology to be used should be established beforehand. There has been a slight improvement over time, but still only a minority of papers gave clear evidence that the study was conducted according to a predetermined protocol or research plan. Many more may have followed protocols but it was not apparent to the reader. If retrospective meta-analyses are to be converted into prospective research, we believe that writing a protocol is an essential first step.

LITERATURE SEARCH

Since a valid meta-analysis should include as many relevant trials as possible, the authors should provide details of their search procedures. At the present time it is insufficient to rely solely on computer literature searches, as they may yield less than two-thirds of relevant trials.[7,8] A computer search can be supplemented by consulting *Current Contents*, reviews, textbooks or experts in the particular field of study, and by reviewing the references of the trials found. The proportion of papers clearly using such exhaustive searching methods has increased.

LIST OF TRIALS ANALYZED AND LOG OF REJECTED TRIALS

The report of a meta-analysis should provide a list of the trials analyzed, and most of the reports in all three periods did so. Just as important is an enumeration of the relevant trials excluded and the reasons why. This is analogous to the log of excluded patients in a clinical trial. We believe it is vital for the reader to be aware of any information that was not used, since the meta-analyst may have had a preconception or bias as to how the result should come out. The proportion of papers reporting excluded trials has steadily increased, but even in the latest period, less than half did so.

TREATMENT ASSIGNMENT

The most important question bearing on the validity of the data pooled is the method of treatment assignment in the primary study. It has been shown that results of trials using historical controls are more likely to favor the new treatment than results of the same therapy tested in randomized control trials.[9] The proportion of papers using data only from randomized trials appears to have declined in recent years.

RANGES OF PATIENTS, DIAGNOSES, AND TREATMENTS

In order for the reader to judge the validity and generalizability of a meta-analysis, data should be provided on the patients, diagnoses, therapies, and end points in the original studies. It is not possible to provide more than broad general outlines here, but specific rules can and should be developed for each particular meta-analysis. The ranges of patient characteristics in all the trials analyzed — age, sex, relevant socioeconomic data, other diseases — should be included. Details on these items were given in only a minority of the meta-analyses in the first two time periods. The ranges of treatments should also be defined. Did all patients in all trials receive the same or similar therapy? In meta-analyses of drug therapies, were trials combined that used the same drug, or the same class of drug? Dosage and route, as well as frequency and duration of therapy in all the trials to be pooled, should be available to the reader. This criterion was variably fulfilled in the papers analyzed. Similarly, data should be presented on the range of diagnoses in the pooled trials. Were diagnostic criteria the same in all trials? What stages or grades of disease were included? Such details on this important question were reported in less than half the meta-analyses in the first two time periods but in more than two thirds in the third time period.

COMBINABILITY

A major issue in pooling data is whether the results of the separate trials can be meaningfully combined. This should be explicitly addressed by the meta-analyst, and in sufficient detail for the reader to determine that a useful and clinically relevant result will be obtained.

CRITERIA

What criteria were used to decide that the studies analyzed were similar enough to be pooled? The meta-analyst should note any differences in the primary studies, and discuss how these differences affect the conclusions of the meta-analysis. Less than half of the meta-analysts in the first samples and 67 percent in the third detailed their criteria for pooling.

MEASUREMENT

Related to the problem of combinability is the statistical issue of heterogeneity. In addressing questions of combining estimates from different studies, statisticians distinguish between two possible models.[10] In Model 1, each study is considered to represent one of the situations in a universe, and all situations in the universe are represented. (For example, we want the average death rate in the three hospitals in a specific city.) In Model 2, each study is considered to be from a different population, the rate varies from study to study, and their differences are due to experimental error and to differences in the populations (between-study variability). There are a variety of methods for deciding which of the above models is more appropriate.[11] In either case, there are tests for homogeneity that help decide the degree of caution with which inferences from the pooled results can be made. Considering the heterogeneity of patients treated by practitioners who might be applying the results of meta-analyses, heterogeneity in the trials may not be so bad. Evidence of a statistical test for homogeneity was found in few of the meta-analyses in the first two time periods, but rose to nearly half in the most recent period.

CONTROL AND MEASUREMENT OF POTENTIAL BIAS

When performing a meta-analysis — as in any scientific endeavor — potential sources of unconscious bias should be controlled for when possible.

SELECTION BIAS

To avoid bias in the selection and rejection of papers, the decision to include a paper should be made by looking only at its methods and not at its results, or looking at the two separately under coded conditions. This important source of bias was not reported in any of the meta-analyses reviewed for our first survey and in only 12 percent in the latest time period.

DATA EXTRACTION BIAS

As with any other data-gathering process that requires interpretation, observers may disagree. When papers list a variety of subgroups, end points, exclusions, and the like, it is quite possible that readers may vary in how they extract the data from a particular study. The ideal way to control for this type of bias is to have the data extracted by more than one observer, each of whom is blinded to the various treatment groups through a coded photocopying process, and then measure the interobserver agreement. In none of the meta-analyses in the first time period was such agreement reported, but this appears to be increasing recently. The data extraction process was not blinded in any of the early papers and in only 12 percent of those in the latest period.

SOURCE OF SUPPORT

We feel that it is useful to the reader to know who financed a study when deciding how much credence to give to its conclusions. Potential conflicts of interest do not necessarily disqualify a study, but they should be clearly acknowledged. The source of support was specified in a minority of papers and, if anything, appears to have declined over time.

STATISTICAL ANALYSIS

METHODS

We evaluated as "adequate" any recognized method of pooling except simple addition of successes across all trials to give an overall average that was rated as "partial." An adequate method was used in well over half of the meta-analyses and appears to have increased over time.

The most commonly used method was the Mantel–Haenszel test or a modification thereof. Other studies combined data by calculating a standardized average effect size; many also performed various types of regression analyses, significance tests, or both. A few papers used various other methods of pooling and five papers did not specify the methods used. For further discussion of these various methods, see the section entitled Remaining Problems, below.

STATISTICAL ERRORS

Less than half of the meta-analyses in the first two time periods showed an awareness of the potential problems of the Type I (concluding that there is a difference when none exists) and Type II (concluding that there is no difference when there is one) statistical errors. There has been some improvement in the recognition of Type II errors in meta-analyses; this contrasts with the survey of randomized control trials also updated in this book (see Chapter 19) which found no improvement over the last ten years.

CONFIDENCE INTERVALS

It is often more useful to the reader to have an estimate with confidence intervals of the difference between the success rates of the modalities being compared than to have only the results of significance tests. Confidence intervals for major outcomes were given in less than half (39 percent) of the meta-analyses in the first two periods, but this has risen to 84 percent recently.

SUBGROUP ANALYSES

One of the purposes of meta-analysis, as previously stated, is to increase the statistical power for subgroup analyses. Relevant subgroups were analyzed in the majority of papers.

SENSITIVITY ANALYSIS

Depending on the test chosen, the same set of data can be combined to give different conclusions.[12] Similarly, the results may vary depending on the overall quality of the primary trials and on whether certain trials, subgroups of patients, or other important variables are excluded or changed.

QUALITY ASSESSMENT

In a meta-analysis, the methodologic rigor and scientific quality of the papers to be combined should be assessed and considered in formulating recommendations.[13,14] If the original methodology is poor, the resulting conclusion will be less reliable. For each primary study, such features as the randomization process, the measurement of patient compliance, the blinding of patients and observers, the statistical analyses, and the handling of withdrawals should be examined. This issue of quality was fully addressed in only a small proportion of the meta-analyses and was not even mentioned in nearly half. There was no evidence of improvement over time.

VARYING METHODS

Each meta-analysis should include in a sensitivity analysis data that show how the results vary through the use of different assumptions, tests, and criteria. This type of analysis was performed in only 15 percent of the meta-analyses reviewed in the first period but increased appreciably in the second and third.

PUBLICATION BIAS

One of the criticisms sometimes made of meta-analysis is that there may be some unpublished studies that would contradict the results of published studies, and there is some evidence that negative studies are less likely to be published than positive ones.[15,16] A simple method has been proposed for calculating the number of unpublished negative studies required to refute the published evidence,[17] which is one possible measure of the strength of the published evidence. In only 8 percent of meta-analyses was publication bias considered in the first period, compared to 17 percent in the second and 41 percent in the third.

APPLICATION OF RESULTS

CAVEATS

Once the results of the pooling process are available, the meta-analyst should attempt to put them into perspective, based on all of the considerations above. Does the new therapy seem to be established as more effective than the old for all patients or for some subgroups? Or should the conclusions be taken only as suggestions for future study? Such caveats were included in the discussion in varying proportions of the papers with, if anything, a decline over time.

ECONOMIC IMPACT

In today's climate of financial constraints on health care expenditure, it is increasingly important to consider the economic impact of adopting new modalities of treatment or diagnosis. Although some may consider this a topic for other studies, we were disappointed that less than 5 percent of the meta-analyses included a thorough analysis of economic impact.

SURVEY CONCLUSIONS

Of the 23 individual items, 7.63 ± 2.84 (mean $\pm$ S.D.) were adequately addressed in the 40 meta-analyses published between 1955 and 1982, 6.80 ± 3.86 were adequately addressed in the 66 published between 1983 and 1986, and 11.91 ± 4.79 were adequately addressed in the 58 published between 1987 and 1990 ($F = 31.3$, $P<0.001$). Twenty-two of the 58 papers published in the last time period referred to our initial survey or to other similar guides for meta-analysis, which suggests that the guidelines are being used.

REMAINING PROBLEMS

What is the role of meta-analysis? When should it be attempted and how should its results be used? We believe that the best way to answer questions about the efficacy of new therapeutic or diagnostic modalities is to perform well-designed, adequately sized randomized control trials. Meta-analysis may have a role when definitive trials are impossible or impractical, when trials have been performed but the results are inconclusive or conflicting, or while awaiting the results of definitive studies. Meta-analysis, like decision analysis, can give quantitative estimates of the weight of available evidence which can be helpful in making clinical decisions. There is, however, a danger that meta-analysis may be used inappropriately or indiscriminately. As with many other types of analysis, the quality of the results depends on the quality of the input. Therefore, the question posed in each meta-analysis should be explicitly stated and clinically relevant.

Beginning in 1988 the National Library of Medicine's bibliographic database, MEDLINE, included "meta-analysis" as a separate term for use in searching the medical literature. This change makes it easier to find more recent meta-analysis articles.

Difficulty still exists, however, in locating both meta-analyses and randomized control trials in the literature because present literature-searching and indexing systems do not always distinguish primary studies and reviews from meta-analyses or randomized control trials from other clinical trials. Thus, it cannot be claimed that the papers found for this analysis are an exhaustive or representative sample. It is also quite likely that there are unpublished meta-analyses. Investigators will facilitate the process of recovery and integration of important clinical information if they insist on inclusion of the terms "randomized" and "meta-analysis" in titles and abstracts so that indexing can be improved.

More attention also needs to be paid to statistical issues and the advantages and disadvantages of the various pooling methods. A variety of statistical techniques have been developed for combining the results of separate studies. For example, there are a number of methods for combining the probability values or test statistics from individual studies.[12] These methods, however, may not distinguish between small studies with large effects and large studies with small effects and do not yield an estimate of the size of the effect. The Mantel–Haenszel method[18] or a modification of that technique for combining separate 2 × 2 tables[19] (Peto method) is becoming increasingly popular. These methods have several useful properties: they compare each treatment only with its own control, and weight studies according to their sample size. They can include a test for homogeneity, as well as an estimate of the effect size, but they may have some undesirable properties (see below). In the psychiatric literature, studies have been combined by computing effect sizes, defined as the mean difference between experimental and control groups, divided by the control group standard deviation.[20] This method allows for the pooling of different end points, since all findings are transformed into common units, but the conclusions may be difficult to interpret clinically; the validity of this process is, thus, open to question. A few papers have used multivariate methods or log-linear models to attempt to adjust for differences between studies.

Since publication of our first paper there have been several important contributions to the statistical methods in meta-analysis. The validity of the Peto one-step method has been challenged.[21] Simulations have confirmed this,[22] and in those experiments the most valid method of determining variances has been that described by Robins, Greenland, and Breslow.[23] When zero observations in one or both groups present a problem, the exact method, as modified and automated by Mehta,[24] is recommended. A Bayesian approach has been advocated.[25] However, we encountered no published meta-analyses that used the last two methods.

Another important development in meta-analysis was a new textbook entitled *Effective Care in Pregnancy and Childbirth*[26] with hundreds of meta-analyses. These have not been included in our survey because they represent a special case — these studies were written on consignment by the editors and based on a database of randomized control trials that had been collected in Oxford.[27] Obviously this approach needs to be replicated by conducting large numbers of meta-analyses in other fields of medicine, but it will first be necessary to collect data on all the randomized control trials available in each field.

Some meta-analysts believe that the only valid pooling of results should include the outcomes for all randomized patients regardless of how long (or even whether) they received the assigned treatment (the intention-to-treat method[19]). The authors of at least one paper apparently felt that dropouts and withdrawals should be excluded (exclusion method[28]). However, the data on withdrawals and dropouts are not always available. We believe that the results should be reported both according to the intention to treat and according to exclusion rules to facilitate evaluation of the differences. If there are none, there is no problem. If clinically important differences exist, the study may be difficult to interpret.[29]

Due to the problem of publication bias, some meta-analysts choose to supplement their published data with unpublished trials, data, or both. Unpublished results may be less reliable, since they have not been found acceptable by peer-reviewers, and may not be collected with the same rigor or accuracy as published results.[30] However, the potential problems inherent in unpublished results make it unclear as to whether both types of data should be given equal weight.

Greater uniformity in reporting meta-analyses would be helpful to readers. Many of the meta-analyses found were written in the standard format of scientific papers with detailed methods and results sections, but several were editorials, leading articles, or letters to the editor with little detail on methodology. We believe that meta-analyses should be presented with sufficient information for readers to draw their own conclusions about the validity of the results.

In conclusion, if meta-analysis is to be accepted as a scientific tool, each meta-analysis should be conducted like a scientific experiment, beginning with a clear plan of the question to be answered and the methodology to be employed. Attention needs to be paid to intraobserver and interobserver variability, and attempts should be made to identify and minimize bias. Concerns have been expressed about the validity of pooling[31,32] but the process is increasingly used and frequently defended.[33-36] We feel that a quantitative synthesis of the data in similar randomized control trials is more useful to the practicing physician than a traditional narrative review article, but such syntheses must be properly performed to warrant serious attention. We hope that the points we have raised here will stimulate discussion that will ultimately lead to better meta-analyses.

REFERENCES

1. Light RJ, Smith PV. Accumulating evidence: procedures for resolving contradictions among different research studies. Harvard Educ Rev 1971; 41:429-71.

2. Glass GV. Primary, secondary and meta-analysis of research. Educational Researcher 1976; 5:3-8.

3. Webster's Seventh New Collegiate Dictionary. Springfield, Mass.: G&C Merriam Co., 1967:532.

4. Louis TA, Fineberg HV, Mosteller F. Findings for public health from meta-analysis. Annu Rev Publ Health 1985; 6:1-20.

5. Sacks HS, Berrier J, Reitman D, Ancona-Berk VA, Chalmers TC. Meta-analyses of randomized controlled trials. N Engl J Med 1987; 316:450-5.

6. Beecher HK. The powerful placebo. JAMA 1955; 159:1602-6.

7. Poynard T, Conn HO. The retrieval of randomized clinical trials in liver disease from the medical literature. A comparison of MEDLARS and manual methods. Controlled Clin Trials 1985; 6:271-9.

8. Dickersin K, Hewitt P, Mutch L, Chalmers I, Chalmers TC. Perusing the literature: comparison of MEDLINE searching with a perinatal clinical trials database. Controlled Clin Trials 1985; 6:306-17.

9. Sacks H, Chalmers TC, Smith H Jr. Randomized versus historical controls for clinical trials. Am J Med 1982; 72:233-40.

10. Laird N, Mosteller F. Some statistical methods for combining experimental results. Int J Tech Assess Hlth Care 1990; 6:5-30.

11. Berlin JA, Laird NM, Sacks HS, Chalmers TC. A comparison of statistical methods for combining event rates from clinical trials. Stat Med 1989; 8:141-51.

12. Rosenthal R. Combining results of independent studies. Psychol Bull 1978; 85:185-93.

13. Chalmers TC, Smith H Jr, Blackburn B, et al. A method for assessing the quality of a randomized control trial. Controlled Clin Trials 1981; 2:31-49.

14. DerSimonian R, Charette LJ, McPeek B, Mosteller F. Reporting on methods in clinical trials. N Engl J Med 1982; 306:1332-7.

15. Dickersin K, Chan S, Chalmers TC, Sacks HS, Smith H Jr. Publication bias and clinical trials. Controlled Clin Trials 1987; 8:343-53.

16. Simes RJ. Publication bias: the case for an international registry of clinical trials. J Clin Oncol 1986; 4:1529-41.

17. Rosenthal R. The "file drawer problem" and tolerance for null results. Psychol Bull 1979; 86:638-41.

18. Mantel N, Haenszel W. Statistical aspects of the analysis of data from retrospective studies of disease. J Natl Cancer Inst 1959; 22:719-48.

19. Yusuf S, Peto R, Lewis J, Collins R, Sleight P. Beta blockade during and after myocardial infarction: an overview of the randomized trials. Prog Cardiovasc Dis 1985; 27:335-71.

20. Glass GV, McGaw B, Smith ML. Meta-analysis in social research. Beverly Hills, Calif.: Sage Publication, 1981:21-56.

21. Greenland S. Quantitative methods in the review of epidemiologic literature. Epidemiol Rev 1987; 9:1-30.

22. Emerson JD. Confidence intervals for the odds ratio in meta-analysis: the state of the art. Cambridge, Mass.: Technology Assessment Group, Harvard University, 1990.

23. Robins J, Greenland S, Breslow NE. A general estimator for the variance of the Mantel-Haenszel odds ratio. Am J Epidemiol 1986; 124:719-23.

24. Mehta CR, Patel NR, Gray R. Computing an exact confidence interval for the common odds ratio in several 2 × 2 contingency tables. J Am Stat Assoc 1985; 80:969-73.

25. Eddy DM, Hasselblad V, Schachter R. A Bayesian method for synthesizing evidence: the confidence profile method. Int J Tech Assess Hlth Care 1990; 6:31-56.

26. Chalmers I, Enkin M, Keirse MJNC. Effective care in pregnancy and childbirth. Oxford: Oxford University Press, 1989.

27. Chalmers I, Hetherington J, Newdick M, et al. The Oxford database of perinatal trials: developing a register of published reports of controlled trials. Controlled Clin Trials 1986; 7:306-24.

28. Levitt SH, McHugh RB, Song CW. Radiotherapy in the postoperative treatment of operable cancer of the breast. Part II. A re-examination of Stjernsward's application of the Mantel-Haenszel statistical method. Evaluation of the effect of the radiation on immune response and suggestions for postoperative radiotherapy. Cancer 1977; 39:924-40.

29. Bhaskar R, Reitman D, Sacks HS, Smith H Jr, Chalmers TC. Loss of patients in clinical trials that measure long-term survival following myocardial infarction. Controlled Clin Trials 1986; 7:134-48.

30. Relman AS. News reports of medical meetings: how reliable are abstracts? N Engl J Med 1980; 303:277-8.

31. Elashoff JD. Combining results of clinical trials. Gastroenterology 1978; 75:1170-4.

32. Goldman L, Feinstein AR. Anticoagulants and myocardial infarction: the problems of pooling, drowning, and floating. Ann Intern Med 1979; 90:92-4.

33. Gerberg ZB, Horwitz RI. Resolving conflicting clinical trials: guidelines for meta-analysis. J Clin Epidemiol 1988; 41:503-9.

34. Wachter KW. Disturbed by meta-analysis? Science 1988; 241:1407-8.

35. L'Abbe KA, Detsky AS, O'Rourke K. Meta-analysis in clinical research. Ann Intern Med 1987; 107:224-33.

36. Thacker SB. Meta-analysis. A quantitative approach to research integration. JAMA 1988; 259:1685-9.

Index

INDEX

Other *NEJM* Books

SI Unit Conversion Guide
By Michael Laposata, M.D., Ph.D.

Law-Medicine Notes: Progress in Medicolegal Relations
By William J. Curran, J.D., LL.M., S.M.Hyg.

By the London Post
By John Lister, M.D.

NEJM Reprint Collections

Breast Cancer, Volume 1

Drug Therapy, Volume 8

AIDS: Epidemiologic and Clinical Studies, Volume 1

AIDS: Epidemiologic and Clinical Studies, Volume 2

Clinical Medical Ethics